Routing Protocols and Architectural Solutions for Optimal Wireless Networks and Security

Dharm Singh
Namibia University of Science and Technology, Namibia

A volume in the Advances in Wireless Technologies and Telecommunication (AWTT) Book Series

DISSEMINATOR OF KNOWLEDGE
www.igi-global.com

Published in the United States of America by
IGI Global
Information Science Reference (an imprint of IGI Global)
701 E. Chocolate Avenue
Hershey PA, USA 17033
Tel: 717-533-8845
Fax: 717-533-8661
E-mail: cust@igi-global.com
Web site: http://www.igi-global.com

Library of Congress Cataloging-in-Publication Data

Names: Singh, Dharm, 1963- editor.
Title: Routing protocols and architectural solutions for optimal wireless networks and security / Dharm Singh, editor.
Description: Hershey, PA : Information Science Reference, [2017] | Includes bibliographical references.
Identifiers: LCCN 2017000832| ISBN 9781522523420 (hardcover) | ISBN 9781522523437 (ebook)
Subjects: LCSH: Routing (Computer network management) | Wireless LANs--Security measures.
Classification: LCC TK5105.5487 .R695 2017 | DDC 004.6--dc23 LC record available at https://lccn.loc.gov/2017000832

This book is published in the IGI Global book series Advances in Wireless Technologies and Telecommunication (AWTT) (ISSN: 2327-3305; eISSN: 2327-3313)

British Cataloguing in Publication Data
A Cataloguing in Publication record for this book is available from the British Library.

All work contributed to this book is new, previously-unpublished material. The views expressed in this book are those of the authors, but not necessarily of the publisher.

For electronic access to this publication, please contact: eresources@igi-global.com.

Advances in Wireless Technologies and Telecommunication (AWTT) Book Series

Xiaoge Xu
The University of Nottingham Ningbo China, China

ISSN:2327-3305
EISSN:2327-3313

Mission

The wireless computing industry is constantly evolving, redesigning the ways in which individuals share information. Wireless technology and telecommunication remain one of the most important technologies in business organizations. The utilization of these technologies has enhanced business efficiency by enabling dynamic resources in all aspects of society.

The **Advances in Wireless Technologies and Telecommunication Book Series** aims to provide researchers and academic communities with quality research on the concepts and developments in the wireless technology fields. Developers, engineers, students, research strategists, and IT managers will find this series useful to gain insight into next generation wireless technologies and telecommunication.

Coverage

- Global Telecommunications
- Wireless Broadband
- Wireless Technologies
- Digital Communication
- Broadcasting
- Mobile Web Services
- Network Management
- Mobile Technology
- Cellular Networks
- Telecommunications

IGI Global is currently accepting manuscripts for publication within this series. To submit a proposal for a volume in this series, please contact our Acquisition Editors at Acquisitions@igi-global.com or visit: http://www.igi-global.com/publish/.

The Advances in Wireless Technologies and Telecommunication (AWTT) Book Series (ISSN 2327-3305) is published by IGI Global, 701 E. Chocolate Avenue, Hershey, PA 17033-1240, USA, www.igi-global.com. This series is composed of titles available for purchase individually; each title is edited to be contextually exclusive from any other title within the series. For pricing and ordering information please visit http://www.igi-global.com/book-series/advances-wireless-technologies-telecommunication-awtt/73684. Postmaster: Send all address changes to above address.

Titles in this Series

For a list of additional titles in this series, please visit: www.igi-global.com/book-series

Emerging Trends and Applications of the Internet of Things
Petar Kocovic (Union – Nikola Tesla University, Serbia) Reinhold Behringer (Leeds Beckett University, UK) Muthu Ramachandran (Leeds Beckett University, UK) and Radomir Mihajlovic (New York Institute of Technology, USA)
Information Science Reference • copyright 2017 • 330pp • H/C (ISBN: 9781522524373) • US $195.00 (our price)

Resource Allocation in Next-Generation Broadband Wireless Access Networks
Chetna Singhal (Indian Institute of Technology Kharagpur, India) and Swades De (Indian Institute of Technology Delhi, India)
Information Science Reference • copyright 2017 • 334pp • H/C (ISBN: 9781522520238) • US $190.00 (our price)

Multimedia Services and Applications in Mission Critical Communication Systems
Khalid Al-Begain (University of South Wales, UK) and Ashraf Ali (The Hashemite University, Jordan & University of South Wales, UK)
Information Science Reference • copyright 2017 • 331pp • H/C (ISBN: 9781522521136) • US $200.00 (our price)

Big Data Applications in the Telecommunications Industry
Ye Ouyang (Verizon Wirless, USA) and Mantian Hu (Chinese University of Hong Kong, China)
Information Science Reference • copyright 2017 • 216pp • H/C (ISBN: 9781522517504) • US $145.00 (our price)

Handbook of Research on Recent Developments in Intelligent Communication Application
Siddhartha Bhattacharyya (RCC Institute of Information Technology, India) Nibaran Das (Jadavpur University, India) Debotosh Bhattacharjee (Jadavpur University, India) and Anirban Mukherjee (RCC Institute of Information Technology, India)
Information Science Reference • copyright 2017 • 671pp • H/C (ISBN: 9781522517856) • US $360.00 (our price)

Interference Mitigation and Energy Management in 5G Heterogeneous Cellular Networks
Chungang Yang (Xidian University, China) and Jiandong Li (Xidian University, China)
Information Science Reference • copyright 2017 • 362pp • H/C (ISBN: 9781522517122) • US $195.00 (our price)

Handbook of Research on Advanced Trends in Microwave and Communication Engineering
Ahmed El Oualkadi (Abdelmalek Essaadi University, Morocco) and Jamal Zbitou (Hassan 1st University, Morocco)
Information Science Reference • copyright 2017 • 716pp • H/C (ISBN: 9781522507734) • US $315.00 (our price)

IGI GLOBAL
DISSEMINATOR OF KNOWLEDGE

www.igi-global.com

701 East Chocolate Avenue, Hershey, PA 17033, USA
Tel: 717-533-8845 x100 • Fax: 717-533-8661
E-Mail: cust@igi-global.com • www.igi-global.com

List of Reviewers

Table of Contents

Detailed Table of Contents

Today's IEEE 802.11n WLAN are capable of delivering the essential bandwidth for video by using MPEG-4 or H.264 codec compression. However, 802.11n devices are popular for delivering wireless video transmission. Even though there are some challenges for these devices specifically severe network congestion that degrade the quality of video transmission. For assessing the video transmission quality during a simulation on 802.11n WLAN technology, we extended a framework and toolset (802.11nMyEvalVid) which can be used to measures the QoS constraints in wireless networks, such as throughput, delays, and end-to-end delay. However, it also supports PSNR, which is a received video quality measuring technique based on a comparison of frame-by-frame. This chapter focused on a framework 802.11nMyEvalVid that can be used for research and evaluating new techniques for MAC-layer optimizations for simulating MPEG-4, H.264/AVC and H.264/SVC video over 802.11n WLAN in a more efficient and reliable way.

There is an increased risk of copyright violation of multimedia data due to the enormous growth of computer networks that provides fast and error free transmission of any multimedia information. A copyright identifier that may contain some information about the lawful owner is inserted in the contents of the image, without sacrificing its quality. The security levels are increased by using a key value and scaling factor for the embedding and extraction process. The dual scrambled watermark using Arnold and Scrambling sequence is embedded by modifying the singular values of the scrambled cover image's

DWT middle frequency sub-band. The simulation was performed on MATLAB 7.7.0 with standard database gray scale images of size 512x512 and watermark of size 64x64 using hybrid dual scrambled watermark schemes. The performance analysis is done on the basis of the degree of scrambling and JPEG compression attack using various parameters. The proposed method achieves better imperceptibility and security for the copyright protection methods.

Mobile communication has various important issues such as congestion control, scarcity of radio spectrum, bandwidth utilization, and security. With increased number of threats in mobile devices, security has become the most important issue now-a-days. Security can be considered primarily at two levels; wireless transmission medium and mobile data level, where malicious applications execute their behavior by exploiting vulnerabilities present in mobile OS. The chapter also focuses on the latter one as these attacks can be easily performed by targeting vulnerabilities present in OS. The content is specifically on Android Operating System security as Android has become the largest installed base of mobile platform. This chapter specifies these vulnerabilities and focuses on malicious intentions of latest and sophisticated malware in detail. The chapter covers static and dynamic analysis tools as well as focuses on limitations of these methods for analyses. The results for malware samples with emulator detection capabilities show that these malware behaves more as benign samples.

Information dissemination in agricultural sector for its growth using information and communication technology (ICT) as a tool is need of the hour. This can be achieved using information systems. ICT benefits are helpful in exchange and dissemination of information among farming stakeholders. By using the latest tool of mobile technology, farmers can get the current information related to their farming jobs around the clock and at any location, as the mobile network have touched every part/location of the India. Using the features of the mobile-phones like GPS etc. one can get the localized information. Only need is to structure the abundant information available across the various organizations. So, a mobile based agricultural information system framework 'mAgIDS' has been proposed employing the hybrid mobile application architecture approach. Client-server architecture using the location Application Programming Interface (API) has been proposed. Inference mechanism of the system has implemented on the basis of improved fuzzy rule promotion technique.

VANET is a type of Mobile Ad-hoc network (MANET) that provides wireless communication between close by moving vehicles as well as between vehicles and road side fixed equipment. It requires the vehicles to be equipped with sensors and wireless devices and fixed roadside wireless infrastructure. There are number of research issues like Highly Dynamic Topology, Signal fading, Limited Bandwidth, Intermittent Connectivity, Mobility Modelling, Propagation Model, Routing which needs attention. The scope of this chapter is focused on various advances in Greedy Forwarding routing techniques. Various VANET routing protocols are reviewed and compared in this chapter.

The main applications of delay tolerant network lie in space research, military, tactical and other security sensitive operations. In these and other applications, secure and energy efficient routing is an important issue in homogeneous and heterogeneous network. This chapter contributes a research work that demonstrates the design, simulation, testing and evaluation of secure, energy efficient and environment independent routing protocol for delay tolerant network. The proposed protocol not only proposed secure approach but also energy efficient approach in homogeneous and heterogeneous environment. The proposed approach utilize sectorization technique to achieve its goals of security, energy efficiency and environment independence. Described in this chapter is a detailed architecture, design, simulation, testing and evaluation of the proposed protocol. Having evaluated these protocols in various simulators, it is concluded that the proposed routing protocol utilized minimum energy and bandwidth and also provided secure transmission.

Establishing trust in Mobile Ad hoc Networks (MANETs) is a challenging task. Trust management consists of various activities like prediction of trust, computation of trust, propagation of trust, aggregation of trust etc. Trust prediction can be done based on past behavior of node, internal parameters of the node, or it can be induction-based or neural-network based. Trust computation can be centralized or distributed. Trust propagation can be social networks based, small world phenomenon based, web of trust based, distributed hash table based. Trust aggregation can be weighted average based, probability based, sequential, conditional-sequential, parallel or subjective logic based. This work presents the review of various activities pertaining to trust management to enable researchers, academicians and practitioners to identify and address trust related issues in a better way.

The work in this chapter mainly focusses on a pseudo-noise (PN) sequence generator which is constructed exploiting the features of one-dimensional chaotic systems such as a logistic map. The use of logistic map to generate strong cryptographic sequences is novel in approach in terms of its use with a range of transmission techniques for wireless communication because it is easy to conceive and requires simple devices to generate the sequence. The generated sequence is used as a spreading sequence in a DS-SS modulation multipath environment with AWGN and Rayleigh fading channel, under various channel fading conditions. The BER curves and channel impulse responses are obtained and compared with the existing linear feedback shift register (LFSR) based PN sequence and Kasami sequence. Results from this comparative analysis indicate that the proposed method generally yields a greater number of reliable, unpredictable and random bits than existing techniques under the same conditions and can be practically implemented for DS-SS scheme as a spreading sequence.

With the wide applications of wireless communication, it becomes an important factor to keep the pace of the spectrum usage for all the new applications. To improve efficiently the spectrum usage in variety of applications, where the primary users are not active all the time, we can introduce the secondary users to exploit the radio frequency spectrum by not creating any interference to the Primary Users (PU). Unlike the pre-assigned spectrum allocation policy that has been adopted by various wireless technologies, in Cognitive Radio (CR), the whole spectrum is divided into many bands and corresponding channels where spectrum holes exist. The main challenge is thus to sense these holes and accommodate the secondary unlicensed users. In this chapter, proposing the Quantized data fusion sensing which is one of the type of cooperative sensing scheme used for unused spectrum sensing and acknowledgement to secondary signals. Simulation results with error rates are improved by the activity of PU and have been presented.

Localization is a fundamental problem in wireless sensor networks. Localization means the determination of geographical locations of sensor nodes, consequently detecting the event location and to initiate a prompt action whenever necessary. The localization process passes with three phases distance and/or angle estimation phase, position phase and algorithm phase. There are many techniques can be used in each phase, some of these techniques may add additional devices, cost, power consumption, or delay to the network. This chapter includes a classification and overview of the most popular localization techniques in each phase, also high lighting the problems of these techniques and their solutions in such a way that helps in understanding localization concepts especially with respect to WSN localization.

Foreword

It is an immense honor to share some preliminary remarks in this foreword to this publication titled *Routing Protocols and Architectural Solutions for Optimal Wireless Networks and Security.*

The wireless networks, Internet, social media, smart devices, smart phones and tablet computers have been playing a larger role in our daily lives. These days the majority of search traffic is a result of video content and social networks need to optimize their platforms. Therefore, the future marketing projects will have to include high investment to manage large files of video and to develop routing protocols and secure wireless networks as these days most of traffic is on wireless networks. Marketers or researchers must understand how each social channel is optimizing its platform to allow for more video content in users' feeds, and how brands are using video content to engage their followers. *This book is* a comprehensive resource on the latest technological advancements in the development of routing protocols and designing secure wireless networks.

Let me congratulate and commend Prof Dharm Singh for the pains he took to bring out a book, which deserves much more than just having a look. I am extremely happy to see the contribution from Namibia University of Science and Technology Exert to Communication Networks and Security.

I hope that this book will serve as a valuable text for students and reference for researchers and practitioners working in the field of Communication Networks and Security.

Anicia Peters
Namibia University of Science and Technology, Namibia

Anicia Peters *is the Dean Faculty of Computing and Informatics at The Namibia University of Science and Technology, Namibia.*

Preface

This publication is an ideal reference source for network designers, industries, researchers, educators, and governments who are involved in designing and implementing wireless technologies for security, wireless Ad hoc networks quality of service (QoS) routing protocols, transmission of data, voice, and video, protocols and algorithms, content networking & multimedia delivery technologies, big data analysis and other innovations.

This book, *Routing Protocols and Architectural Solutions for Optimal Wireless Networks and Security,* is a comprehensive resource on the latest technological advancements in designing secure wireless networks. With the proliferation of emerging wireless technologies and electronic networks devices, there is a fast growing interest in multimedia services and future networks. In the days to come, we expect that the wireless communication and networking technologies will become the emergence of many future wireless networking technologies. The future wireless network will offer security, multiservice, multimedia services, mobility, quality of service, variable connectivity, spontaneous networking, autonomic networking and other capabilities as the norm.

IEEE 802.11 standard based wireless local area networks (WLAN) have become popular wireless access technology. In addition, nowadays, users expect their laptops and other portable communication devices to deliver services like Video over IP, VoIP, video conferencing and media streaming. Such services are time sensitive and their delivery is greatly affected by delay and jitter introduced in the network. Hence, these applications demand strict guarantees for Quality-of-Service (QoS) from the underlying network.

Unfortunately, 802.11 based WLAN does not incorporate any mechanism to ensure QoS on WLAN. The standard, IEEE 802.11e has been introduced to provide QoS support. However, channel overloading still remains a major problem as the network performance is degraded under heavy load. When a new flow is admitted while the network is operating close to full capacity, it may not achieve its required QoS, further it may also jeopardize the QoS experienced by other already admitted flows.

Failure in assuring QoS support will pose the challenge to the survival for 802.11 based WLAN. To address this challenge, it is necessary to incorporate suitable mechanism in the WLAN to ensure the QoS support. One of the mechanism is controlling admission of new flows in resource constrained WLAN and protecting the existing sessions. However, imperfect and poorly designed admission control can underutilize the network resource.

Every day cities worldwide suffer from serious organizational and logistic difficulties due to overcrowding and adverse weather conditions, traffic congestion and pollution issues, limits in the quality of service offered to senior and disabled citizens, safety and security problems in peripheral areas and new challenges continuously arise. Smart environment rely on wireless and mobile technologies for providing services such as healthcare assistance, security and safety, real-time traffic monitoring, and managing the environment, to name a few. Such applications have been a main driving force in the development of smart environment using wireless technologies.

These wireless and mobile technologies enable several new services that result in better decision making and actions made by enterprises and governments. Without the appropriate communication networks, it is really difficult for a governments to facilitate its citizens in sustainable, efficient, and safer manner or environment. Considering the significance of mobile and wireless technologies for realizing the vision of smart cities, there is a need for conducting research to further investigate the standardization efforts and explore different issues or challenges in the wireless technologies, mobile computing and smart environments. Providing secure, reliable, and dependable wireless services is the primary objective of modern data networks. While enabling technology for "on-demand" services through any of the common wireless architectures, such as WiFi, ad hoc networks, sensor networks, and vehicular networks, has made large progresses,

Current information age is witnessing a dramatic use of digital and electronic devices in the workplace and beyond. Wireless, Mobile Networks & its applications had received a significant and sustained research interest in terms of designing and deploying large scale and high performance computational applications in real life. Many challenges remain to be overcome in wireless networks such as the integration of mobile and WiFi, infrastructure-based and ad hoc networks, reconfigurable wireless networks, robust algorithms for self-organizing, on-demand service models and their performance in highly-volatile interconnect topologies as well as the smooth interoperability of different architectures. The growing number of interactions and wireless communications between surveillance devices, vehicle-to-vehicle communications, smart meters, intelligent recognition systems,, traffic management technologies, and a large variety of sensors exposes the need for solutions that provide both security and privacy guaranties for our citizens.

In particular, the topics of the book are the following: Video quality of service in wireless networks, dual scrambled image watermarking, mobile security, mobile networks, wireless vehicular area networks, direction aware routing protocol, trust management in mobile ad hoc networks, stochastic wireless channels, reconfigurable wireless transceiver, wireless honeypots, wireless sensor networks localization and self-organizing network solutions.

Dharm Singh
Namibia University of Science and Technology, Namibia

Acknowledgment

Thank you to each and every individual that was involved in this project. Without their support, this book would not have become a reality.

This book came to light due to the direct and indirect involvement of many researchers, engineers, academicians, designers, developers, and industry practitioners. Therefore, I acknowledge and thank the contributing authors, research institutions, and companies whose papers, reports, notes, websites, and study materials have been referred to in this book. I also wish to acknowledge the valuable contributions of the reviewers regarding the improvement of quality, coherence, and content presentation of chapters. Most of the authors also served as referees; I highly appreciate their double task.

First of all I would like to sincerely thank our Vice-Chancellor Prof. Tjama Tjivikua, The Namibia University of Science and Technology for his encouragement and support.

I extend my appreciation to Dr. Anicia Peters, the Dean Faculty of Computing and Informatics, Namibia University of Science and Technology for providing me a warm academic environment for completing this book. I appreciate her help and support very much.

I am grateful to the editorial team at IGI Global for their support through the stages of preparation and production. Their positive attitude and patience are greatly appreciated. Thanks to Jan Travers who closely worked with me to make the process smooth and efficient.

Finally, I would like to acknowledge with gratitude, the support and love of my family: my daughter, Charu Singh; my son, Chirag Singh; and my wife, Dr. Poonam Dhaka. The all kept me going, and this book would not have been possible without them.

I hope that this book will serve as a valuable text for students and reference for researchers and practitioners working in the field of Communication Networks and Security.

Dharm Singh
Namibia University of Science and Technology, Namibia

Introduction

The Institute of Electrical and Electronics Engineers, Inc. (IEEE) developed a standard for wireless local area network (WLAN) and named as IEEE 802.11 WLAN (IEEE Std 802.11-Part 11, 1999) in 1997. It operates on 2.4 GHz frequency with 2Mbps data rate. Soon after in 1999 the 802.11 standards modified and two standard IEEE 802.11b and IEEE 802.11a were developed. The 802.11b operates on 2.4 GHz frequency and Industrial Scientific Medical (ISM) band. It provides 11 Mbps data rate. The 802.11a operates on 5GHZ frequency and Unlicensed National Information Infrastructure band. In 2005, the new standard IEEE 802.11e (IEEE Std 802.11e-Part 11, 2005) was developed. This standard supports Quality of Service (QoS) for multimedia traffic. The QoS was achieved by updating the medium access control (MAC) layer through traffic differentiation. IEEE 802.11e is efficiently used in the video, audio, and real-time voice over IP (VoIP) types of applications.

By using spread-spectrum or orthogonal frequency-division multiplexing (OFDM) technologies, the user remains linked with the network while moving around within a local coverage area. The products using the IEEE 802.11 WLAN standards are promoted under the Wi-Fi brand name. Various 802.11 wireless networking protocols jointly are known as Wi-Fi.

In 2009, WLAN standard 802.11e updated and enhanced MAC and Physical (PHY) layers, also developed new standard for IEEE 802.11n (IEEE Std 802.11n-Part 11, 2009). It operates on 20 and 40 MHz radio frequency (RF) band. It supports multiple antenna by using multiple input and multiple output (MIMO) techniques with maximum data-transfer speed is 300 Mbps. These new improvements in 802.11n are an MIMO-OFDM physical layer, frame aggregation, Block Acknowledgement (BA) and Reverse Direction (RD).

Networking capabilities have been significantly enhanced in recent years. With emerging advancements in technology, wireless communication has increased exponentially. IEEE 802.11 WLAN is the most common wireless communication technology used on premises for data, voice and video communication. Simplicity in deployment and high bandwidth capacity has led to the rapid deployment of WLANs almost everywhere. Its popularity is further driven by the ability to use portable mobile handheld devices that support applications with increasing multimedia content. The popular applications are digital video, voice over IP (VoIP), video-conferencing, networked games and other time-sensitive critical applications. Such services are time sensitive and their delivery is greatly affected by delay and jitter introduced in the network. Hence, these applications demand strict guarantees for Quality-of-Service (QoS) from the underlying network.

The study observed (1) the ever-increasing number and diversity of wireless devices (2) running applications that become more critical and bandwidth-hungry, and having very diverging QoS requirements, and (3) the presence of multiple heterogeneous wireless technologies, often integrated in the same device, sharing the same physical environment and in many cases also sharing the same spectral bands ("Softwarization of radio and wireless networks - IEEE software defined networks," 2017b).

The original 802.11 WLAN is capable of provisioning only best-effort services and does not provide any service guarantees. In order to ensure QoS on WLAN, IEEE 802.11e standard introduced. It offers improved performance over traditional WLAN; however, it does not always offer desired QoS guarantees, especially under heavy network load conditions.

MULTIMEDIA COMMUNICATION OVER WIRELESS NETWORKS

In the last few years the interactive multimedia services have gained popularity in almost all areas. It became even more buzzword when personal computers were connected via networks.

One the most important application of multimedia services is Video on Demand (VoD), where the user chooses a movie from the already stored in the video server and can access instantly when user want to watch via a wired, wireless or hybrid networks. VoIP telephony, which is highly interactive, is now widely used to exchange information between people.

Computer simulation is the research tool of choice for a majority of the wired and wireless network research community and these days most of research works verify their novel ideas by adopting of network simulators. For measurement of quality of video over wireless networks various studies and frame works are available for the research community and can be extended for the upcoming QoS support IEEE 802.11 standard. Received video quality can measured by QoS parameters, delay and packet loss rate.

The framework can be used for development of new mechanism at MAC layer of wireless networks. Researchers can evaluate new techniques for MAC-layer optimizations in IEEE 802.11 WLAN as MAC controls the flow of different type of applications by allocating the different resources based on the characteristics of applications.

However, because of the bandwidth constraints that come with using wireless, video-over-wireless was not really feasible until about seven years ago, at that time 802.11n and 3G HSPA technologies were mature enough for use with commercial applications. IEEE 802.11n WLAN are capable of delivering the essential bandwidth for video by using MPEG-4 or H.264 codec compression. However, 802.11n devices are popular for delivering wireless video transmission. Even though there are some challenges for these devices specifically severe network congestion that degrade the quality of video transmission.

In this book an overview of the framework for evaluating the quality of received video transmitted over a wireless networks is presented and also provides the idea to develop new framework for upcoming compression and WiFi technologies.

VIDEO OVER WIRELESS NETWORKS

With the aid of this evaluation framework, researchers or practitioners have greater freedom to analyse their proposed network designs for video transmission without considering an appropriate tool-set.

The Network Simulator 2 (NS2) is freeware simulation tool used for wired and wireless network research. NS-2 is developed by the Virtual Inter Network Testbed (VINT) project in order to reduce duplication of effort within the network research and develop community ("Nsnam web pages," n.d.).

Initially EvalVid and MyEvalvid framework and tool-set for evaluation of the quality of video transmitted over a real or simulated communication network using NS2 have been developed. EvalVid present a complete framework and tool-set for evaluation of the quality of video transmitted over a real or simulated communication network using NS2 Jirka Klaue, Berthold Rathke, and Adam Wolisz (2003). But the simulated environment provided by EvalVid is simply an error model to represent corrupted or missing packets in the real network. MyEvalvid has done to build the connecting interfaces between Evalvid and NS2. Then make it be more complete and useful to network researchers (C. H. Ke, C. K. Shieh, W. S. Hwang, A. Ziviani, 2008). The framework can be installed in Windows XP using Cygwin version 2.4 or 2.5 and it is difficult to use Upgraded version of Cygwin and Windows.

In Chapter 1 titled "A Framework for Evaluation of Video Quality of Service in Wireless Networks," the author presents extended a framework and toolset (802.11nMyEvalVid) which can be used to measures the QoS constraints in wireless networks, such as throughput, delays, and end-to-end delay. However, it also supports PSNR, which is a received video quality measuring technique based on a comparison of frame-by-frame. This chapter focuses on a framework 802.11nMyEvalVid that can be used for research and evaluating new techniques for MAC-layer optimizations for simulating MPEG-4, H.264/AVC and H.264/SVC video over 802.11n WLAN in a more efficient and reliable way. This chapter also includes the important aspects and some challenges for 802.11n enabled devices specifically severe network congestion that degrades the quality of H.264 video transmission.

DUAL SCRAMBLED IMAGE WATERMARKING

Copyright violation of multimedia traffic over wired or wireless network has become an important issue for content owners and service providers. The process known as watermarking is the process of embedding information into the multimedia content to prove the "ownership" authentication and also minimised the quality degradation and imperceptible level is also maintained.

In Chapter 2 titled "Dual Scrambled Image Watermarking Algorithm in DWT-SVD Composite Domain," the authors explain an increased risk of copyright violation of multimedia data due to the enormous growth of computer networks that provides fast and error free transmission of any multimedia information. A copyright identifier that may contain some information about the lawful owner is inserted in the contents of the image, without sacrificing its quality. The security levels are increased by using a key value and scaling factor for the embedding and extraction process. The dual scrambled watermark using Arnold and Scrambling sequence is embedded by modifying the singular values of the scrambled cover image's DWT middle frequency sub-band. The simulation was performed on MATLAB 7.7.0 with standard database gray scale images of size 512x512 and watermark of size 64x64 using hybrid dual scrambled watermark schemes. The performance analysis is done on the basis of the degree of scrambling and JPEG compression attack using various parameters. The proposed method achieves better imperceptibility and security for the copyright protection methods.

WIRELESS AND MOBILE SECURITY

These days' mobile devices are essential part of our life and society. Every year several new wireless devices with improved capabilities and intelligence are accessible through mobile networks. In 2015 more than half a billion (563 million) mobile devices and network connections were added and these devices are one of the main contributors to global mobile traffic growth (affiliates, 2016).

The device and connection are not only having smarter capabilities but also developing from lower version generation network connectivity (2G) to higher generation network connectivity (3g, 3.5G, 4G or LTE and 5G) (affiliates, 2016). Device computing capabilities and higher bandwidth networks support leads to adoption of high bandwidth required multimedia applications that also contribute to increased global mobile data and Wi-Fi traffic. Worldwide, mobile devices and network connections will grow to 11.6 billion by 2020 at a compound annual growth rate (CAGR) of 8 percent including M2M modules (e.g., GPS systems in car, asset tracking systems in shipping and manufacturing sectors, or medical applications making patient records and health status more readable available, etc.,) that are exceeding the worlds projection population at that time (7.8 Billion). Today most of mobile user using their Mobile devices for accessing video and as per CISCO visual networking index three-fourth of the world's mobile data traffic will be video by 2020 (affiliates, 2016).

Most of the people rely on their mobile devices in their personal lives as well as their businesses. Various mobile device security threats including mobile malware, eavesdropping, unauthorized access, theft and loss, and unlicensed and unmanaged applications pose a significant risk to corporate data. Mobile devices like laptop, palmtop, WiFi enabled desktops, smartphones and WiFi enable tablet PCs are vulnerable to digital attacks, but they are also highly susceptible to physical attacks given their portability (2017, 2003).

Mobile device security threats: Mobile malware – Similar to desktops smartphones and tablets are also susceptible to viruses, worms, trojans and spyware. Mobile malware can steal corporate or individual sensitive data, and also collect user data from long distance. These days' high-profile mobile malware attacks are few, but that can be changed in near future. In addition, attackers can also use mobile malware to carry out targeted attacks against mobile device users (2017, 2003).

Eavesdropping: Eavesdropping is one of typical security threats in terrestrial wireless ad hoc networks (Wang, 2013). It is difficult to detect the eavesdroppers in wireless networks especially in Ad hoc network since they just passively tap the transmissions without disclosing their existence in the networks. Carrier-based wireless networks lack end-to-end upper-layer security even though such networks have good link-level security (2017, 2003). Data traffic including video and voice is often unencrypted during transmission from client to server and allowing intruders to eavesdrop on users' sensitive communications.

In Chapter 3 "Mobile Security: Attacks and Prevention," the authors consider important issues in mobile communication such as congestion control, scarcity of radio spectrum, bandwidth utilization, and security. With increased number of threats in mobile devices, security has become the most important issue now-a-days. Security can be considered primarily at two levels; wireless transmission medium and mobile data level, where malicious applications execute their behavior by exploiting vulnerabilities present in mobile operating system (OS). The chapter also focuses on the latter one as these attacks can be easily performed by targeting vulnerabilities present in OS. The content is specifically on Android operating system security as Android has become the largest installed base of mobile platform. This chapter specifies these vulnerabilities and focuses on malicious intentions of latest and sophisticated malware in detail. The chapter covers static and dynamic analysis tools as well as focuses on limitations of these methods for analyses. The results for malware samples with emulator detection capabilities show that these malware behaves more as benign samples.

Lately mobile devices have become full-fledged computing platforms as they have started supporting operating systems, complex applications, Internet, data transfer and software development kits. This has also increased the necessity for human beings to become aware of various threats which are rapidly growing with this revolution. This chapter explore the various security issues and their analyses for mobile devices. Most of the market in mobile operating system is covered by Android. Because of this, the chapter was primarily focused on security issues related to android devices. By studying the intentions of various attacks, analysis methods like static analysis and dynamic analysis have been developed to detect malicious applications in Android Operating System. Researchers can use these analysis methods to provide better security solutions for mobile devices. Applying these solutions, users would probably feel more comfortable in storing private information in their mobile devices.

MOBILE NETWORKS AND AGRICULTURAL SECTOR

In earlier years, the print media played vital role in transforming the agricultural productivity from extremely low to reasonably good point. Information dissemination using the CDs and DVDs has further strengthened agricultural output. These days' mobile and Internet based communication are precise, fast and economical mode to transfer the information from research farms and laboratories for disseminating agriculture related information to farmers.

Chapter 4 titled "Mobile Networks and Indian Agricultural Sector" presents information dissemination in agricultural sector for its growth using information and communication technology (ICT) as a tool as it is need of the hour. This can be achieved using information systems. ICT benefits are helpful in exchange and dissemination of information among farming stakeholders. By using the latest tool of mobile technology, farmers can get the current information related to their farming jobs around the clock and at any location, as the mobile network have touched every part/location of the India. Using the features of the mobile-phones like GPS etc. one can get the localized information. Only need is to structure the abundant information available across the various organizations. So, a mobile based agricultural information system framework 'mAgIDS' has been proposed employing the hybrid mobile application architecture approach. Client-server architecture using the location Application Programming Interface (API) has been proposed. Inference mechanism of the system has implemented on the basis of improved fuzzy rule promotion technique.

VEHICULAR AD-HOC NETWORK (VANET)

In recent years, advances in wireless communication technologies and motor vehicle industry a vehicular ad-hoc network (VANET) has triggered a significant research interest in the field of VANETs to improve road safety and motor traffic efficiency through the development of Intelligent transport systems (ITS).

VANET consists of vehicle-to-infrastructure (V2I) in which a vehicle can communicate directly with another vehicle and vehicle-to-vehicle (V2V) communications where vehicle can communicate to an infrastructure at road side supported by wireless access technologies. V2V communication in VANETs is one of the important networks establishing a network of vehicles for a specific need or situation where vehicles receive relevant traffic information using wireless communication technologies from their peer.

VANETs applications such as traffic information dissemination and vehicle collision warning are using for communication purpose on urban or highway environment without using any central base controller. By sending traffic information and warning to vehicle drivers through wireless communication system can assist with the tasks of avoiding traffic accidents and congestion on transport system. One of the main applications of VANET is in the critical medical emergency situations where there is no network infrastructure while it is critical to pass on the information to the expert for saving human lives. Now a day VANETs is a reliable network but still there are number of challenges such as provisioning of Quality of Service (QoS), high bandwidth network connectivity and security to vehicle and individual privacy (Rehman et al., 2013).

Though VANETs is an established network it is still scope of research in VANETs and include unique characteristics like High mobility with the constraint of road topology, rapidly changing network topology, frequently disconnected network, Potentially unbounded network size, Geographical type of communication, Time-sensitive data exchange, Potential support from infrastructure, Abundant energy and storage and Better physical protection (Kuk-Hyun Cho and Min-Woo Ryu 2012).

Rapid topology changes and frequent wireless link disconnection In vehicular network makes it difficult to design an efficient routing protocol for routing data between vehicles and roadside wireless network infrastructure. In VANET, routing algorithms based on greedy routing mechanism, which does not require building and maintaining routing tables such as greedy perimeter stateless routing. Greedy routing algorithms require a geometric coordinate assignment such as greedy embedding, where messages are transmitted over a distance decreasing path from the source to the destination. Various researches have been done on computation of greedy embedding but robustness of greedy coordinates in dynamic topologies has not studied thoroughly (Kuk-Hyun Cho and Min-Woo Ryu 2012).

This chapter presents state-of-the-art of VANET and discusses the related issues like network architecture; routing protocols and network security are discussed in detail. This chapter also provides an overview on current research state, challenges, potentials of VANETs as well the way forward to achieving the long awaited ITS.

Chapter 5 titled "Wireless Networks: Greedy Routing in Vehicular Area Networks" addresses VANET is a type of Mobile Ad-hoc network (MANET) that provides wireless communication between closely moving vehicles as well as between vehicles and road side fixed equipment's. It requires the vehicles to be equipped with sensors and wireless devices and fixed roadside wireless infrastructure. There are number of research issues like highly dynamic topology, signal fading, limited bandwidth, intermittent connectivity, mobility modelling, propagation model, and routing which needs attention. The scope of this chapter is focused on various advances in greedy forwarding routing techniques. Various VANET routing protocols are reviewed and compared in this chapter.

DIRECTION-AWARE ROUTING PROTOCOL FOR DELAY-TOLERANT NETWORK

Delay Tolerant Network (DTN) is evolved from the Mobile Ad Hoc Network (MANET). It is sparse and intermittently connected mobile adhoc network where reliable communication and end-to-end connectivity is not available for message transmission. In DTN, the main principle for routing message is "Store and Forward". Each node in DTN stores incoming message in the buffer and then delivered it to the other desirable node towards destination whenever contact is initiated. The Intermediate node copies incoming message and work as a relay node in order to increase the probability of message delivery. DTN is called opportunity network because the intermediate node is always searches opportunity to relay message from source to destination. So DTN are mostly vulnerable to many security threats and also gave birth to new security challenges.

Designing a fool proof routing protocols which are capable of working in homogeneous and heterogeneous environment is an important issue. During designing process for DTN some of constraints shall be considered like routing objective, resource allocation, effective buffer space utilization and energy constraints.

Routing protocols in Delay Tolerant Network: Routing in Delay Tolerant Network in a broad sense, categorize into Flooding and Forwarding protocols. Flooding protocols are based on the principal of replicating messages to enough nodes in order to deliver to the targeted destination. Forwarding protocols make use of predicted and heuristic knowledge about network to select the best path to the destination.

In Chapter 6 titled "Direction-Aware Routing Protocol for Delay-Tolerant Network", authors described a detailed architecture, design, simulation, testing and evaluation of the proposed protocol. Having evaluated these protocols in various simulators, chapter conclud that the proposed routing protocol utilized minimum energy and bandwidth and also provided secure transmission. The main applications of delay tolerant network lie in space research, military, tactical and other security sensitive operations. In these and other applications, secure and energy efficient routing is an important issue in homogeneous and heterogeneous network. This chapter contributes a research work that demonstrates the design, simulation, testing and evaluation of secure, energy efficient and environment independent routing protocol for delay tolerant network. The proposed protocol not only propose secure approach but also energy efficient approach in homogeneous and heterogeneous environment. The proposed approach utilizes vectorization technique to achieve its goals of security, energy efficiency and environment independence.

TRUST MANAGEMENT IN MOBILE AD HOC NETWORKS

The ability of communication between two nodes and also to transmit traffic beyond its radio transmission range using these intermediate nodes using wireless network technologies led to the growth of Mobile Ad Hoc NETwork (MANET). In an increasingly wired and wireless networked traffic, increased network connectivity could lead to facilitate collaboration, information sharing and enable distributed decision making. In MANETs, the distributed decision making which is the underlying concept in Network Centric Operations, should take into account trust in the elements: the sources of information, the processors of information, the elements of the communications network across which the information is transmitted, etc. Trust management is very important in MANETs and is needed when participating nodes, desire to establish a network without any previous interactions with an acceptable level of trust relationships among themselves (Jin-Hee Cho, Ananthram Swami, and Ing-Ray Chen, 2011).

In Chapter 7 titled "A Review of Trust Management in Mobile Ad Hoc Networks," authors introduce a generalized trust management framework which evaluates trust of a node on the basis of selected parameters. Authors also presents experimental results and the outcome indicates that the approach is workable and can be used to evaluate trustworthiness of a node in MANET environment. Establishing trust in Mobile Ad hoc Networks (MANETs) is a challenging task. Difficulty comes from the fact that trustworthiness depends on several factors beyond the control of environment and other participants.

Trust management consists of various activities like prediction of trust, computation of trust, propagation of trust, aggregation of trust etc. Trust management activities need to be selected in way that they ultimately yield an accurate and precise value of trust. This work presents the review of various activities pertaining to trust management to enable researchers, academicians and practitioners to identify and address trust related issues in a better way.

STOCHASTIC WIRELESS CHANNELS

In wireless networks content privacy of user data is important issues against illegal access. Various techniques are developed to secure user data using cryptographic techniques. One such cryptography technique is introduced to generate a pseudo-random noise (PN) sequence which is used for various encoding processes. Good statistical properties of PN sequence can be used to determine the strength and stability of advanced cryptographic methods.

In Chapter 8 titled "A Logistic-Map-Based PN Sequence for Stochastic Wireless Channels," authors mainly focusses on a pseudo-noise (PN) sequence generator which is constructed exploiting the features of one-dimensional chaotic systems such as a logistic map. The use of logistic map to generate strong cryptographic sequences is novel in approach in terms of its use with a range of transmission techniques for wireless communication because it is easy to conceive and requires simple devices to generate the sequence. The generated sequence is used as a spreading sequence in a DS-SS modulation multipath environment with AWGN and Rayleigh fading channel, under various channel fading conditions. The BER curves and channel impulse responses are obtained and compared with the existing linear feedback shift register (LFSR) based PN sequence and Kasami sequence. Results from this comparative analysis indicate that the proposed method generally yields a greater number of reliable, unpredictable and random bits than existing techniques under the same conditions and can be practically implemented for DS-SS scheme as a spreading sequence.

RECONFIGURABLE WIRELESS TRANSCEIVER

In Chapter 9 titled "Sensing and Performance Issues With Reconfigurable Wireless Transceiver," authors proposes the quantized data fusion sensing which is one of the type of cooperative sensing scheme used for unused spectrum sensing and acknowledgement to secondary signals. With the wide applications of wireless communication, it becomes an important factor to keep the pace of the spectrum usage for all the new applications. To improve efficiently the spectrum usage in variety of applications, where the primary users are not active all the time, secondary users can be introduce to exploit the radio frequency spectrum by not creating any interference to the Primary Users (PU). Unlike the pre-assigned spectrum allocation policy that has been adopted by various wireless technologies, in Cognitive Radio (CR), the whole spectrum is divided into many bands and corresponding channels where spectrum holes exist. The main challenge is thus to sense these holes and accommodate the secondary unlicensed users. Simulation results with error rates are improved by the activity of PU and have been presented.

WIRELESS HONEYPOT

Honeypots are trap networks used to attract attackers to understand their motives and the attacking techniques used. Honeypots have been around for number of years and are mainly used in wired environments. Wireless networks, on the other hand, are gaining popularity because of their mobility, scalability, and ease of integration with existing networks. However, due to the broadcast nature of radio frequency signals and ease of intercepting them, maintaining data security within wireless environments is a serious challenge.

WIRELESS SENSOR NETWORKS LOCALIZATION

Wireless sensor network (WSN) consists of many sensors deployed in a certain area to monitor or detect an event or some events depending on the application and sensor specifications; sensors can detect temperature, sound, pressure, etc. Wireless sensor networks (WSNs) are composed of hundreds, possibly thousands, of tiny low-cost and smart devices called sensor nodes that are capable of measuring various physical values, this sensors are deployed in large numbers, and they are communicating with each other and organizing themselves in order to cooperatively achieve a desired task.

Localization is an important problem in wireless sensor networks (WSN) which determine the geographical locations of sensor nodes, consequently detecting the event location and to initiate a prompt action whenever necessary. The localization process passes with three phases distance and/or angle estimation phase, position phase and algorithm phase.

In Chapter 10 titled "Wireless Sensor Networks Localization," authors includes a classification and overview of the most popular localization techniques in each phase, also high lighting the problems of these techniques and their solutions in such a way that helps in understanding localization concepts especially with respect to WSN localization. Localization is a fundamental problem in wireless sensor networks. Localization means the determination of geographical locations of sensor nodes, consequently detecting the event location and to initiate a prompt action whenever necessary. The localization process passes with three phases distance and/or angle estimation phase, position phase and algorithm phase. There are many techniques which can be used in each phase, some of these techniques may add additional devices, cost, power consumption, or delay to the network.

SELF-ORGANIZING NETWORK

Due to latent complexity of cellular mobile communication systems, operating radio networks is a difficult task. This complexity arises from the number of network elements and interconnections between their configurations. In a heterogeneous network, it is difficult to handle the variety of technologies and their precise operational paradigms. Before Self-organising networks (SON) planning and optimisation tools are typically semi-automated and management tasks need to be tightly supervised by human operators. This manual effort by the human operator is time-consuming, expensive, error prone and requires a high degree of expertise (Xin Su, Jie Zeng and Chiyang Xiao, 2014).

Self-organizing networks (SON) concept has established only after the evolution 4G and can be used to reduce operating costs by reducing tasks at hand and to protect proceeds by minimising human error. SON is to realize the autonomic management function of the wireless network by self-configuration, self-optimization and self-healing, which reduces the human intervention and improves the user experience. Small and moving cells will increase due to the frequency property and data offload in 5G networks. These changes will cause neighbour cell lists to be changed more frequently.

In Chapter 11 titled "Self-Organizing Network Solutions: A Principal Step Towards Real 4G and Beyond," authors presents a detailed discussion in self-organizing network (SON) methodology, as a novel solution to deal with network challenges, e.g. inter-cell interference coordination (ICIC), mobility, power control, etc. to improve the network quality and capacity. Quality of service (QoS) and network capacity are being insisted as the two dominant factors for the utmost network satisfaction within any mobile network contracts. On the other hand, the heterogeneous network (HetNets), which are constructed based on sub-network layers' cooperation between macrocell and shorter-range applications like micro, femto and relay nodes, are also introduced as an open door to the recent researches towards the desired network satisfaction for the recently upgraded networks like LTE-Advanced (LTE-A). Nevertheless, since any network cooperation is expected to include a number of challenges; this cooperation is not excluded of dealing with degrading effects, such as interference, among the sub-network elements.

Dharm Singh
Namibia University of Science and Technology, Namibia

REFERENCES

Cho, J.-H., Swami, A., & Chen, I.-R. (2011). A Survey on Trust Management for Mobile Ad Hoc Networks. *IEEE Communications Surveys and Tutorials*, *13*(4), 562–583. doi:10.1109/SURV.2011.092110.00088

Cho, K.-H., & Ryu, M.-W. (2012). A Survey of Greedy Routing Protocols for Vehicular Ad Hoc Networks. *Smart Computing Review*, *2*(2), 125–137.

Cisco visual networking index: Global mobile data traffic forecast update, 2015–2020. (2016). Retrieved from http://kom.aau.dk/group/16gr1022/Data_Stuff/Cisco_Mobile-growth.pdf

IEEE. (2009). IEEE Std 802.11n - Part 11. Wireless LAN Medium Access Control (MAC) and Physical Layer (PHY) specifications, Amendment 5. *Enhancements for Higher Throughput*. IEEE.

IEEE. (1999). *IEEE Std 802.11 - Part 11, Wireless LAN Medium Access Control (MAC) and Physical Layer (PHY) Specifications*. IEEE.

IEEE. (2005). *IEEE Std 802.11e - Part 11 (2005): Wireless LAN Medium Access Control (MAC) and Physical Layer (PHY) specifications, Amendment 8: Medium Access Control (MAC) Quality of Service Enhancements*. IEEE.

Ke, C. H., Shieh, C. K., Hwang, W. S., & Ziviani, A. (2008). An Evaluation Framework for More Realistic Simulations of MPEG Video Transmission. *Journal of Information Science and Engineering*, *24*(2), 425–440.

Klaue, J., Rathke, B., & Wolisz, A. (2003). EvalVid - A Framework for Video Transmission and Quality Evaluation. *Proc. of the 13th International Conference of modelingtechnques and tools for computer performenceevalution.*

Mobile device protection and security threat measures. (2003). Retrieved February 21, 2017, from http://searchmobilecomputing.techtarget.com/guides/Mobile-device-protection-and-security-threat-measures

Nsnam Web Pages. (n.d.). Retrieved February 27, 2017, from http://www.isi.edu/nsnam

Rehman, Khan, Zia, & Zheng. (2013). Vehicular Ad-Hoc Networks (VANETs) - An Overview and Challenges. *Journal of Wireless Networking and Communications, 3*(3), 29-38. DOI: 10.5923/j.jwnc.20130303.02

Su, Zeng, & Xiao. (2014). Key Technologies for SON in Next Generation Radio Access Networks. *Journal of Communication and Computer, 11*, 448-459. doi: 00510.17265/1548-7709/2014.05

Wang, Q., Dai, H.-N., & Zhao, Q. (2013). Eavesdropping Security in Wireless Ad Hoc Networks with Directional Antennas. *Proceedings of IEEE WOCC*. doi:10.1109/WOCC.2013.6676462

Chapter 1
A Framework for Evaluation of Video Quality of Service in Wireless Networks

Dharm Singh
Namibia University of Science and Technology, Namibia

Lal Chand Bishnoi
Uttrakhand Technical University, India

ABSTRACT

Today's IEEE 802.11n WLAN are capable of delivering the essential bandwidth for video by using MPEG-4 or H.264 codec compression. However, 802.11n devices are popular for delivering wireless video transmission. Even though there are some challenges for these devices specifically severe network congestion that degrade the quality of video transmission. For assessing the video transmission quality during a simulation on 802.11n WLAN technology, we extended a framework and toolset (802.11nMyEvalVid) which can be used to measures the QoS constraints in wireless networks, such as throughput, delays, and end-to-end delay. However, it also supports PSNR, which is a received video quality measuring technique based on a comparison of frame-by-frame. This chapter focused on a framework 802.11nMyEvalVid that can be used for research and evaluating new techniques for MAC-layer optimizations for simulating MPEG-4, H.264/AVC and H.264/SVC video over 802.11n WLAN in a more efficient and reliable way.

INTRODUCTION

These days' video traffic is rapidly increasing in wireless networks, and video traffic is different from ordinary data traffic that has higher data rates and tighter delay constraints. According to Cisco's annual Visual Networking Index Forecast, it is estimated that internet video streaming and downloads will soon take over three-fourth of consumer traffic by 2019 ("Cisco Visual Networking Index: Forecast and Methodology, 2014-2019 White Paper," 2014). In 2019, IP video traffic will be 80 percent of every customer's Internet traffic globally, which was 67 percent in 2014. Video exchanged through peer-to-peer (P2P) file sharing is not in this percentage. It is expected that in 2019, the Internet traffic linked through wireless

DOI: 10.4018/978-1-5225-2342-0.ch001

networks may reach up to 75 percent, most of this traffic (66 percent) may depend on Wi-Fi networks. Network management will be even more complex due to anticipated growth in online video streaming, video conferencing, and other IoT equipment's, such as health observations.

Heavy load video data rate over the wireless link cannot support continuous video playback. Therefore, more research needed to be done on video traffic over wireless networks. Nowadays most organizations are using IEEE 802.11n access points (APs). Even though IEEE 802.11n provides more than 300mbps connectivity but we need some mechanism for video transmission over IEEE 802.11n as there are still some challenges for these devices specifically severe network congestion that degrade the quality of video transmission. The 802.11n is backward compatible with earlier versions also.

IMPORTANCE OF SIMULATORS

Nowadays, in advanced network research, the simulation software and tool-set are necessary and useful. In eminent journals and conferences, the published research papers are using computer simulation framework getting new techniques for network performance improvement. As a survey of 2246 research papers published in the eminent IEEE journals, and conferences says that computer simulation was used by more than 51 percent of all published papers. (Pawlikowski, 2003).

The (Pawlikowski, 2003) reports say that improvement in latest video communication over an IP network simulation framework that may be applied to multimedia research and its applications such as video course network and digital libraries. The framework supports delivery of MPEG-4, H.264/AVC, and H.264/SVC video that is accessed from video server through various type of networks i.e.; Ethernet and wireless. It also concentrates about dissimilar QoS facility over IP hybrid networks during video transmission. The main aim of the simulation framework is to improve interoperability.

The developed framework will help the researchers to do more research on video transmission over IEEE 802.11n and earlier versions also.

QOS ON AN IP-NETWORK

The services in which current Internet system assigned similar priorities to all packets during the network are known as the best-effort service. Differentiated Services (DiffServ) model RFC 2475 for the Internet (Blake et al., 1998) has been defined by the Internet Engineering Task Force (IETF) to support the multimedia traffic over the Internet and facilitate true end-to-end QoS on an IP-network as shown in Figure 1. In this model, IP field based type of services (ToS) and traffic is considered by intermediate systems with a comparative priority.

Figure 1. Differentiated service code point (DSCP)

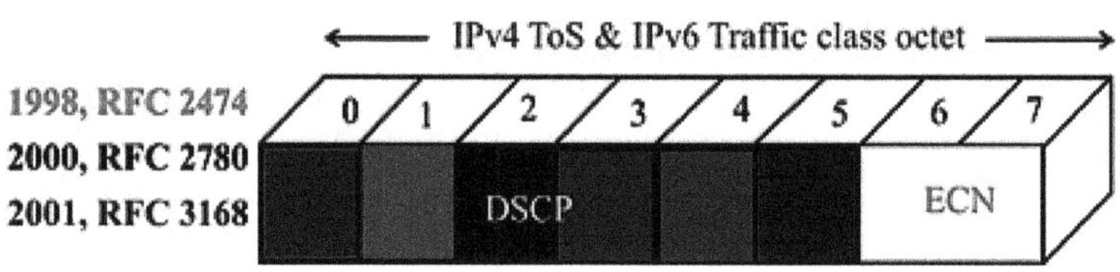

RFC 2474 defined a differential service (DS) field in the lower 6 bits of the ToS field, and the encoding of the DS field is called differentiated service code point (DSCP). In the DiffServ field architecture, the last two Currently Unused (CU) bits in the DiffServ field were not defined. As defined in RFC3168 (Ramakrishnan, Floyd, & Black, 2001), DSCP and CU are now used as Explicit Congestion Notification (ECN) bits. In a DiffServ network, classification of packets is marked by either DSCP value or IP priority value through the router at the boundary of the network.

The Network Simulator

The Network Simulator 2 (NS2) ("Network Simulator (NS2.29.3)," 2011) is freeware simulation tool used in network research both in wired and wireless. For removing the manpower effort duplication during network research and new mechanism investigation, the NS-2 was released by the Virtual Inter-Network Testbed (VINT) project ("The Network Simulator - ns-2," 2009).

The NS2 simulation framework is commonly used in network research work. Still NS2 is very popular among network researchers as we can view from Figure 2. These days over 2846 NS2 users downloaded NS2 in a week ("SourceForge: NS2 downloader stats," 2014). The 33% of downloaders are from India, and 46% of downloaders are Linux users. Therefore, we have chosen NS2 simulator tool for our research work. The framework is implemented in Window environment as well as Linux environment. On window operating system the Cygwin ("Cygwin: Get that Linux feeling - on Windows," 2010) software is widely used for getting that Linux feeling on Windows environment. The framework can be installed in Window XP using Cygwin version 2.4 or 2.5 and it does not support upgraded version of Cygwin. It would not contain full facility, which is provided by Linux environment. However, Cygwin based simulation framework is slow compared to Linux. Therefore, we have chosen Fedora for getting Linux platform.

By using NS2, some of the developed complete framework and toolsets for examining the transmitted video quality during network simulation are described as follows.

Figure 2. NS2 downloader during 12.11.2014 to 18.11.2014
"SourceForge: NS2 downloader stats," 2014.

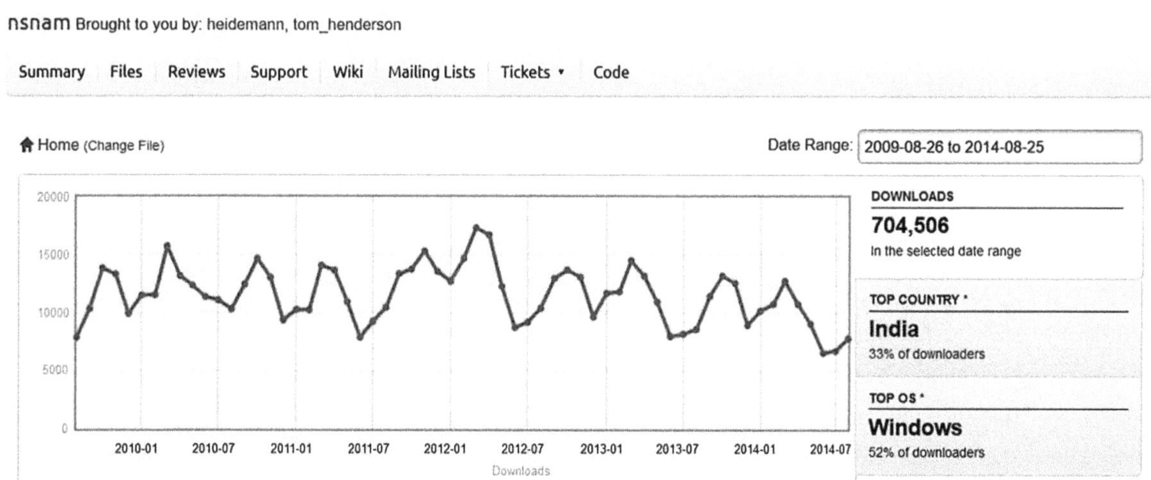

EvalVid FRAMEWORK

EvalVid introduces a network simulation framework with the tool set for examining the video quality during transmission by using NS2 (Klaue, Rathke, & Wolisz, 2003). Its simulation methodology describes the error model for finding lost or damaged packets for a real network. For getting the satisfactory quality of received video at the end user, performance metrics like packet loss or jitter are insufficient. They might arise from 30% frame error probability by the loss of 3% packets (Boyce & Gaglianello, 1998). Therefore, better-delivered video quality cannot be indicated by lower packet loss rate.

myEvalvid FRAMEWORK

The myEvalvid framework is built by integration of Evalvid and NS2 with video transmission facility. It is a complete framework for doing network researches in the field of video transmission (Ke, Shieh, Hwang, & Ziviani, 2008). The framework can be installed in Window XP using Cygwin version 2.4 or 2.5 and it is not supported by the Upgraded version of Cygwin. It would not contain all facility, which is provided by Linux. This framework designed for MPEG-4 and H.264/AVC video simulation for 802.11e technology (Ke, 2013).

myEvalSVC Framework

The myEvalSVC is integrated framework designed for H.264/SVC video over 802.11e WLAN (Ke, 2012a). It is installed in Fedora environment. It is used for SVC video simulation for temporal scalability. The streaming Evaluation Framework (SVEF) toolset used for H.264 Scalable Video coding and integrated with NS2 for simulation.

802.11nMyEvailVid FRAMEWORK

We upgraded the NS2 framework ("Network Simulator (NS2.29.3)," 2011) with the integration of myEvailvid (Ke, 2013), and myEvalSVC (Ke, 2012a) and finally proposed a new framework for IEEE 802.11n WLAN standard under Fedora ("Fedora 16 (32 bit)," 2011) environment with the help of VMware Workstation virtual machine ("VMware Workstation," 2011) framework called **802.11nMyEvalVid**. This framework is capable of simulating MPEG-4, and H.264/AVC video for IEEE 802.11e/n WLAN standards. We also made necessary enhancement in this framework for H.264/SVC video simulation.

SIMULATION STEPS OF FRAMEWORK

Due to rising demand video transmission over the wireless network and rapid growth of the WLAN has experienced magnificent attention from industry, hotels, and academic. During communication of real-time video generally has the larger bandwidth requirements, and it produces the delay during communication. Video transmission over a WLAN puts many QoS challenges. To overcome these challenges extensive research has been conducted in the various areas of wireless video applications, but still it requires more research to do. Therefore, we require a framework for simulating our research proposed algorithms by using latest available WLAN and video technologies. Nowadays, the latest popular WLAN technology (e.g. IEEE 802.11n) for most-recent video compression (e.g. H.264/SVC) are market demands.

The NS2 is most popular and demanding open source for simulating network's problems either wired or wireless. During our research, We have developed a new simulation framework by using Network Simulator 2 (NS2) ("Network Simulator (NS2.29.3)," 2011). We upgraded this framework ("Network Simulator (NS2.29.3)," 2011) and proposed a new framework for the IEEE 802.11n WLAN standard under Fedora ("Fedora 16 (32 bit)," 2011) environment with the help of VMware Workstation virtual machine ("VMware Workstation," 2011) framework called as **802.11nMyEvalVid**. This framework is capable of simulating MPEG-4, and H.264/AVC video for IEEE 802.11e/n WLAN standards. In further enhancement also incorporated necessary updating for H.264/SVC video simulation.

Steps for Developing Simulation Framework Video Transmission

This section is aimed at dissemination of the contributions a simulation environment for video transmission over the IEEE 802.11n wireless network in Fedora's environment. This section includes major steps for installing and preparing the simulation framework.

Pre-Installation Work for NS2

After installing Fedora 16 (32 bit) ("Fedora 16 (32 bit)," 2011), require following packages to be installed before NS2 installation.

- Packages for compilation NS2 source code, graphics display: gcc, gcc-c++, compat-gcc-34-c++, python, tcl-devel, libX11-devel, libXt-devel, libXmu-devel, xorg-x11-proto-devel, gtk2, gtk2-devel, tcpdump, automake, autoconf, gnuplot, make automake.
- Packages required for X264/AVC Encoder: yasm.
- Packages for MP4BOX Toolset: freeglut, freeglut-develd.
- Packages for running MS Window Exe Tool Set files: apt, wine

Installation of NS2.29

Network simulator installation requires following steps:

- Download NS2.29 ("Network Simulator (NS2.29.3)," 2011) and extract Tar file.
- Set environment variable as: export CC=gcc34 CXX=g++34
- Set path library and another application path in ***.bashrc*** Profile file
- Install NS2.29 on home directory.
- Download and install patch file for NAM ("Patch file for NAM,").
- Download and install patch file for the wireless application ("Patch file for Wireless Application," 2011).
- Replace the ns-2.29 from WMNLAB (Wang & Wei, 2013) for converting 802.11e technology simulation into 802.11n technology simulation.

Installation of Software Toolset Used for Video Processing

After NS2.29 installations, we require install other simulation toolset software's those are used for video processing are described as follows:

- **wine:** Download and install Microsoft window environment application packages for running **.EXE* files in fedora environment. Use two fedora packages (*apt* and *wine*) for this step.
- **FFmpeg:** Download and install FFmpeg tool-set software from ("FFmpeg (GitHub)," 2011) for video coder and decoder.
- **MP4Box:** Download and install MP4Box tool-set software from ("MP4Box," 2014) for MPPEG-4 raw video coding and decoding.
- **EvalVid:** Download and install video quality tool-set from (Klaue, Rathke, & Wolisz, 2014) for processing and examining video quality.
- **myEvalvid:** Download and install necessary steps for integrating MPEG-4 video transmission with NS2 from (Ke, 2013).
- Download and install necessary steps for integrating H.264/AVC video transmission with NS2 from (Ke, 2006).
- Download and install SVC reference software JSVM from ("JSVM Reference Software," 2013; VCEG, 2007) for encoding and decoding SVC video. It also used for getting video quality parameters such as Y-PSNR ("JSVM Software Manual," 2013).
- Download and install necessary step for integrating H.264/SVC video transmission with NS2 from myEvalSVC (Detti et al., 2009; Ke, 2012a; "SVEF: Scalable Video-streaming Evaluation Framework," 2012).

Video Test Sequences for Simulation

After preparing simulation framework as per above two steps. We require video YUV test sequences for simulation in proposed simulation framework. Most of the Video sequence are available in two formats QCIF (Quarter Common Intermediate Format) (176x144), CIF (Common Intermediate Format) (352x288). These sequences are available in 15 frames per second (15fps) as well as 30 frames per second (30fps) recorded samples. The various video test sequences are downloaded from as follows:

MPEG-4 and H.264/AVC Video Test Sequences

The MPEG-4 and H.264/AVC videos YUV test sequences download from ("MPEG-4 video trace file," 2010; "YUV test sequences," 2014; "YUV Video Sequences," 2014). The raw video in YUV format (QCIF or CIF) available here in 30 fps and directly used to our stimulating framework process as a source video.

H.264/SVC Video Test Sequences

The H.264/SVC video YUV test sequences downloaded from ("SVC: Original YUV sequences in 4CIF and CIF resolution," 2014; "Video Library and Tools - NSL (Raw Sequences from the H.264/SVC Group)," 2015). The raw video in YUV format (QCIF or CIF) available here in 15/30 fps and directly used to our stimulating framework process as a source video.

Simulation Framework Architecture

There are various components of 802.11nMyEvalVid Framework Architecture, as shown Figure 3. The simulation requires various fixed steps in a particular order during simulations. The details of framework components are described as follows:

Video Source

The video source generates video streams in YUV format. This YUV format is two types: CIF (352x288), and QCIF (176x144).

Video Encoder and Decoder

It is used for video encoding and decoding purpose. The video converted into YUV format to H.264/MPEG-4 format at the sender side. For this, purpose video encoder is used. The video converted into H.264/MPEG-4 to YUV format at the receiver this purpose video; decoder is used. ("FFmpeg," 2011).

Video Sender (VS)

It reads the compress video file from the output of the video encoder as an input. After inputting the file, it checks the size of video frames and compares with maximum transmission unit (MTU). If video

Figure 3. 802.11nMyEvalVid framework architecture

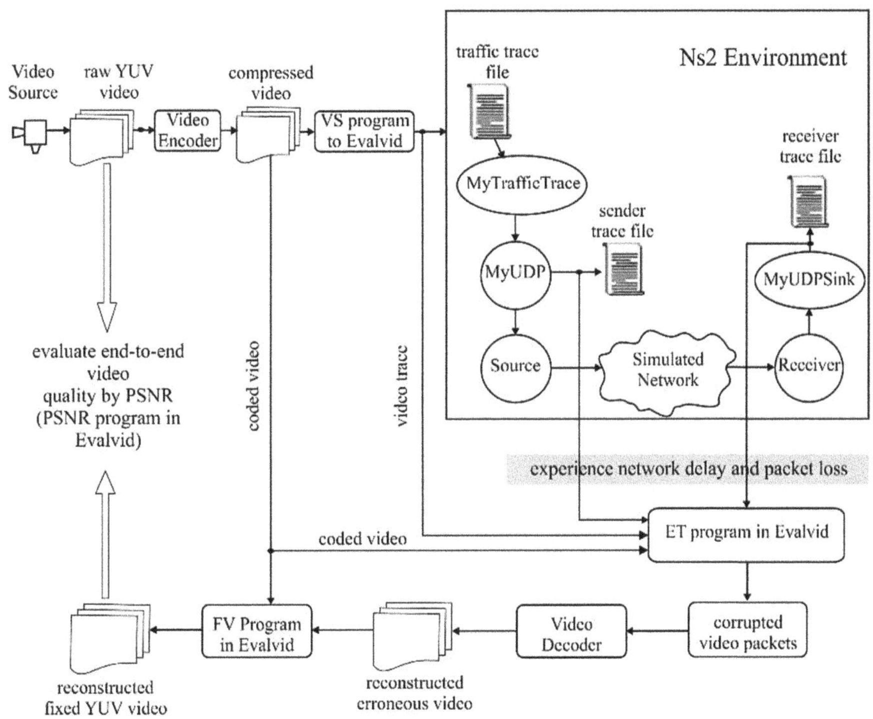

frames are larger than MTU, it split them into small segments. After that, it transmits split frames over a simulated network in the form of packets by using User Datagram Protocol (UDP) or Internet Protocol (IP) protocols. This framework gives video trace file that contained the information about all transmitted UDP packets such as packet id, timestamp, and packet payload size. Such video traced file is known as the sender trace file (*.st file).

Video Reconstruction

After the completion of video communication over a simulated network; the performance rating begins by using evaluate trace (ET) program.

Evaluate Trace (ET)

It reconstructs the video by using four input files; coded original video, video-trace, sender-trace, and receivers trace. It generates the report that contains the information about packet/frame such as loss, delay, and jitter. Other than this information, it also generates the reconstructed video file. There is a possibility of corrupt video may be generated on the receiver side then ET omitted the lost or corrupted frames by copying original file frame by frame.

Fix Video (FV)

The video quality (such as PSNR) is assessed by comparing received video with transmitted video frame by frame. During transmission if the video decoder not able to manage the lost frame, then such problem can be solved by replacing each missing frame with inserting previous successfully decoded frame.

SIMULATION STEPS FOR MPEG-4 VIDEO

The simulation step for Running MPEG-4 video shown in Fig. 4 and batch file of Fedora operating system described in Table 1.

The raw (*.yuv) given to video encoder. The *xvid_encraw.exe* or *mpeg4Encoder.exe* utility used for encoding the input video. The output of video encoder given to *MP4Box.exe* or *MP4Box.exe* utility for generating **.mp4* file. In the last step of pre-processing use mp4trace.exe utility, it gives NS2 send trace file. The format of Forman video trace file given in Table 2. These trace files are given as input of NS2. It defined frame number, frame type, frame length, total packet, and sending start time. NS2 give video data file as shown in Table 3.

In post-processing *et.exe* utility used for converting received data into video format. In next step *mp4decoder.exe* or *etmp4.exe* utility, convert received video into *yuv* format. The ffmpeg.exe utility also used in *yuv* conversion.

After post-processing step, we can use video quality measurement tools such as psnr.exe, an awk script, etc.

Figure 4. Pre-processing work on raw video

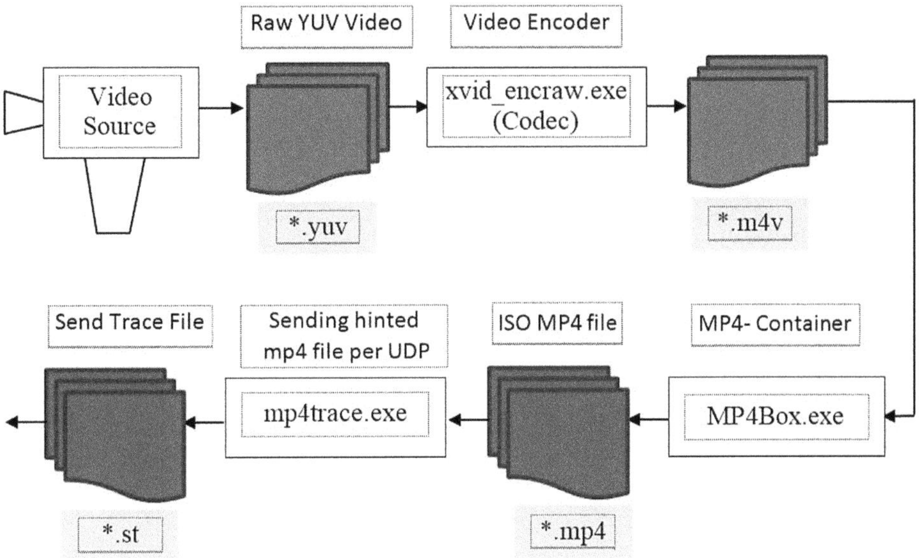

Table 1. Batch file for Testing MPEG-4 Video simulation

```
#
# Batch File of Fedora Command
# Includes Steps for simulating MPEG-4 Video
#
if [ $# -lt 5 ] ; then
 echo "It needs 5 Arguments (1:AggSize 2:Choice
3:noVoice 4:noCBR/TCP 5:noBackground/UPD)"
 echo " "
 exit
fi

#
# Adaptive Croslayer Mapping [802.11n] for MPEG-4 Video
#

./xvid_encraw.exe -i foreman_qcif.yuv -w 176 -h 144 -framerate 30 -
 max_key_interval 30 -o foreman_qcif.m4v

 ./MP4Box -hint -mtu 1024 -fps 30 -add foreman_qcif.m4v foreman_qcif.mp4

./mp4trace.exe -f -s 192.168.0.2 12346 foreman_qcif.mp4 >foreman_qcif.st
```

```
# -------- TCL Program Command Line Argumets List --------------------
# [80] CBR Interval (CBR traffic flow with 80us interval time)
# [0] Set PacketSize in bytes
# [0] Value between 0 to 3 (RD-Reverse Direction)
# [0] Block Acknoledge type (0-for None, 1-for Normal, 2-for Compressed)
# [16383] Byte, Aggregating Size (16383 Byte-Optimal Aggr Size,
# 0-means not Aggr)
#
# Choice 0-EDCA-802.11n, 1-Static, 2-ACMA, 3-DACMMD, 4-DLDCA
# NoOfVoiceFlow No of Voice traffic flow
# NoOfVideoFlow No of Video traffic flow
# NoOfTCP_Flow No of TCP (Best Effort) traffic flow
# NoOfCBR_Flow No of CBR (Background) traffic flow
```

continued on following page

Table 1. Continued

```
ns crosslayer802_11n.tcl $1 $2 $3 $4 $5

awk -f replaceHwithI.awk foreman_qcif.st > modified.st

export MALLOC_CHECK_=0

#max=$(($4 - 1))
max=0
for i in `seq 0 $max`
do
 ## It calculate Packet & Frame delay as well losses too.
 ./et.exe sd_foreman_$i rd_foreman_$i modified.st 1 1 >frameLosses$i.tr
 mv delay_1.txt delay_frame$i.tr
 mv delay_pkt.txt delay_pkt$i.tr

 #./etmp4.exe sd_foreman_$i rd_foreman_$i modified.st foreman_qcif.mp4
foreman_qcife$i
 etmp4 -F -0 sd_foreman_$i rd_foreman_$i modified.st foreman_qcif.mp4
foreman_qcife$i

 #./ffmpeg -i foreman_qcife$i.mp4 -y foreman_qcife$i.yuv
 ./ffmpeg -i foreman_qcife$i.mp4 -vcodec rawvideo -pix_fmt yuv420p -an -y
foreman_qcife$i.yuv
done
```

```
echo "***************** R E S U L T S ***************"
 echo "----------------- Throughput -----------------"
 awk -f newTP.awk out.tr > tp.tr
 gnuplot plotTP.plt

 awk -f throughputAverage.awk out.tr

max=$(($4 - 1))
for i in `seq 0 $max`
do
 echo " "
 echo "---- Frame Losses for :: video($i)----"
 awk -f frameLosses.awk frameLosses$i.tr

 echo " "

 echo "---- Average PSNR for :: video($i)----"
 ./avgpsnr 176 144 420 foreman_qcif.yuv foreman_qcife$i.yuv

./psnr.exe 176 144 420 foreman_qcif.yuv foreman_qcife$i.yuv >
psnr.tr
 gnuplot plotPSNR.plt

 #echo "---------- Packet & Frame Send-Receive Delay----------------"
 gnuplot plotPacketDelay.plt
 gnuplot plotFrameDelay.plt

 #echo "----------- Average & Maximum Packet Delay------------------"
 awk -f packetDelay.awk delay_pkt.tr

 #echo "---------- Average & Maximum Frame Delay -------------------"
 awk -f frameDelay.awk delay_frame.tr

 # average AckDelay of ACs
 awk -f avgAckDelay.awk recvAckDelay

#echo "---------- Queue Space Utilization -----------------------"
 awk '{print $1, $2}' queueLengthAC.txt > qsAC0.tr
 awk '{print $1, $3}' queueLengthAC.txt > qsAC1.tr
 awk '{print $1, $4}' queueLengthAC.txt > qsAC2.tr
 awk '{print $1, $5}' queueLengthAC.txt > qsAC3.tr
 gnuplot plotQueue.plt
done
echo " "
```

Table 2. Format of video trace file (foreman_qcif.st)

$frame_{no}$	$frame_{type}$	$frame_{length}$	$total_{packet}$	$sendingStart_{time}$
1	H	6865	7	0.042
2	P	5446	6	0.135
3	B	1442	2	0.136
4	B	1453	2	0.136
5	P	5047	5	0.245
6	B	1541	2	0.245
7	B	1552	2	0.245
8	H	9966	10	0.339
9	B	1256	2	0.339
10	B	1359	2	0.339
----	----	----	----	----

Table 3. Format of sender trace file (video.dat)

Time (msec)	$frame_{length}$	$type_{video}$	$prio_{p}$	maxFramentSize (bytes)
42000	6865	1	1	1024
93000	5446	2	1	1024
1000	1442	3	1	1024
0	1453	3	1	1024
108999	5047	2	1	1024
0	1541	3	1	1024
0	1552	3	1	1024
94000	9966	1	1	1024
----	----	-----	-----	-----

Simulation Steps for H.264/AVC Video

The simulation of H.264/AVC video almost similar as MPEG-4 video simulation. The details of all steps are described in Table 4 batch file.

Simulation Steps for H.264/SVC Video

The framework for video transmission over the wireless network in NS2 ("The Network Simulator - ns-2," 2009) on Fedora 16 (32 bit) desktop operating system ("Fedora 16 (32 bit)," 2011) environment integrated with EvalSVC (Le, Nguyen, & Zhang, 2010) and myEvalSVC (Ke, 2012a, 2012b) framework used for the simulation for this study. The jsvm9.19.14 ("JSVM Reference Software," 2013) and svef-1.5 ("SVEF: Scalable Video-streaming Evaluation Framework," 2012) reference software used for coding, decoding SVC video data during simulation. The details of all steps are described in Table 5 batch file. It uses JSVM utility tools for pre-processing and post-processing of H.264/SVC video.

Table 4. Batch file for testing MPEG-4 video simulation

```
#
# Adaptive Cross-layer Mapping [802.11n] for H.264/AVC Video
#
if [ $# -lt 5 ] ; then
echo "It needs 5 Arguments (1:Choice 2:noOfVoice
3:noVideo 4:noCBR/TCP 5:noBackground/UPD)"
echo " "
exit
fi
./lencod.exe
./parser.exe config
ns crosslayer802_11n.tcl 80 0 0 0 16383 $1 $2 $3 $4 $5
echo "-------------------------------------"
max=$(($3 - 1))
for i in `seq 0 $max`
do
./errinsert foreman_qcif.264 distorted.264 sd_foreman_$i rd_foreman_$i >
packet_losses
awk -f packetLosses.awk packet_losses
done
./ldecod.exe decoder_distorted.cfg
echo " ----------------------- PSNR & AvgPSNR --------------------------------"
./psnr.exe 176 144 420 foreman_qcif.yuv foreman_qcif_distorted.yuv > psnr.tr
gnuplot plotPSNR.plt
awk -f avgPSNR.awk psnr.tr
echo "------------------------- Throughput --------------------------"
awk -f newTP.awk out.tr > tp.tr
gnuplot plotTP.plt
awk -f throughputAverage.awk out.tr
exit
#echo "------------------- Queue Space Utilization ----------------"
awk '{print $1, $2}' queueLengthAC.txt > qsAC0.tr
awk '{print $1, $3}' queueLengthAC.txt > qsAC1.tr
awk '{print $1, $4}' queueLengthAC.txt > qsAC2.tr
awk '{print $1, $5}' queueLengthAC.txt > qsAC3.tr
gnuplot plotQueue.plt
echo " "
awk -f converCIF-ST.awk foreman_qcif.txt > modified.st
./et.exe sd_foreman_0 rd_foreman_0 modified.st 1 1
mv delay_1.txt delay_frame.tr
mv delay_pkt.txt delay_pkt.tr
./etmp4 -F -0 sd_foreman_0 rd_foreman_0 modified.st foreman_qcif.yuv
foreman_qcife
#echo "------------ Packet & Frame Send-Receive Delay----------"
gnuplot plotPacketDelay.plt
gnuplot plotFrameDelay.plt
#echo "------------ Average & Maximum Packet Delay----------"
awk -f packetDelay.awk delay_pkt.tr
#echo "------- Average & Maximum Frame Delay ---------"
awk -f frameDelay.awk delay_frame.tr
        echo " "
```

Table 5. Batch file for testing MPEG-4 Video simulation

```
#===============================================
if [ $# -lt 3 ] ; then
echo "It requires 3 Arguments as follows:"
echo " Argument-1 [Aggregation Size]::
0, 1023, 2047, 4095, 8191, 16383, 32767, 65535"
echo 0 - No Aggregation or used when Dynamic Aggregation ON"
echo " "
echo " Argument-2 [Cross layer mapping type]::
0 - Default 802.11n EDCA"
echo " 1 - LCB Cross-Layer Algorithm"
echo " "
echo " Argument-3 [Dynamic Aggregation ON/OFF]::
0 - OFF-No Dynamic Aggr."
echo " 1 - ON -Start Dynamic Aggr."
echo " "
exit
fi
```

```
#======================================================
echo "Step 07 ....[Output File(s)=> sd, rd,out.tr, video1.dat, queuelength.txt]"
ns test_svc_802_11n.tcl $1 $2 $3
```

```
#======================================================
#-------- Queue Space Pringitng ----------
awk '{print $1, $2}' queuelength.txt > AC_VO.tr
awk '{print $1, $3}' queuelength.txt > AC_VI.tr
awk '{print $1, $4}' queuelength.txt > AC_BE.tr
awk '{print $1, $5}' queuelength.txt > AC_BK.tr
gnuplot plotQueue.plt
echo "Step 08 ....[Output File(s)=> Pe2edelay.txt] Packet End 2 End Delay"
awk -f Pe2edelay.awk rd > Pe2edelay.txt
gnuplot plotPe2eDelay.plt
echo "Step 09 ....[Output File(s)=> ns2received]"
awk -f prepare_receivedtrace1.awk rd > ns2received
echo "Step 10 ....[Output File(s)=> received.txt]"
./prepare_receivedtrace2 ns2send ns2received originaltrace-frameno.txt > received.txt
echo "Step 11 ....[Output File(s)=> filteredtrace.txt]"
#Usage: nalufilter <sent stream trace file>
# <received trace file>
# <play out buffer in milliseconds>
# <frames per second> > <filtered trace file>
#
# 5000 means 5000 miliseconds for play out buffer.
# 30 means 30 frames per second for the processed video.
nalufilter originaltrace-frameno.txt received.txt 5000 30 > filteredtrace.txt
echo "Step 12 ....[Output File(s)=> test-filter.264]"
# Genrate Received Video H.264 "
BitStreamExtractorStatic test.264 test-filter.264 -et filteredtrace.txt > packetLoss.txt
echo "Step 13 ....[Output File(s)=> test-filter.yuv]"
# Generate YUV file from Received H.264 Video"
H264AVCDecoderLibTestStatic test-filter.264 test-filter.yuv >
frameRecived.txt
echo "Step 14 ....[Output File(s)=> test-conceal.yuv]"
#./myfixyuv cif 300 test-filter.yuv test-conceal.yuv
#Usage: framefiller <filtered received stream's trace>
# <bytes per frame>
# <total frames>
# <distorted YUV>
# <concealed YUV (output)>
#
#<byte per frame>:: Width*height*1.5. i.e. 38016 for QCIF(176x144),
# 152064 for CIF(352x288),
# 608256 for 4CIF(704x576),
# 4866048 for HD.
framefiller filteredtrace.txt 152064 300 test-filter.yuv test-conceal.yuv
echo "Step 15 ....[Output File(s)=> psnr.txt, psnr.tr]"
#Usage: PSNRStatic <w> <h> <org> <rec> [<t> [<skip>
[<strm> <fps> [strg]]]] [-r]
# w : original width (luma samples)
# h : original height (luma samples)
# org: original file
# rec: reconstructed file
#
PSNRStatic 352 288 foreman_cif_30.yuv test-conceal.yuv > psnr.txt
awk -f prepare_psnr.awk psnr.txt > psnr.tr
gnuplot plotPSNR.plt
# Plot Dynamic Aggregation Queue size with time varriying
gnuplot plotDyAggrSize.plt
echo ""
```

continued on following page

Table 5. Continued

```
echo "------------------------ R E S U L T S -------------------------------"
echo " "
awk -f displayAvgPSNR.awk psnr.txt
awk -f frameSendCalculation.awk encoding.txt
awk -f frameRecivedCalculation.awk frameRecived.txt
# -- Calculate frame Losses ----
awk -f frameSendCalculation.awk encoding.txt >frameLoss.txt
awk -f frameRecivedCalculation.awk frameRecived.txt >>frameLoss.txt
awk -f frameLossCalculation.awk frameLoss.txt
#-----------------------------
awk -f avgE2Edelay.awk Pe2edelay.txt
awk -f packetStatistcs.awk out.tr
awk -f packetLossCalculation.awk packetLoss.txt
#echo "------------------------------ Throughput -----------------------"
awk -f newTP.awk out.tr > tp.tr
gnuplot plotTP.plt
awk -f throughputAverage.awk out.tr
echo "-----------------------------------------------------------------------"
echo " "
echo "-----------------------------Queue Utilization----------------------"
echo "AC_VO"
awk -f quePrtUtilize.awk AC_VO.tr
echo "AC_VI"
awk -f quePrtUtilize.awk AC_VI.tr
echo "AC_BE"
awk -f quePrtUtilize.awk AC_BE.tr
echo "AC_BK"
awk -f quePrtUtilize.awk AC_BK.tr
```

VIDEO QUALITY ASSESSMENT METHODS FOR MULTIMEDIA APPLICATIONS

Nowadays videos are a significant part of media communication since most of the information are relayed via television, The Internet, etc., and video is the easiest and quickest way to make information understandable. There have been huge developments in multimedia since there is a need for improving the quality of video and the transfer speed. Increasing prevalence of mobile devices forces many stakeholders to put a lot of efforts to increase the quality of their multimedia services.

Video Compression

In the last decade, the digital video communication becomes more useful and play an important role in various prime fields of our life. Video files are transferred in real or non-real time, stored on data storage devices, etc. They differentiate from each other according to the transfer speed demanded or required an amount of free space on displayed. By the end of 2014 mobile video traffic was 67% and mobile video predicted to achieve 80% of total mobile data traffic by the end 2019 ("Cisco Visual Networking Index: Forecast and Methodology, 2014-2019 White Paper," 2014), and this is a reason to make it faster and easier by decreasing file sizes to a minimum.

For this very reason, the idea of video compression was established which is intended to adjust video file to required method of data transfer by decreasing the number of bits in an original file and expressing the exact same information without minimum or no loss. Delivery of high-quality video to the end user has been made possible by the significant advancement in compression and transmission techniques. Especially by these new technologies upcoming in the market, makes us develop lots of modern telecommunication services (e.g. digital television, direct broadcast satellite, video teleconferencing, high-definition TV, and Internet video). Digital video communication system performance can be measured by examining the changes in video quality at each stage.

The very important factor is to maintain a high quality of compressed video file. Video quality is defined as a similarity measurement of original and compressed video file, where the highest possible similarity is demanded. Video quality examination is done by directly or indirectly, comparing the test video with the original video.

Video compression is divided into lossless compression, which allows exact reconstruction of original video file after compression and lossy one, which cannot reconstruct it identically, but significant information and features are maintained. However in lossy compression, the problem of video distortions and compression artifacts occurs. These effects are caused by discarding some of the data, in order to adjust a signal to the limitation of file size or data transfer. In the situation when established or demanded limitation is not sufficient to recreate original file, then diminishing of quality, or introduction of artifacts appear. This degradation of video quality may occur during capturing, compression, transmission and displaying (Shahid, Rossholm, Lövström, & Zepernick, 2014). In order to minimize distortion effects, established video coding standards were there, which ensure effective compression performance.

Advanced Video Coding (H.264/MPEG-4 AVC) is the most popular video coding mechanism used during the video communication. AVC standard was very innovative compared to previous standards (Ries, Nemethova, & Rupp, 2007). The final draft was provided in 2003, and then many extensions have been added in subsequent editions (Ries et al., 2007). Nowadays H.264 is the most common video standard, which is used in web software like the Adobe Flash Player and Microsoft Silverlight and also in High Definition TV. All video files saved on Blu-ray Discs are decoded in this standard. Also, the most important video-sharing websites use this standard in theirs video files, e.g. YouTube, Vimeo and iTunes.

Video Quality Assessment

As high perceptual video quality is demanded nowadays, many efforts were put to develop Video Quality Assessment (VQA) methods. Presented below is an overview of quality metric methods, which allow understanding basis of the proposed prediction model.

Overview of Video Quality Assessment Methods

The quality of a video can be most accurately determined by computing it using psychophysical examine, related with a human.

Assessment of video quality conducted in two ways, subjective and objective. Subjective assessment is mostly conducted in a laboratory experiment, where human viewers assess different videos by scoring each of them on a questionnaire. Results are computed as mean opinion score (MOS). The Video Quality Assessment methods showed Figure 5.

Subjective Video Quality Assessment

Testing of video quality procedures was recommended by the International Telecommunication Union (ITU). Recommendation ITU-T P.910 ("Recommendation ITU-T P.910 (04/2008): Subjective video quality assessment methods for multimedia applications," 2008) has proposed assessing non-interactive subjective method for examining video quality for multimedia application such as storage and retrieval applications, video conferencing, telemedical applications, etc. During video connection, these methods are mainly used for grading video system performance, algorithm selection and measuring video quality

Figure 5. Video quality assessment methods

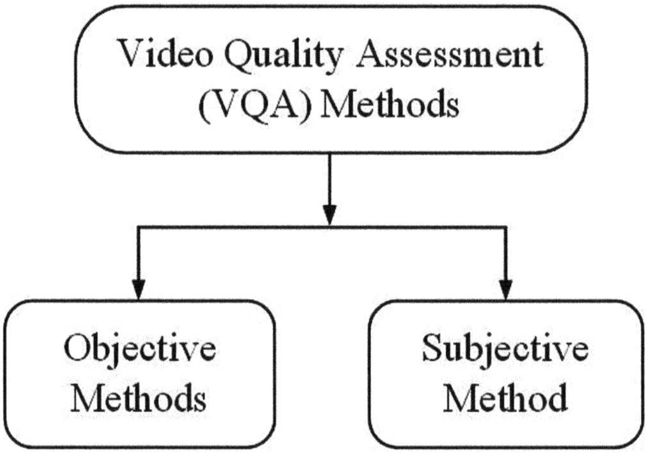

level. The subjective experiments are used for measuring video quality more precisely. These experiments use a set of video test sequence for examining about its quality. For each set of sequence, the average value collected known as Mean Observer Score (MOS).

In general, these psychophysical experiments are very complex, time-consuming and expensive, and this is a reason for using objective methods instead. A great amount of the experimenter's time is consumed in the execution, design and data analysis. During experiment execution, it requires the accessibility of equipment, subjects, and physical space. Therefore, the ability to efficiently and accurately measuring of video quality, and considering the experiment number limitation are also considered. Thus, for getting the most out of the resources without using human observers, an appropriate methodology should be used.

The fast algorithms are used for physical measuring of video quality when videos transmitting, receiving or viewing methods are performed. A few objective quality measures applied on video quality measuring techniques such as mean square error (MSE), the mean absolute error (MAE), and the peak signal-to-noise ratio (PSNR). Results of these methods are always not a perfect judgment by a human.

Objective Video Quality Assessment

The video quality metrics are mainly applied for; observing the quality of the video, comparison of Video algorithms and processing systems performance, and dynamically improving parameters' settings and algorithm for video processing system. By using the application necessities and restrictions, the metric type is decided. By enlarging the quality of video metrics, categories in three types depends upon availability of reference (original) video signal which are:

The new qualities of service are considered in digital television, with complex associations between picture quality for objective and subjective parameter measurements. For getting optimal quality in the cable television system, there should be a good correlation between the measurement of objective and subjective quality of the video. It is also considered that the measurements of objective quality cannot be replaced by subjective quality judgment. Subjective quality assessments are considered according to the requirement of human viewers, and consequently particular video set of sequences are applied to an application ("Recommendation ITU-T J.144 (03/2004): Objective perceptual video quality measurement techniques for digital cable television in the presence of a full reference," 2004).

Therefore, subjective and objective quality assessment measurements are opposite rather than exchangeable. Objective measurements have equipment specification demands and require monitoring of system performance measurement day-to-day, whereas subjective assessments are suitable for research areas. The assessment of objective video quality mechanism can be further divided into three groups: reduced reference, full reference, and no reference metrics as shown in Figure 6 (Shahid et al., 2014).

For compressed image transmission, especially in broadcasting the QoS is controlled by objective video quality assessment metrics. There are three basic Image quality methods to perform objective measurements:

Full Reference (FR) Methods

On the availability of full reference video signal the full reference mechanism can be applied. It is a measuring method with full reference, which has similarity on both sides. It objectively measures perceptual video quality. During primary input of the system, the original video signals are compared with undistorted input or reference input and degraded video signals at the output of the system are obtained. By this process, we can be evaluating the performance of systems. It works through the comparison of frame-by-frame among the referenced input video and received video. For the quality assessment, the original video is needed to be in an unimpaired and uncompressed form. Furthermore, due to the expectation of two videos' alignment by precise spatial and temporal property and subsequently in real-

Figure 6. Objective video quality assessment methods

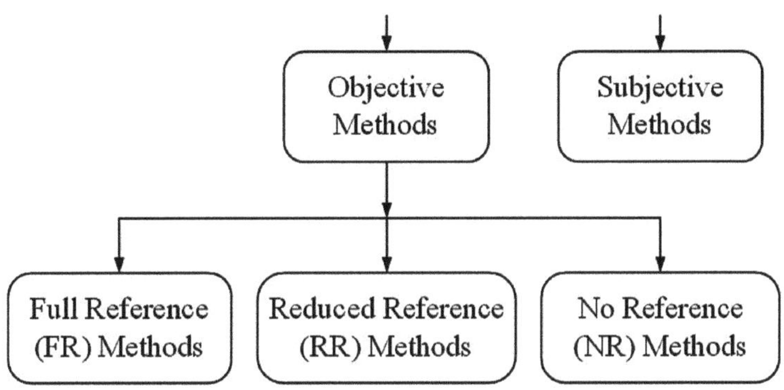

Figure 7. Full reference metric

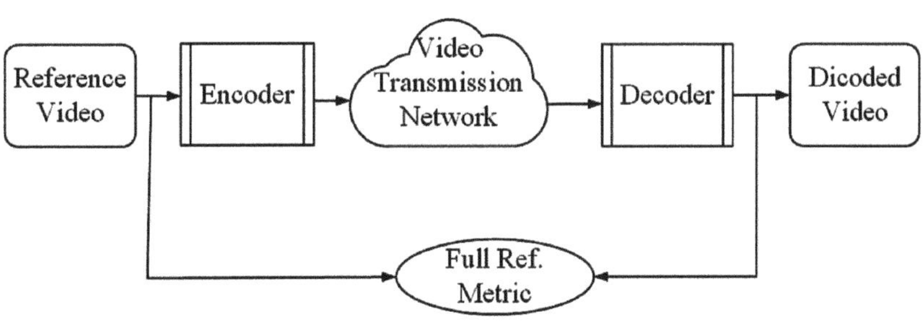

time applications it also becomes impractical. The Full Reference methods based applications are MSE/PSNR and HVS as shown in Figure 7.

Reduced-Reference (RR) Methods

On the availability of reduced video reference data, the reduced-reference mechanism can be applied. It is also a double-ended method and has no access to original video; use only reduced information about the original videos, such as some features values or characteristics. The overview of RR metrics as shown in Figure 8. It trusts on back channel access to reference video even though it has benefit of partially arranged requirements.

For evaluating the performance of video quality metrics on an objective basis the RR metric commonly preferred metrics used are Spearman Rank Order Correlation Coefficient (SROCC), Root Mean Square Error (RMSE) (Chikkerur, Sundaram, Reisslein, & Karam, 2011), Pearson Correlation Coefficient (PCC) and Outlier Ratio (OR).

No Reference (NR) Methods

A mechanism observes the test video (degraded video signal or coded video) with no reference video. This method ends on a single point and has no access to original video; use only information of distorted

Figure 8. Reduced reference metric

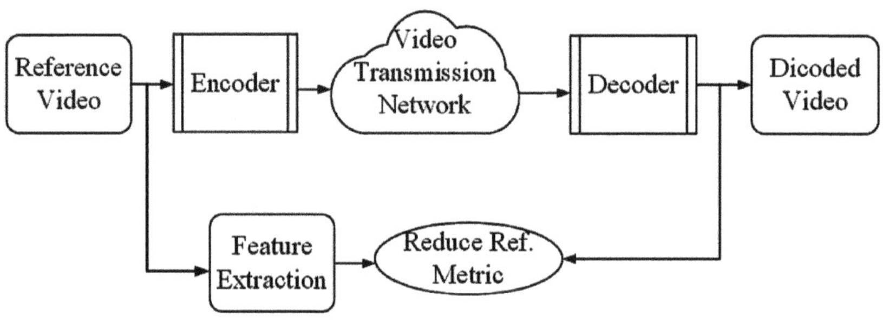

Figure 9. No reference metric

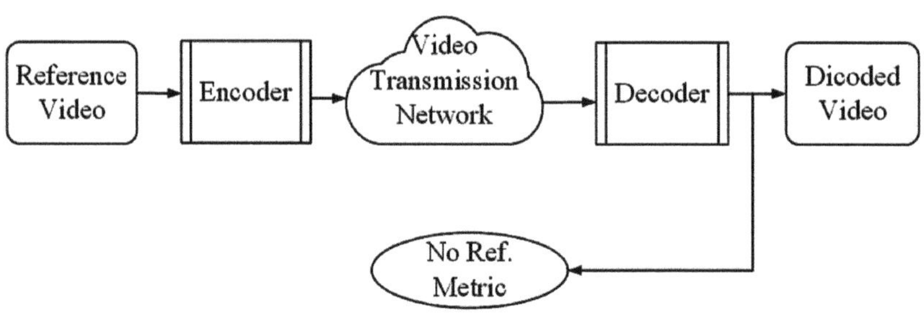

video file. In comparison to FR method, such method gives more flexibility and is free from positioning issues. NR method each time makes predictions about the distortions and video images. The overview of NR Metric shown in Figure 9.

This method is useful for examining for real-time measurement demands for the in-service video system.

- **Pixel-Based Methods:** Estimation is performed on the basis of statistics derived from pixel-based features.
- **Bitstream Based Methods:** Estimation quality of the video is made with the ground from bit-stream parameters and achieved features values such as motion vectors, coding modes, QP values.
- **Hybrid Methods:** Combination of two previous methods. By selecting characteristics from the bitstream using extraction and pixel domain depend on requirements of the specific case.

All above methods are used by various applications for giving several levels of finding correctness of the application, that expressed in terms of subjective rating results by correlation.

Video Quality Assessment Methods

Distortion produced in processed video during video coding and transmission. MSU is an important video quality measurement tool. Due to simplicity, the Peak Signal to Noise Ratio (PSNR) is commonly used for video quality measurement. However, it is known by each that PSNR does not essentially correctly model for perceptual quality. This section also discusses video quality measuring latest approaches proposed in the last decade and provides a brief explanation of each video quality method and the main video components it examined by them.

The video quality measuring metrics includes Video Quality Metric, PSNR, Single Scale Structural Similarity, and Mean Square Error. The wireless network quality measured by using Throughput, end-to-end delay, and Packet loss metrics.

Figure 10. Overview of VQM
Xiao, 2000.

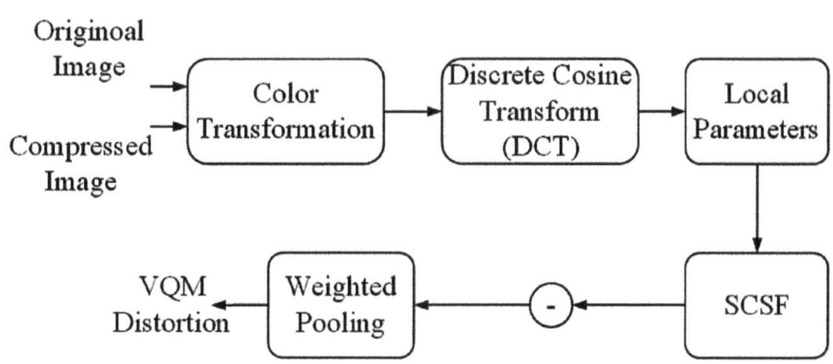

Video Quality Metric (VQM)

VQM metric developed in 1991 by using Digital Video Quality (DVQ) and Discrete Cosine Transform (DCT) measurements. These DVQ and DCT indicate a 95% correlation to the particular conditions. The overview of VQM shown in Figure 10.

As an input original image and compressed image are given. The Color transformation used raw YUV data from the input section. DCT transform divides input images into various spatial frequency components. After that by using local contrast (LC) it converts all DCT coefficients as shown in following Eqn. (1).

$$LC(i,j) = DCT(i,j)\frac{\left[\frac{DC}{1024}\right]^{0.65}}{DC} \tag{1}$$

where DC is 8-bit image DC component of each block and 1024 is DCT value. Afterward, this maximum values remains between [-1, 1]. All above steps are equal to Watson's DVQ model (Xiao, 2000).

In next step, the static spatial contrast sensitivity function (SCSF) matrix is calculated. The DCT coefficients are multiplying with the corresponding element in default image quantization matrix. All entry in the static SCSF matrix is generated by a power of account for the temporal characteristics of SCSF, and finally, power is determined by the frame rate of input video sequences. Two sequences are subtracted first for weighted pooling, and then VQM calculated by using formula mention in Eqn. (2).

$$VQM(x,y) = 1000\left|\overline{diff}\right| + 0.005\left(1000\max\left(\left|diff\right|\right)\right) \tag{2}$$

where maximum distortion weight 0.005 and standardization ratio is 1000.

Mean Square Error (MSE)

The MSE is the cumulative square among real and the compressed video. For a given matrix, size $m \times n$, which is both input, frame "I" and its coded representation. The MSE is a simple metric for calculating the difference among I and K. The MSE equation shown in Eqn. (3).

$$MSE = \frac{1}{m.n}\sum_{i=0}^{m-1}\sum_{j=0}^{n-1}\left[I(i,j), K(i,j)\right]^2 \tag{3}$$

where $I\left(i,j\right)$ is the actual input signal of the pixel $\left(i,j\right)$, $K\left(i,j\right)$ is the regenerated signal, and picture size is $M \times N$. The MSE calculated by using MSU tool set (Skupin, Hellge, Schierl, & ThomasWiegand, 2011).

Peak Signal-to-Noise Ratio (PSNR)

PSNR is measured by calculating the MSE in the context of the highest possible value of the luminance (Rossholm & Lovstroem, 2008), which detained by matches the error energy with signal energy. It matches the highest signal energy MAX_I with receiver end calculated error energy $\sqrt{MSE}$. It can be representation shown in Eqn. (4) (Skupin et al., 2011).

$$PSNR = 10.\log_{10}\left(\frac{MAX_I}{\sqrt{MSE}}\right) \tag{4}$$

The result denoted by decibels (dB). The low to high-quality video PSNR ranges between 30-40. Even though several video quality measuring models designed in past years, but PSNR is most popular and still continuing for evaluating video quality.

An alternative option for calculating the Mean Opinion Score (MOS) is that first determining the frames' percentage with an MOS worse than the sent video. The distortion caused by the network can be displayed by this method at a single look. The MOS values are calculated based on the PSNR above Eqn. (4) for MOS translation from Evalvid (Klaue et al., 2003) are given in Table 6.

The precise method to measure average PSNR for a sequence is to compute all frames average MSE (where average MSE is the mathematics mean of the MSE values for frames) and after that PSNR calculated by using the ordinary method used for PSNR. The PSNR calculated by using MSU tool set (Skupin et al., 2011).

Structural Similarity Index Measure (SSIM)

SSIM is a technique of likeness computation between two images. SSIM uses structural distortions to estimate perceptual distortion on the basis of luminance comparisons. To find the likeness between an input image and the relative reference, the SSIM method calculates the contrast $c(x,y)$, structure $s(x,y)$,

Table 6. PSNR to MOS translation

PSNR (dB)	MOS
> 37	5 (Very Good)
$31-37$	4 (Good)
$25-31$	3 (Not Good)
$20-25$	2 (Poor)
< 20	1 (Bad)

Klaue et al., 2003.

and luminance $l(x,y)$ of the input image y and the relative reference image x, described in following Eqn. (5) (Chikkerur et al., 2011):

$$SSIM(x,y) = \left[l(x,y)\right]^{\alpha} \cdot \left[c(x,y)\right]^{\beta} \cdot \left[s(x,y)\right]^{\gamma} \tag{5}$$

In above equation β, and γ are used as importance coefficients that describe the comparative important status of the contrast, structure, and luminance components.

$$l(x,y) = \frac{2\mu_x\mu_y + c_1}{\mu_x^2 + \mu_y^2 + c_1'} \tag{6}$$

where μ_x, μ_y are mean values of luminance samples of comprising signals x and y as Eqn. (6).

$$c(x,y) = \frac{2\sigma_x\sigma_y + c_2}{\sigma_x^2 + \sigma_y^2 + c_2'} \tag{7}$$

where σ_x, σ_y are standard deviation values of luminance samples of two images x and y as Eqn. (7).

$$s(x,y) = \frac{\sigma_{xy} + c_3}{\sigma_x\sigma_y + c_3'} \tag{8}$$

where σ_{xy} is a covariance of luminance samples of two images x and y, and where small constants denoted by C1, C2, and C3 as Eqn. (8). These constants given by

$$C1 = 2 * (K1.L), C2 = 2 * (K2.L)$$

and $C3 = \frac{C_2}{2}$. Here L is 8 bits/pixel grayscale images dynamic range equal to 255. When α=β=γ=1, then formula has a form as Eqn. (9):

$$SSIM(x,y) = \frac{(2\mu_x\mu_y + C_1)(2\sigma_x\sigma_y + C_2)}{(\mu_x^2 + \mu_y^2 + C_1)(\sigma_x^2 + \sigma_y^2 + C_2)} \tag{9}$$

The local quality value is obtained by using following formula as Eqn. (10):

$$SSIM_{ij} = W_y.SSIM_{ij}^{Y} + W_{cb}.SSIM_{ij}^{cb} + W_{cr}.SSIM_{ij}^{Ycr} \quad (10)$$

where W_Y, W_{Cb}, W_{Cr} are weighted for Y, Cb and Cr color components.

The SSIM values are varying the range of '0' and '1'. Where value '1' is the best possible value. The VQM calculated by using MSU tool set (Skupin et al., 2011).

Luminance PSNR (Y-PSNR)

In H.264/SVC video the PSNR calculated by utility provided in JSVM ("JSVM Reference Software," 2013; "JSVM Software Manual," 2013; VCEG, 2007) toolset. The SVEF supports only single layered pendency. It calculates the PSNR values: Y-PSNR, U-PSNR, and V-PSNR. The Y-PSNR denotes luminance components of the video. The U-PSNR and V-PSNR denote chrominance video component. Since the luminance (brightness) are more sensitive to the human eye in comparison to chrominance (color), the PSNR is typically evaluated and examined for the luminance (Y) component. The following Eqn. (11) shows the definition of the Y-PSNR among luminance component of the source and destination image denoted as Y_S and Y_D respectively.

$$Y - PSNR(n)_{dB} = 20\log_{10} * V_{peak} \left[\frac{1}{\sqrt{\frac{1}{N_{col}N_{row}} \sum_{i=0}^{N_{col}} \sum_{j=0}^{N_{row}} \left[Y_S(n,i,j) - Y_D(n,i,j) \right]^2}} \right] \quad (11)$$

where

V_{peak} presented by $2^N - 1$ and N represents pixels' size in bits,
N_{col} Represents total columns available in SVC image,
N_{row} Represents total rows available in SVC image.

IEEE 802.11 Performance Evaluation

The performance of IEEE 802.11e/n wireless network can be calculated from the receiving traffic between the wireless mobile node and multimedia server. There are several major metrics for measuring and comparing the performance of IEEE 802.11e/n WLAN. In NS2, output trace is used for calculating these performance metrics. A sample of NS2 trace file output shown in Table 6 and new Wireless Trace file format shown in Table 7. Generally, small AWK programs are used to process these trace files for finding.

Table 7a. NS2 output trace file content

```
s 0.100000000 _0_ AGT --- 1 cbr 1000 [0 0 0 0] ------- [0:1 1:1 32 0] [0] 0 0
s 0.100000000 _0_ AGT --- 2 cbr 1000 [0 0 0 0] ------- [0:3 2:1 32 0] [0] 0 0
s 0.100000000 _0_ AGT --- 3 cbr 1000 [0 0 0 0] ------- [0:4 2:2 32 0] [0] 0 0
s 0.100000000 _2_ AGT --- 4 tcp 40 [0 0 0 0] ------- [2:0 0:2 32 0] [0 0] 0 0
s 0.108000000 _0_ AGT --- 5 cbr 1000 [0 0 0 0] ------- [0:1 1:1 32 0] [1] 0 0
s 0.108000000 _0_ AGT --- 6 cbr 1000 [0 0 0 0] ------- [0:3 2:1 32 0] [1] 0 0
s 0.108000000 _0_ AGT --- 7 cbr 1000 [0 0 0 0] ------- [0:4 2:2 32 0] [1] 0 0
r 0.111256000 _1_ AGT --- 1 cbr 1020 [14a 1 0 0] ------- [0:1 1:1 32 1] [0] 1 0
s 0.116000000 _0_ AGT --- 8 cbr 1000 [0 0 0 0] ------- [0:1 1:1 32 0] [2] 0 0
s 0.116000000 _0_ AGT --- 9 cbr 1000 [0 0 0 0] ------- [0:3 2:1 32 0] [2] 0 0
s 0.116000000 _0_ AGT --- 10 cbr 1000 [0 0 0 0] ------- [0:4 2:2 32 0] [2] 0 0
D 0.116000000 _0_ RTR IFQ 2 cbr 1020 [0 0 0 0] ------- [0:3 2:1 32 0] [0] 0 0
r 0.120520667 _1_ AGT --- 5 cbr 1020 [14a 1 0 0] ------- [0:1 1:1 32 1] [1] 1 0
s 0.124000000 _0_ AGT --- 11 cbr 1000 [0 0 0 0] ------- [0:1 1:1 32 0] [3] 0 0
s 0.124000000 _0_ AGT --- 12 cbr 1000 [0 0 0 0] ------- [0:3 2:1 32 0] [3] 0 0
D 0.124000000 _0_ RTR IFQ 3 cbr 1020 [0 0 0 0] ------- [0:4 2:2 32 0] [0] 0 0
s 0.124000000 _0_ AGT --- 13 cbr 1000 [0 0 0 0] ------- [0:4 2:2 32 0] [3] 0 0
D 0.124000000 _0_ RTR IFQ 6 cbr 1020 [0 0 0 0] ------- [0:3 2:1 32 0] [1] 0 0
r 0.129805333 _1_ AGT --- 8 cbr 1020 [14a 1 0 0] ------- [0:1 1:1 32 1] [2] 1 0
s 0.132000000 _0_ AGT --- 14 cbr 1000 [0 0 0 0] ------- [0:1 1:1 32 0] [4] 0 0
s 0.132000000 _0_ AGT --- 15 cbr 1000 [0 0 0 0] ------- [0:3 2:1 32 0] [4] 0 0
D 0.132000000 _0_ RTR IFQ 7 cbr 1020 [0 0 0 0] ------- [0:4 2:2 32 0] [1] 0 0
s 0.132000000 _0_ AGT --- 16 cbr 1000 [0 0 0 0] ------- [0:4 2:2 32 0] [4] 0 0
D 0.132000000 _0_ RTR IFQ 9 cbr 1020 [0 0 0 0] ------- [0:3 2:1 32 0] [2] 0 0
r 0.139270000 _1_ AGT --- 11 cbr 1020 [14a 1 0 0] ------- [0:1 1:1 32 1] [3] 1 0
s 0.140000000 _0_ AGT --- 17 cbr 1000 [0 0 0 0] ------- [0:1 1:1 32 0] [5] 0 0
..........................................................................................
..........................................................................................
```

Table 7b. AWK script for throughput calculation

```
# AWK Script for calculating the Average Throughput
BEGIN { recvdSize = 0 startTime = 400 stopTime = 0}
{
event = $1 ;#e.g . s time = $2 ;#e.g. 5.000000000 node_id = $3 ;#e.g. _0_ level = $4 ;#e.g. AGT flow_id = $7 ;#e.g. cbr or video or tcp etc. pkt_size = $8 ;#e.g. 160
# assume MTU size is 200 # Store start time if (level == "AGT" && event == "s" && pkt_size >= 200) { if (time < startTime) {
startTime = time }
} # Update total received packets' size and store packets arrival time if (level == "AGT" && event == "r" && pkt_size >= 200) { if (time > stopTime) { stopTime = time } # Rip off the header hdr_size = pkt_size % 200 pkt_size -= hdr_size # Store received packet's size recvdSize += pkt_size
pkt_size = $3 hdr_size = pkt_size % 200
pkt_size -= hdr_size currTp = recvdSize/(time-startTime)*(8/1000)
}
}
END {printf("Average Throughput[kbps]:%.2f\tStartTime=%.2f\
StopTime=%.2f\n", (recvdSize / (stopTime-startTime)) *
(8/1000), startTime,stopTime)
}
```

Throughput

Throughput is the rate at which successful message/video transferred in a wireless network. The throughput of a wireless network can be calculated by using the traffic size through a link in a selected range of time. Higher throughput indicates better performance of the wireless network. The formula of throughput calculation shown in Eqn. (12) (Arif & Sari, 2010).

$$Throughput = \frac{\sum packet}{\sum time} \times \frac{8}{1000} Kbps \tag{12}$$

The awk script for throughput calculation from NS2 simulation trace file shown in Table 7b.

Table 8. AWK script for packet statics and end-to-end-delay calculation

```
## AWK Script for calculating the
# 1. Send, Received, Dropped Packets, Received Packets,
# Packet Delivery Ratio
# 2. Average end to End Delay#BEGIN {
seqno = -1;
droppedPackets = 0; receivedPackets = 0; count = 0;
sendPackets=0}
{ #packet delivery ratio
if($4 == “AGT” && $1 == “s” && $7 == “video”) { if (seqno < $6) { seqno = $6; }
sendPackets++;
} else if($4 == “AGT” && $1 == “r” && $7 == “video”) {
receivedPackets++;
} else if ($1 == “D” && $7 == “video” && $8 > 512){ droppedPackets++;
}
#end-to-end delay if($4 == “AGT” && $1 == “s” && $7 == “video”) {
start_time[$6] = $2;
} else if($4 == “AGT” && $1 == “r” && $7 == “video”) {
end_time[$6] = $2;
} else if($1 == “D” && $7 == “video”) {
end_time[$6] = -1;
}}END { for(i=0; i<=seqno; i++) { if(end_time[i] > 0) { delay[i] = end_time[i] - start_time[i];
count++;
} else {
delay[i] = -1; } } for(i=0; i<count; i++) {
if(delay[i] > 0) {
n_to_n_delay = n_to_n_delay + delay[i];
}
} n_to_n_delay = n_to_n_delay/count;
print “----------- Packet Statistics for Video Data -------------”; print “Generated Packets = ” sendPackets; print “Received Packets = ”
receivedPackets;
print “Packet Delivery Ratio = ” receivedPackets/
(sendPackets)*100 ” %”;
print “Total Dropped Packets = ” droppedPackets; print “Packet Drop Ratio = ” droppedPackets/
(sendPackets)*100 ” %”;

    print “Average End-to-End Delay = ” n_to_n_delay ” sec”;
}
```

End-to-End-Delay

The mean time of packet send and packet received is known as Delay. Less delay indicates better performance of the wireless network. The mathematical equation for delay calculation shows in Eqn. (13) (Arif & Sari, 2010).

$$Delay = \sum_{i=1}^{m} \frac{\left(t_i^{received} - t_i^{send}\right)}{m} ms \tag{13}$$

AWK script for end-to-end delay calculation from NS2 simulation trace file shown in Table 8.

Packet Loss

The comparison of total packet send and number of successful packets received at other end are used for the percentage of packet loss calculation. The less packet loss indicates better performance of the wireless network. The mathematical formula for packet loss shown in Eqn. (14) (Arif & Sari, 2010).

$$Packetloss = \left[\frac{\sum packet_{send} - \sum packet_{received}}{\sum packet_{send}}\right] \times 100\% \tag{14}$$

AWK script for packet generated, send, drop, loss ration calculations from NS2 simulation trace file shown in Table 8.

CONCLUSION

Computer simulation is the research tool of choice for a majority of the wired and wireless network research community and these days most of research works on the network adopt computer simulation to verify novel ideas. The framework and a tool-set is useful for evaluating the quality of video transmitted over a simulated wireless networks. Besides measuring the quality of service (QoS) parameters of the underlying network, such as loss rates, delays and jitter, frameworks also supports a video quality evaluation of the received video.

On wired as well as wireless networks, the Medium Access Control (MAC) plays the important role in the performance of video transmission. MAC handles allocating the resources to the different type of applications or wireless stations. Many types of research conducted for video quality measurement on network and application layer. The framework can be used for research and evaluating new techniques for MAC-layer optimizations. In this chapter an overview of the framework and a tool-set for evaluating the quality of video transmitted over a simulated wireless networks is presented. This chapter also provides the idea to develop new framework for upcoming compression technology.

REFERENCES

Arif, T. Y., & Sari, R. F. (2010). *Performance Comparison of Video Traffic Over WLAN IEEE 802.11e and IEEE 802.11n*. Paper presented at the UBICOMM 2010: The Fourth International Conference on Mobile Ubiquitous Computing, Systems, Services and Technologies, Florence, Italy. Retrieved from http://www.thinkmind.org/index.php?view=article&articleid=ubicomm_2010_13_30_10179

Blake, S., Black, D., Carlson, M., Davies, E., Wang, Z., & Weiss, W. (1998). *An Architecture for Differentiated Services (RFC-2475)*. Academic Press.

Boyce, J. M., & Gaglianello, R. D. (1998). Packet Loss Effects on MPEG Video Sent Over the Public Internet. *Proceedings of the sixth ACM International Conference on Multimedia.*

Chikkerur, S., Sundaram, V., Reisslein, M., & Karam, L. J. (2011). Objective Video Quality Assessment Methods: A Classification, Review, and Performance Comparison. *IEEE Transactions on Broadcasting*, *57*(2), 165–182. doi:10.1109/TBC.2011.2104671

Cisco Visual Networking Index. Forecast and Methodology, 2014-2019 White Paper. (2014). Retrieved from http://www.cisco.com/c/en/us/solutions/collateral/service-provider/ip-ngn-ip-next-generation-network/white_paper_c11-481360.html

Cygwin: Get that Linux feeling - on Windows. (2010). Retrieved from https://www.cygwin.com/

Detti, A., Bianchi, G., Pisa, C., Saverio, F., Loreti, P., Kellerer, W., . . . Widmer, J. (2009). *SVEF: An Open-Source Experimental Evaluation Framework for H.264 Scalable Video Streaming.* Paper presented at the IEEE Symposium on Computers and Communications, 2009 (ISCC 2009), Sousse.

Fedora 16 (32 bit). (2011). Retrieved from https://fedoraproject.org/

FFmpeg. (2011). Retrieved from https://www.ffmpeg.org/

FFmpeg (GitHub). (2011). Retrieved from https://github.com/FFmpeg/FFmpeg

JSVM Reference Software. (2013). Retrieved from http://www.hhi.fraunhofer.de/en/departments/video-coding-analytics/research-groups/image-video-coding/research-topics/svc-extension-of-h264avc/jsvm-reference-software.html

JSVM Software Manual. (2013). Retrieved from http://evalsvc.googlecode.com/files/SoftwareManual.doc

Ke, C.-H. (2006). *A prototype for H.264 evaluation framework using NS2*. Retrieved from http://140.116.72.80/~smallko/ns2/h264.htm

Ke, C.-H. (2012a). *How to do H.264 SVC transmission simulations?* Retrieved from http://csie.nqu.edu.tw/smallko/ns2/svc.htm

Ke, C.-H. (2012b). myEvalSVC: An Integrated Simulation Framework for Evaluation of H.264/SVC Transmission. *Transactions on Internet and Information Systems (Seoul)*, *6*(1), 379–394. doi:10.3837/tiis.2012.01.021

Ke, C.-H. (2013). *myEvalvid: An Evaluation Framework for MPEG video transmission in NS2 environment*. Retrieved from http://csie.nqu.edu.tw/smallko/ns2/myEvalvid.htm

Ke, C.-H., Shieh, C.-K., Hwang, W.-S., & Ziviani, A. (2008). An Evaluation Framework for More Realistic Simulations of MPEG Video Transmission. *Journal Of Information Science And Engineering, 24*(2), 425–440.

Klaue, J., Rathke, B., & Wolisz, A. (2014). *EvalVid - A Video Quality Evaluation Tool-set.* Retrieved from http://www2.tkn.tu-berlin.de/research/evalvid/fw.html#bin

Klaue, J., Rathke, B., & Wolisz, A. (2003). *EvalVid - A Framework for Video Transmission and Quality Evaluation.* Paper presented at the 13th International Conference, TOOLS 2003, Urbana, IL.

Le, T. A., Nguyen, H., & Zhang, H. (2010). *EvalSVC - An evaluation platform for Scalable Video Coding transmission.* Paper presented at the IEEE 14th International Symposium on Consumer Electronics (ISCE 2010), Braunschweig. doi:10.1109/ISCE.2010.5523712

MP4Box. (2014). Retrieved from http://gpac.wp.mines-telecom.fr/mp4box/

MPEG-4 video trace file. (2010). Retrieved from http://www2.tkn.tu-berlin.de/research/trace/trace.html

MSU Video Quality Measurement tools. (2014). Retrieved from http://compression.ru/video/quality_measure/index_en.html

Network Simulator (NS2.29.3). (2011). Retrieved from http://sourceforge.net/projects/nsnam/files/allinone/ns-allinone-2.29/ns-allinone-2.29.3.tar.gz/download

Patch file for NAM. (n.d.). Retrieved from http://www.linuxquestions.org/questions/attachment.php?attachmentid=3326&d=1271149430

Patch file for Wireless Application. (2011). Retrieved from http://www.telematica.polito.it/fiore/ns2_wireless_update_patch.tgz

Pawlikowski, K. (2003). *Do Not Trust All Simulation Studies of Telecommunication Networks.* Paper presented at the International Conference, ICOIN 2003, Cheju Island, Korea.

Ramakrishnan, K., Floyd, S., & Black, D. (2001). *The Addition of Explicit Congestion Notification (ECN) to IP (RFC-3168).* Retrieved from.

Recommendation ITU-T J.144 (03/2004): Objective perceptual video quality measurement techniques for digital cable television in the presence of a full reference. (2004). (pp. 1-156).

Recommendation ITU-T P.910 (04/2008): Subjective video quality assessment methods for multimedia applications. (2008). (pp. 1-42).

Ries, M., Nemethova, O., & Rupp, M. (2007). *Motion Based Reference-Free Quality Estimation for H.264/AVC Video Streaming* Paper presented at the 2nd International Symposium on Wireless Pervasive Computing, 2007. ISWPC '07, San Juan.

Rossholm, A., & Lovstroem, B. (2008). *A new low complex reference free video quality predictor.* Paper presented at the IEEE 10th Workshop on Multimedia Signal Processing, 2008, Cairns, Qld.

Shahid, M., Rossholm, A., Lövström, B., & Zepernick, H.-J. (2014). No-reference image and video quality assessment: A classification and review of recent approaches. *EURASIP Journal on Image and Video Processing*, *1*(40), 1–32. doi:10.1186/1687-5281-2014-40

Skupin, R., Hellge, C., & Schierl, T., & Wiegand, T. (2011). Packet level video quality evaluation of extensive H.264/AVC and SVC transmission simulation. *Journal of Internet Services and Applications*, *2*(2), 129–138. doi:10.1007/s13174-011-0025-z

SourceForge. NS2 downloader stats. (2014). Retrieved from http://sourceforge.net/projects/nsnam/files/stats/timeline

SVC. Original YUV sequences in 4CIF and CIF resolution. (2014). Retrieved from ftp://ftp.tnt.uni-hannover.de/pub/svc/testsequences/

SVEF. Scalable Video-streaming Evaluation Framework. (2012). Retrieved from http://svef.netgroup.uniroma2.it/

The Network Simulator - ns-2. (2009). Retrieved from http://www.isi.edu/nsnam/ns/

VCEG. J. V. T. J. o. I. I. M. I.-T. (2007). JSVM 9.8 Software. Geneva, Switzerland.

Video Library and Tools - NSL. (Raw Sequences from the H.264/SVC Group). (2015). Retrieved from https://cs-nsl-wiki.cs.surrey.sfu.ca/wiki/Video_Library_and_Tools

Video Sequences, Y. U. V. (2014). Retrieved from http://trace.eas.asu.edu/yuv/

VMware Workstation. (2011). Retrieved from http://www.vmware.com/in/products/workstation

Wang, C.-Y., & Wei, H.-Y. (2013). *WMNLAB IEEE NS-2 802.11n Module*. Retrieved from http://wmnlab.ee.ntu.edu.tw/a/ns-2.29-11n.tar.gz

Xiao, F. (2000). *DCT-based Video Quality Evaluation*. Retrieved from http://compression.ru/video/quality_measure/vqm.pdf

YUV test sequences. (2014). Retrieved from http://videocoders.com/yuv.html

Chapter 2
Dual Scrambled Image Watermarking Algorithm in DWT-SVD Composite Domain

Dharm Singh
Namibia University of Science and Technology, Namibia

Madhuri Agarwal
College of Technology and Engineering, India

Charu Singh
VIT University, India

ABSTRACT

There is an increased risk of copyright violation of multimedia data due to the enormous growth of computer networks that provides fast and error free transmission of any multimedia information. A copyright identifier that may contain some information about the lawful owner is inserted in the contents of the image, without sacrificing its quality. The security levels are increased by using a key value and scaling factor for the embedding and extraction process. The dual scrambled watermark using Arnold and Scrambling sequence is embedded by modifying the singular values of the scrambled cover image's DWT middle frequency sub-band. The simulation was performed on MATLAB 7.7.0 with standard database gray scale images of size 512x512 and watermark of size 64x64 using hybrid dual scrambled watermark schemes. The performance analysis is done on the basis of the degree of scrambling and JPEG compression attack using various parameters. The proposed method achieves better imperceptibility and security for the copyright protection methods.

DOI: 10.4018/978-1-5225-2342-0.ch002

INTRODUCTION

Digital multimedia technology has gradually developed and increased the ease of access to digital information. As digital multimedia (video, audio and images) become available for retransmission, reproduction, and publishing over the Internet, a real need for protection against unauthorized copy and distribution has increased, thus gives rise to the digital watermark technique for copyright protection of digital contents. A digital watermark is a visible or invisible identification code that may contain some information about the intended recipient, the lawful owner or author of the original data in the form of textual data or image. On the modification of any data content, could lead to absence or degradation of the watermark.

Thus, an important role can be played by the digital watermarking in providing the evidence of copyright infringements and thus making it traceable for the improper use of protected multimedia. Digital watermarking has a wide-span of practical applications such as digital cameras, medical imaging, image databases, video-on-demand systems, broadcast monitoring, transaction tracking, authentication, copy control, and device control among many others.

Generally, watermark techniques can be broadly divided into:

- **Spatial Domain:** The watermark data is embedded in the pixel value by doing changes in the pixel value intensity in this approach (Schyndelet al., 1994).
- **Transform Domain:** The watermark is added in the transform domain in order to have imperceptibility as well as robustness. In this method, transform coefficients are modified for embedding the watermark (Nikolaidis and Pitas, 2003).

BACKGROUND

In this chapter proposed watermarking method is based on DWT and SVD techniques.

1. Discrete Wavelet Transform (DWT)

Wavelet Transform uses wavelets of finite energy. Wavelets are used as basal functions for representing signals in the DWT transform. Wavelets can be realized by iteration of filters with rescaling. They are local in both frequency/scale (via dialations) and in time (via translations). The resolution of the signal is determined by the filtering operations, and the upsampling and downsampling (subsampling) operations determine the scale of an image.

- **Haar Wavelet:** It is a step function taking values 1 and -1, on [0; 1/2] and [1/2; 1], respectively. These are good for edge detection and reconstructing binary pulses.

The sequence x[n] denotes the signal in the figure, where n is an integer. G_0 denote the low pass filter while H_0 denotes the high pass filter. At each level, the high pass filter produces detail information, d[n], while the low pass filter associated with scaling function produces coarse approximations, a[n] (Dhubkarya and Sonam, 2009).

Figure 1. Three level wavelet decomposition tree

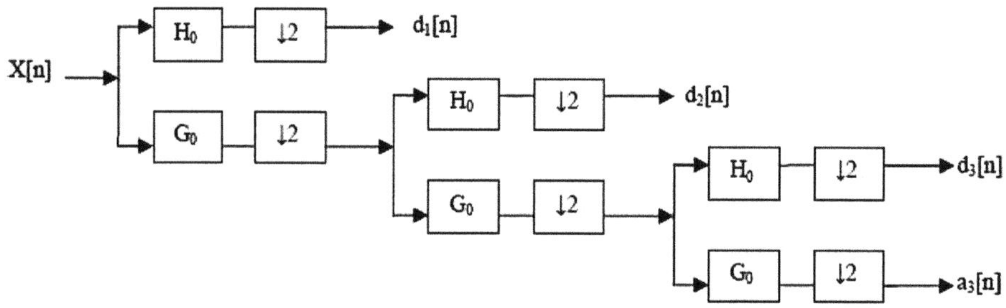

After passing the signal through the half band low pass filter, half of the frequencies get removed and the resolution gets halved but leaves the scale unchanged. This signal is then subsampled by 2 which doubles the scale and the final wavelet coefficients can be obtained after the recursive iterations on the subsampled output of the low pass filter. In other words the DWT can be thought of as consisting of two processes. The filtering is associated with retrieving of the wavelet coefficients corresponding to the current scale (i.e. high pass filter) and the scaling corresponds to sub sampling of the remaining data (i.e. low pass filter sub sampling). Generally the lower frequency sub-bands (LLX) contains most of the image energy and the high frequency sub-bands (HHX) have less energy and cannot be perceived by the human eye but is less prone to attacks.

2. Singular Value Decomposition (SVD)

The Singular Value Decomposition for square matrices was discovered by Beltrami in 1873 and Jordan in 1874. Then it was extended to rectangular matrices by Eckart and Young in the 1930s (Lalitha, Suresh and Sailaja, 2011). The discrete image can be regarded as a matrix which is an array of nonnegative scalar entries. Let such an image be denoted as A. Given a real matrix A which represents an input image of dimension mxn can be decomposed as:

$$A = USV^T$$

where U and V are mxm and nxn orthogonal matrices resp and nonnegative diagonal entries of S are called the singular values (SVs) of A.

The three main properties of SVD are (Liang and Sun,2006):

- The singular values do not change considerably on the addition of small perturbation to an image, that is, they have very good stability.
- The luminance of an image layer is specified by each singular value while the geometry of the image is specified by the corresponding pair of singular vectors (Sverdlov et al., 2005).
- Singular values represent intrinsic algebraic image properties.

PROPOSED METHOD

In most of the research work done, the watermark is scrambled before the embedding procedure and is generally a binary message of small size as a watermark [Gunjal et al., 2010, 2011).

This chapter proposed a technique using two different kinds of scrambling method in wavelet and singular domain for embedding and extraction of a digital image watermark. The cover image and the watermark image both are scrambled and the host image is scrambled using another scrambling sequence before the watermark is inserted.

1. Arnold Scrambling

Arnold Transform of an N x N matrix image is:

$$\begin{pmatrix} a_n \\ b_n \end{pmatrix} = \begin{bmatrix} 1 & 1 \\ 1 & 2 \end{bmatrix} \begin{pmatrix} a \\ b \end{pmatrix} \left(\bmod Q\right) \qquad (1)$$

Figure 2. Embedding after scrambling

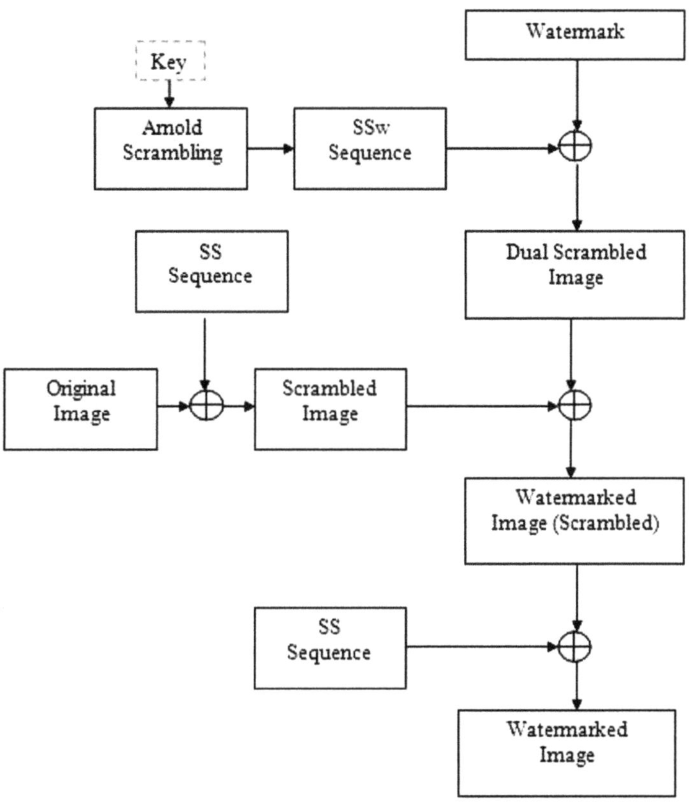

Figure 3. Extraction after scrambling

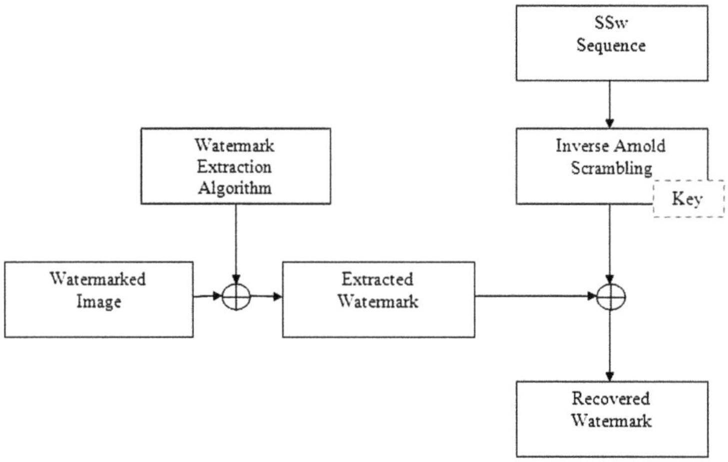

where (a, b) = {0, 1 … N} are pixel coordinates of an original image and (a_n, b_n) are corresponding pixel coordinates after the transformation. The layout of gray values is changed by changing the coordinates of the pixels.

- **Periodicity of Arnold Transform (P):** It is the 'number of iterations' that make up the specific property of Arnold Transform. This P is dependent on the size of given image. From equation 3.3 we have,

$$a_n = a + b$$

$$b_n = a + 2 * b$$

If $\left(\mathrm{mod}\left(a_n, Q\right) == 1 \,\&\&\, \mathrm{mod}\left(b_n, Q\right) == 1\right)$ then P = Q.

2. Scrambling Methodology

The scrambling sequence is generated using equation 3.1. The size of this SS sequence is equal to the host image

SS = {S || S Є (0, 1)} (2)

For the sake of result analysis and to study the effect of scrambling, the number of binary bits (either 0s or 1s) can be controlled in the SS sequence.

- **The Degree of Scrambling:** It is the number of pixels of an image gets interchanged with it corresponding diagonal counterpart. Image Scrambling is generally used as a preprocessing or post-processing treatment for information hiding. It is used to transform a meaningful image into an insignificant or chaotic image to enhance the security by increasing the power to resist various attacks (Ting and Ling, 1990).

3. Procedure for Embedding and Extraction

The work assumes that the dimension of the cover image (I) is 512x512 gray scale 8-bit image and the dimension of the visual watermark(W) is 64 x 64 gray scale 8-bit image. The main idea of the proposed algorithm is as follows:

a. Watermark Embedding Procedure

Step 1: Read_Cover Image(I)
Step 2: Generate_Scramble Sequence (SS)
Step 3: Loop
If SS=1, Scramble the particular positions
Else, No change
End
Step 4: Decompose_scrambled cover image (I') by DWT using Haar wavelet

LL1, HL1, LH1 and HH1

Step 5: Re-Decompose_LL1 using DWT to get four smaller sub bands

LL2, HL2, LH2 and HH2

Step 6: Re-Decompose_LL2 using DWT to get four smaller sub bands

LL3, HL3, LH3 and HH3

Step 7: Perform SVD_HL3 sub-band according to equation (3.4) and obtain U, S and V

$cH1=USVT$Step 8: **Read**_Watermark Image(W)

Step 9: Arnold Scramble_get 'Scrambled Watermark Image' (W') with key value
Step 10: Generate_Scramble Sequence (SSw)
Step 11: Loop
If SSw=1, Scramble the particular positions
Else, No change
End
Step 12: Perform SVD_Final Scrambled Watermark (W'')
Step 13: Add_watermark using scaling factor
Loop

$$U'S'V'=S+\alpha W'' \quad (4)$$

End
Step 14: Inverse SVD (ISVD)_HL3 sub-band

$$HL3*=U\ S'VT \quad (5)$$

Step 15: Inverse DWT (IDWT)_level3, level2 and level1 sequentially
Step 16: Rescramble_SS sequence
Step 17: Loop
If SS=1, Rescramble the particular positions
Else, No change
End
Step 18: Get_Final Watermarked Image (I*).

b. Watermark Extraction Procedure

Step 1: Read _Final Watermarked Image(I*)
Step 2: Decompose_upto three level DWT

$$LL3, HL3^{*}, LH3 \text{ and } HH3$$

Step 3: Perform_SVD to HL3* sub-band

$$HL3^{*}=U'\ S'^{*}V^{*}T \quad (6)$$

Step 4: Compute

$$D = U'S'^{*}V'^{T} \quad (7)$$

Step 5: Get_Scrambled watermark using scaling factor
Loop

$$W^{**}=D-S/\alpha \quad (8)$$

End
Step 6: Rescramble_SSw sequence
Step 7: Loop
If SSw=1, Rescramble the particular positions
Else, No change
End
Step 8: Perform_Inverse Arnold Scrambling using the key value
Step 7: Inverse SVD (ISVD)_HL3 band using equation **(6)**
Step 8: Inverse DWT (IDWT)_level3, level2 and level1 sequentially
Step 9: Get_Final Extracted Watermark (W*).
Step 10: Get_Final Recovered Image(I†).

EVALUATION PARAMETERS

1. Imperceptibility

Imperceptibility means that the perceived quality of the cover image should not be distorted by the presence of the watermark. As a quantitative measure, Mean Square Error (MSE) and PSNR metrics are used (Singh, Choudhary and Agrawal, 2012).

2. Peak Signal to Noise Ratio (PSNR)

PSNR is calculated between the original and the watermarked image. Larger the PSNR value, more similar is watermarked image to the original image. This image quality metric is defined in decibels as:

$$MSE = \sum_{i=0}^{n-1} \sum_{j=0}^{m-1} \left(\left(a_{i,j} - b_{i,j} \right) \right)^2 / \left(n * m \right)$$

$$PSNR = 10 \log_{10} \left(\frac{255}{MSE} \right)^2 dB$$

where m × n is the image size, $a_{i,j}$ and $b_{i,j}$ are the corresponding pixel values of two images.

3. Robustness

Robustness is a measure of the immunity of the watermark against attempts to remove or degrade it, internationally or unintentionally, by different types of digital signal processing attacks.

a. Normalized Correlation (NC)

If the extracted watermark image is absolutely tally with the origin watermark image, the Normalized Correlation (NC) =1. Otherwise the NC is between 0 and 1. When the extracted watermark image is more tally with the origin watermark image, the NC is bigger (Singh, Choudhary and Agrawal, 2012).

$$NC = \frac{\sum_{i=0}^{N} \sum_{j=0}^{M} W\left(i, j\right) * W^{*}\left(i.j\right)}{\sum_{i=0}^{N} \sum_{j=0}^{M} \left[W\left(i, j\right) \right]^2}$$

where,

W (i, j) = Original watermark image,

W*(i, j) = Extracted watermark image,
N and M = Width and height of the watermark image.

b. Bit Error Rate (BER)

Bit error rate refers to the amount of watermark data that may be incorrectly embedded within a host signal per unit of time or space, such as error bits per second or error bits per pixel. For the cover image and watermarked image of length L bits, the BER (in percent) is given by the expression:

$$BER = \frac{100}{L} \sum_{n=0}^{L-1} \begin{cases} 1, & (n) \neq c(n) \\ 0, & (n) = c(n) \end{cases}$$

EXPERIMENTAL SETUP

The Matlab software has been used for implementation as it provides an easy and user friendly environment for numerical computing and programming. For analysis of invisible watermark data, cover images with different formats like bmp, tif, png, gif, jpg, etc. have been used. Five images provided at CIPR Still Images at Rensselaer Polytechnic Institute(including the Kodak set) and test images by Hlevkin as a standard evaluation database for watermarking algorithms along with our image as the logo of the College (College of Technology and Engineering, Udaipur) where work have been done is used. The PSNR and NC parameters are used to measure the quality of the watermarked image.

The JPEG compression is a lossy compression and its compression rate depends on the quantization level used. Quantization eliminates the high frequency data that contains image details, noise and embedded watermark. JPEG compression attack with quality: 30, 40, 50, 60, 70, 80, 90 is applied to the test images. Hence, higher compression ratio corresponds to lower image fidelity.

PERFORMANCE ANALYSIS

1. On the Basis of Degree of Scrambling (DOS)

Degree of scrambling can be defined as the number of bits in an image that gets exchanged with its diagonal counterpart. Figure 4 shows the effect of degree of scrambling on PSNR and BER values using the test images and the logo image as watermark.

In the scrambling methodology, as the degree of scrambling increases, the no. of bit error increases. This increase in the bit error leads to the rise of noise in the signal which finally results in the reduction of the PSNR value. In the proposed strategy the scrambling of the image leads to decrease in its quality but it makes the watermarking more robust.

Figure 4. PSNR values and bit error rate at different degrees of scrambling

Cover Image	DOS	PSNR (db)	BER	Watermark Image
	200	64.4849	0.0020	
	400	50.5837	0.0040	
	600	44.3839	0.0077	
	200	63.4545	0.0021	
	400	59.7390	0.0051	
	600	58.6415	0.0075	
	200	59.2816	0.0018	
	400	54.5280	0.0041	
	600	54.9114	0.0056	
	200	69.9820	0.0017	
	400	69.6580	0.0040	
	600	68.3252	0.0056	

2. On the Basis of JPEG Compression Attack

The JPEG compression standard is considered as the most significant attack and is a lossy type of compression. Its compression rate depends on the quantization level used. The quantization eliminates the high frequency data that consist of image details, noise and embedded watermark.

Figure 5 shows the NC values corresponding to the various JPEG compression degrees respectively. This shows a slight decrease in the values as the degree increase but the values can be compensated for the introduction of security purposes.

Figure 6 shows the PSNR values corresponding to the various JPEG compression degrees respectively. The robustness of the proposed method can be clearly observed with the graph as the value of the parameters increases with the increase in the compression degrees. Security is considered as having the most financial implication with a high demand in the commercial market, thereby considering the security as the prime concern.

CONCLUSION

The proposed algorithm mainly focuses on gray scale digital images using the wavelet and singular domain with a different dual scrambling methodology to increase the robustness of the watermarked

Figure 5. NC values at different JPEG compression degrees

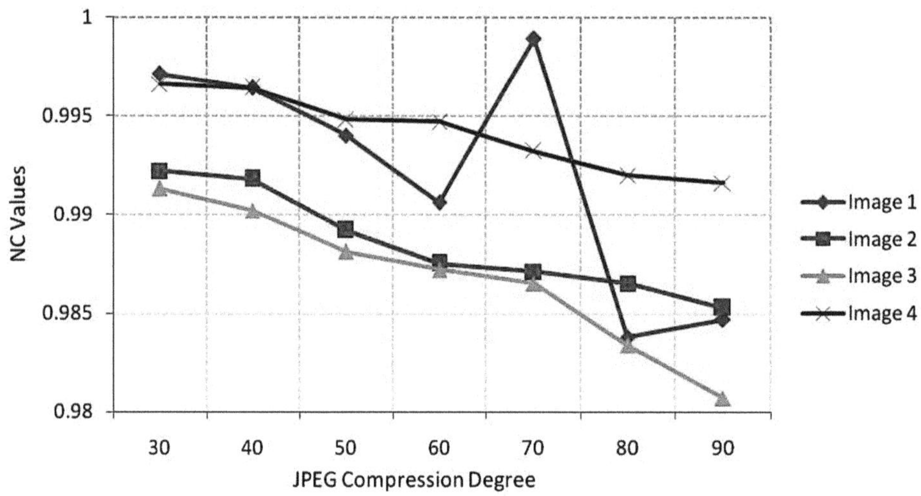

Figure 6. PSNR values at different JPEG compression degrees

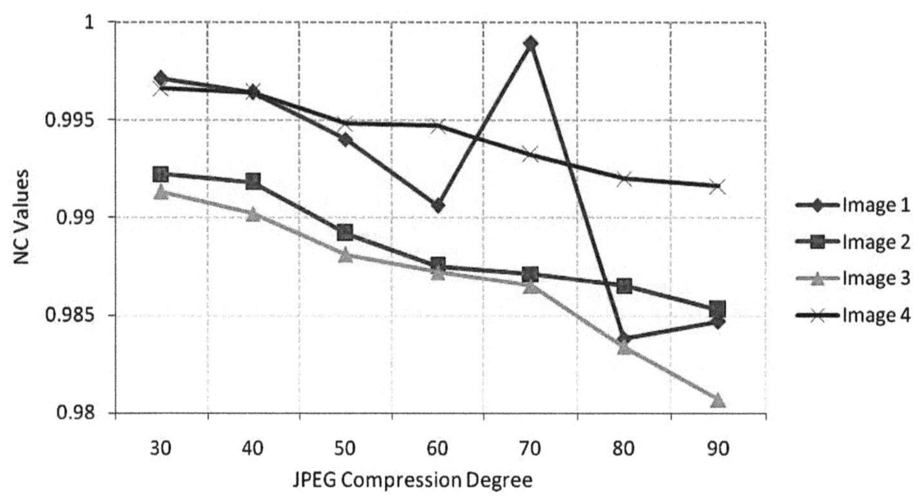

image. This chapter observes the characteristics of the proposed watermarking scheme which can be summarized as follows:

- The varied applicability of the proposed method over different types of images is shown as the method fully exploits the features of both the DWT and SVD transform domains by implementing on several images having different properties.
- Dual scrambling methodology having different characteristics is implemented resulting in the high level of security. The cover image is also scrambled along with the watermark image. It is able to accept JPEG lossy compression upto 90% where the compression level can be set according to the application.

- Various degrees of scrambling and Bit Error Rate are also presented with different images that can be applied with the proposed method to achieve better imperceptibility and security for the copyright protection methods.

Thus, algorithm used in this paper chapter can be successfully implemented for the Copyright Protection purpose.

- **Scope for Future Enhancements:** The method can be combined with some Visual Cryptographic (VC) methods along with the encryption and decryption at different levels to further enhance the security of watermarks. Further, research is needed to make it work in the audio and video systems.

REFERENCES

Dhubkarya & Sonam. (2009). High Quality Audio Coding At Low Bit Rate Using Wavelet And Wavelet Packet Transform. *Journal of Theoretical and Applied Information Technology, 6*(2), 194-200.

Gunjal, B. L., & Manthalkar, R. R. (2010). Discrete Wavelet Transform based Strongly Robust Watermarking Scheme for Information Hiding in Digital Images. *IEEE 3rd International Conference on Emerging Trends in Engineering and Technology*, 124 – 129.

Gunjal Baisa, L., & Mali Suresh, N. (2011). Secured Color Image Watermarking Technique In DWT-DCT Domain. *International Journal of Computer Science, Engineering and Information Technology*, 36-44.

Lalitha, N. V., Suresh, G., & Sailaja Dr, V. (2011). Improved Audio Watermarking Using DWT-SVD. *International Journal of Scientific & Engineering Research, 2-6*, 1–7.

Liu, L., & Sun, Q. (2006). A new SVD-DWT composite watermarking. *Proceeding 8th International Conference on Signal Processing*, 4.

Nikolaidis, A., & Pitas, I. (2003). Asymptotically optimal detection for additive watermarking in the DCT and DWT domains. *IEEE Transactions on Image Processing, 12*(5), 563–571. doi:10.1109/TIP.2003.810586 PMID:18237932

Singh, Naveen, & Madhuri. (2012). Spatial and Frequency Domain for Grey level Digital Images. *IJCA*, 16-20.

Singh, D., Naveen, C., & Madhuri, A. (2012). Scrambled Hybrid Wavelet-Singular Domain Digital Image Watermarking. *CSI 6th International Conference on Software Engineering (CONSEG)*, 219-224.

Sverdlov, A., Dexter, S., & Eskicioglu, A. M. (2005). Robust DCT-SVD domain image watermarking for copyright protection: embedding data in all frequencies. *Multimedia Computing and Networking Conference*, 16-20.

Ting & Ling. (1999). Hidden digital watermarks in Image. *IEEE Transactions on Image Processing, 8*(1), 58-68.

van Schyndel, R. G., Tirkel, A. Z., & Osborne, C. F. (1994). A Digital Watermark. *Proceedings of the IEEE International Conference on Image Processing*, 2, 86-90. doi:10.1109/ICIP.1994.413536

Chapter 3
Mobile Security:
Attacks and Prevention - Security in Mobile Communication

Meenakshi Tripathi
Malaviya National Institute of Technology Jaipur, India

Jyoti Gajrani
Malaviya National Institute of Technology, India

Vinesh Kumar Jain
Government Engineering College Ajmer, India

ABSTRACT

Mobile communication has various important issues such as congestion control, scarcity of radio spectrum, bandwidth utilization, and security. With increased number of threats in mobile devices, security has become the most important issue now-a-days. Security can be considered primarily at two levels; wireless transmission medium and mobile data level, where malicious applications execute their behavior by exploiting vulnerabilities present in mobile OS. The chapter also focuses on the latter one as these attacks can be easily performed by targeting vulnerabilities present in OS. The content is specifically on Android Operating System security as Android has become the largest installed base of mobile platform. This chapter specifies these vulnerabilities and focuses on malicious intentions of latest and sophisticated malware in detail. The chapter covers static and dynamic analysis tools as well as focuses on limitations of these methods for analyses. The results for malware samples with emulator detection capabilities show that these malware behaves more as benign samples.

INTRODUCTION

Wireless communication is playing a vital role in our daily life due to the portability of wireless devices. Nearly every person is using many types of the wireless devices such as mobile phones, pagers, Global Positioning System (GPS), Remote controls of Home-entertainment-system, Remote door openers for Vehicles (WIRELESS Types). As large number of competing users transmit their data through wireless

DOI: 10.4018/978-1-5225-2342-0.ch003

medium, stringent laws are mandatory to circumvent transmissions by interfering with the each other. Governments are spending a considerable amount of time in allocating specific frequencies for specific services to avoid interferences. These allocations are agreed world wide to minimize interferences.

According to applications and usages, wireless communication technology is further classified as Radio Communication, Cellular or Mobile Communication, Satellite Communication and Wi-Fi Communication. Among these, Cellular and Wi-Fi technologies are the most common technologies used in our daily life. Also, Cellular (mobile) technology is used by almost every person. Starting with simple regular handsets which were used just for phone calls, mobiles have turned our lives and have become an essential part in our lives. Mobiles have vast applications such as camera, music player, video calling, document readers, web browsing, chats, etc. The mobile phone market, which has 4.33 billion users in 2013 is likely to reach 5.13 billion users globally by 2017 (Internet Usage, 2014). The average mobile connection speed will increase 2.4-fold, from 1.7 Mbps in 2014 to 4.0 Mbps by 2019 (VNI). According to CNN Global News of January 2014, 55% of Internet usage was from Mobile devices and only 45% from PC in the United States. Out of 55% of mobile device Internet usage, mobile applications uses 47% of Internet traffic and remaining 8% is of mobile browsers (CNN, 2014).

There are many issues need to be focused in mobile communication such as congestion control, scarcity of radio spectrum, bandwidth utilization, security. Among these security issue is most important looking at cases of attacks happening on mobile devices. Security goals of Mobile Communication are also same as other networks: authentication, confidentiality, integrity, availability, and non-repudiation. Authentication is the process of verification of the identity of a user. Confidentiality means that read, write and execute of protected data or programs is allowed only by authorized people. In mobile communication, confidentiality of data is more important than any other requirement as it contains private data. Integrity ensures that the information or data is not altered or damaged by unauthorized users. Availability means the network is able to provide services as and when required (Sen et al., 2010).

In Mobile communication, one of the classification of attacks can be according to network protocol stack. Even a single attack can occur at any of the layer of protocol stack. Same DoS attack can occur in different layers in different ways, e.g. jamming at physical layer, hello flood attack at network layer, and SYN flood attack at transport layer (Sen et al., 2010). Figure 1 gives some examples of attacks implemented at various layers.

The wireless medium is used by mobile devices for the purpose of communication with other mobile devices through base station. Both active and passive attacks can be performed on wireless medium. In an active attack, communication between the two innocent communicating entities can be corrupted by injecting some additional data, deleting some part of data, modifying the communication, or by replaying previous messages. In a passive attack, communication between two entities can be eavesdropped on the communication link by the passive attacker. Wireless medium is primarily a broadcast based medium(IMAI). Most of the carrier-based 3G wireless networks provide good link-level security but they lack end-to-end upper-layer security and leave it to the user. Data going from the client to server is usually unencrypted which allow attacker to eavesdrop it by placing him anywhere within the area of the cell. Visiting hosts can overload the network with unreasonable transmissions which is also difficult to control. This results in an abrupt decrease in network performance and may lead to DoS(denial of service) attack due to the congested network. Therefore, Businesses cannot rely only on link encryption for transmitting their important data over the network. This leads to the need of providing a good end-to-end security.

Figure 1. Attacks on protocol stack

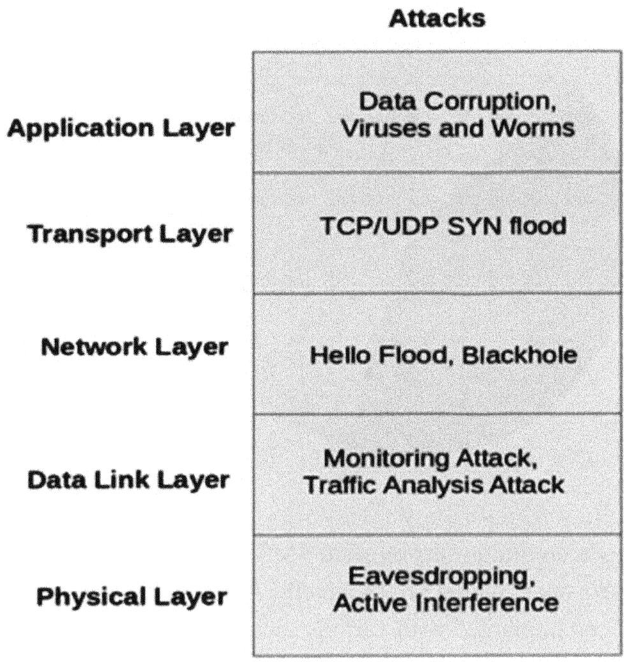

Mobile devices are designed in such a way to make them portable, light weighted and battery powered. Not only operating system services and various running applications consume battery but also data transmission consumes battery power. Therefore, power consumption should also be reduced to avoid frequent recharging. Due to this, complex encryption algorithms cannot be used and also the number of messages involved in security protocols must be constrained. Additionally, enriching mobile devices with more secure features must consider additional constraints like short packet size, limited bandwidth, higher transmission costs, less processing power and storage resources.

MOBILE OPERATING SYSTEMS

Present Mobile Operating Systems include Android, IOS, Java ME, Symbian, Windows Phone, and BlackBerry. Figure 2 shows share of mobile OS as of January, 2015 (MOBILE Share, 2015). This clearly shows that Android is the mostly used OS. Android provides world class platform for creating applications and games for the users and provide an open marketplace for deploying them easily.

Mobile Data Security

As smart-phone is a personal device and manages variety of user's private and sensitive information such as personal information, contacts, SMS's, documents, pictures, call recordings, bank account information, various user-names and passwords and email accounts, the security of this information is the responsibility of the underlying operating system. Apart from loss of information smart-phone malware

Figure 2. Share of mobile operating systems

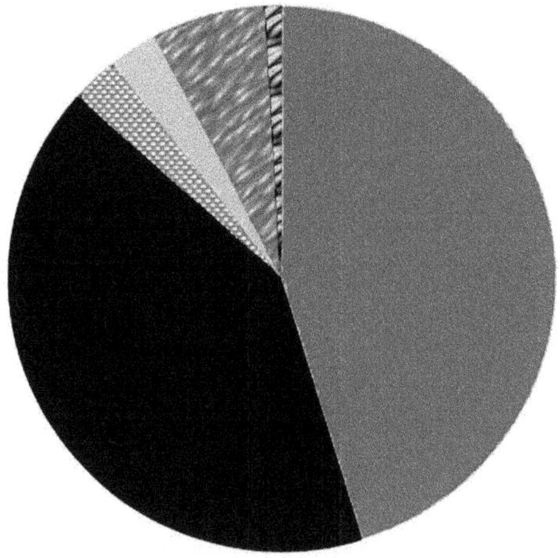

can cost monetary charges accounting for premium SMS and calls or even frame a person for a crime. So security issues related to smart-phone OS especially Android is the prime concern of this chapter.

Though Android has been implanted with various security mechanisms, malware authors are able to find vulnerabilities and exploit them for their use. As Android is strengthening itself, malware authors too are creating smart malware making their detection difficult. Advanced malware with various survival techniques such as emulation or virtualization detection, code obfuscation, encrypted transmission are already found on Android platform. Thus the task of analyzing these apps is challenging. Android users simply download the applications from Google playstore or third party stores without having knowledge that the application they are downloading is malicious application or benign application. Innocently, they make their phones target for attack. Before getting into detail of android security model, it is essential to get familiar with Android architecture. The next section covers the Android architecture in brief.

ANDROID ARCHITECTURE

Android is an open-source project of Google under the Apache License. Android supports a complete software package for mobile devices which includes an operating system, middle-ware and key mobile applications. It is based on custom Dalvik Virtual Machine (DVM) which is designed to optimize memory and hardware resources in a mobile environment. It has a multi-layered architecture with its core consisting of modified Linux kernel. It is basically a four layered architecture having: Linux Kernel layer, Native Userspace layer, Application Framework layer and Applications (Zhauniarovich, Y., 2014) as shown in Figure 3.

- **Linux Kernel Layer:** The kernel is tuned for embedded environment to perform core tasks of operating system such as memory management, process management, device management (through drivers) and networking along with additional components such as power management module and IPC binder (a driver, which provides the support for custom inter-process communication in Android).

Figure 3. Android software stack

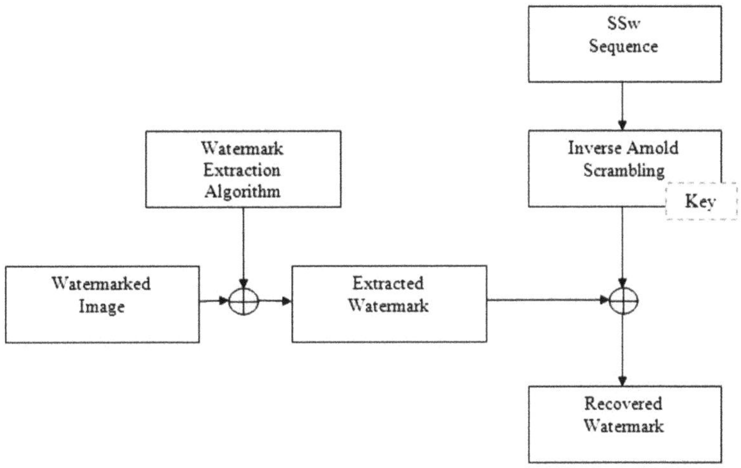

- **Native Userspace Layer:** All the user-space components not belonging to the Linux Kernel layer and running outside Dalvik Virtual Machine comes under this layer. This layer has four subcomponents.
- **Hardware Abstraction Layer (HAL):** The first subcomponent at this layer. Android defines an API for every hardware type. These APIs are used by upper layers to interact with various hardware components. A software module implementing APIs defined in Android for that particular hardware must be provided by the suppliers of hardware . Therefore, this solution permits Android to avoid embedding all probable drivers into the kernel and thus reducing the kernel size.
- The next subcomponent is init process. This process starts all other processes and services in Android.
- Another important component is Daemons which start executing with system startup and keeps running till the system is running.
- Last subcomponent is Native libraries which are compiled to native machine code for the device. These are loaded at run time and make various functionalities available to Android processes. Native libraries enable to call code written in non-java languages within the Android application. This native code is executed outside DVM in contrast to Java/Dex code which runs under DVM.
- **Application Framework Layer:** Application framework layer provides uniform and concise view of Java libraries which are used by applications. Android runs all its sensitive functionality like telephony, GPS, network, power-management, radio and media as system services under this layer.
- **Android Applications:** These are the set of all software applications that run on Android OS and provide various functionalities to the user. Many build-in applications are delivered by default in Android.

ANDROID APPLICATION STRUCTURE

Android applications are available as apk files (android package file). This file contains several files and folders as follows:

- assets folder,
- META-INF folder,
- res folder,
- AndroidManifest.xml file,
- classes.dex file,
- resources.arsc file.

AndroidMainfest.xml file contains the meta-data like package name, required permissions, minimum and maximum version support, definition of various components such as Activities, Services, Broadcast Receivers, Content Providers, and libraries to be linked etc.

Executable file classes.dex stores the Dalvik bytecode to be executed on Dalvik Virtual Machine. Dex files are used to hold a set of class definitions and their associated adjunct data.

Resources.arsc or application resource files, are used by programs developed for the Google Android mobile operating system. They contain compiled resources in a binary format, and may include images, strings, or other data used by the program. Resource files enable developers to access resources in a standard way in the program source code. They also can be used for protecting proprietary data.

Res folder contains classes for accessing application resources, such as raw asset files, colors, drawables, media or other files in the package, plus important device configuration details (orientation, input types, etc.) that affect how the application may behave.

META-INF folder stores the certificate of the app's developer to verify the third party developer identity. This folder contains three files CERT.RSA, CERT.SF and MANIFEST.INF folder assets contains non-compiled resources.

ANDROID INBUILT SECURITY MODEL

Android OS provides following inbuilt security features:

- Android take advantage of DAC (Discretionary Access Control) feature of Linux at kernel level. It assigns every application process a unique UID to avoid interference of app with other apps or system services.
- Android provides permission-based security model at application framework level to restrict apps from accessing important functionality of a smartphone such as telephony, network, contacts, SMS, sdcard and GPS location. To access any functionality of these components, first the related permission must be written in AndroidManifest.xml file and user must accept that permission before installing the app.
- System partition of smartphone contains Android's kernel, system libraries, runtime, framework and applications (ANDROID Security, 2014). System partition's are made read-only to protect unauthorized access and modification. Android secure some part of file-system like application cache and sdcard with proper privileges to avoid damage from adversary during the period device is connected with the PC using USB port.
- Google verifies all applications with Bouncer before making them open to users for download. Bouncer is a dynamic analysis framework that executes application in a sandboxed environment to confirm that it is a benign application.

- Android has started integrating SELinux in Android Jelly Bean 4.3 and later versions to provide better security (SE-Linux, 2013). It enforces Mandatory Access Control (MAC) policies upon the traditional Discretionary Access Control (DAC) on the device.

VULNERABILITIES IN ANDROID OS

Following vulnerabilities are still present in Android OS and these are exploited by attackers in many ways:

- In Android architecture, application request for all permissions it requires at installation time and if the user has accepted for these permissions then at run time there will be no control on the resources accessed and operation performed by the application. The application can do any malicious activity with the component, if it has permission of that component.
- Before Android version 2.3, many vulnerabilities like allowed default access to system property space, correct SSL displays were present. Patches for all these are coming in recent versions but if user does not apply patches then their mobiles will remain vulnerable.
- Third party application stores are not protected and if the user installs application from third party then the system becomes vulnerable.
- Social-engineering techniques are used many times to lure people to install malware apps.
- Recently performed collusion attacks have shown that poorly written apps can be exploited by interacting with them by malicious apps and making use of their permissions.
- Users of smart-phones are not that much aware of installing anti-virus applications and also Android does not provide any default anti-virus app.
- Many users are unaware of the risk of connecting to untrusted Wi-Fi networks and websites. These free and open Wi-Fi networks always attracts people and they try to connect to these networks without knowing that their data can be leaked during communication
- Browser Application in Android makes in vulnerable to all kind of Web security attacks.

MALICIOUS INTENTIONS

Malicious intentions of attackers are classified in following ways in (Masud et al., 2014):

- **Stealing Device and User Credential Information:** Many Android malware gather information regarding International Mobile Subscriber Identity (IMSI) number, International Mobile Equipment Identity (IMEI) number, Phone number, GPS Location, Installed packages and many other private information of users phone. Android malware also grab user activities like SMS received or send and outgoing or incoming call information.
- **Communicating With Command and Control (C&C) Server:** Similar to Botnet, Android malware try to communicate to a C&C server through HTTP or SMS. These connections are built to send captured information from the device. These can also be used for updating the malware packages or obtaining any other malicious information like SMS premium number or any other malicious URL from the server.

- **Sending Premium Rate SMS and Spam:** Android malware are able to automatically send SMS without the user awareness. This can be used for gathering illegal income by charging user after sending SMS to the premium number from his phone.
- **Search Engine Optimization:** Android malware can have an automatic script to visit advertisements, for increasing the number of visitor to a website or for the purpose of generating a charge per click without having actual knowledge of user. Even though, it does not cause any harm to users but this activity can increase mobile data consumption and thus they have to pay more charge of Internet usage.
- **Updating and Downloading Packages:** Android malware can communicate to the C&C server to update the existing package and add malicious intention to it or even they can download a new package and install it in the user devices without his knowledge.
- **Draining Resources:** Android malware can execute a malicious payload which can take over all the disk storage or memory (RAM) quotas and monopolize the CPU.

MAJOR ATTACKS IN ANDROID

Android malware are implemented using either .dex/.java code or using native code libraries.

- **Attack From Java/JavaScript:** Android applications are developed in java language and these applications runs in Virtual Machine to avoid compiling applications on different platform again. There are large number of development tools available for java language, several mobile phones are already using java ME. Due to these reasons, attackers attract to java language and try to obfuscate malicious code into it.
- **Attack From Java Native Interface (JNI):** All the attacks which are implemented using native code written in languages such as C, C++ comes under this category. These are due to improper use of the JNI in Android applications. Here attacker can take benefit of loopholes present in other languages. Native code is more flexible when working with manipulation of memory or when operating with hardware. Some exploits rely on excessive number of processes thus new processes are quickly launched to increase the maximum process number, and overwriting other processes memory. Also, header files exist for different data structures used by the Android operating system, which can conveniently be used in native code implementations. Native Code is convenient for reliable and effective exploit development (Shih et al., 2008).

Attacks Classification

Following are the major attacks exploited using android architecture vulnerabilities:

- **Collusion Attacks:** These attacks are due to the collusion of two applications which are signed by same certificate. As these apps share UID and address space, they can take benefit of it and use permissions of each other which lead to some malicious activity as shown in Figure 4. Individually each of these apps may be benign.
- **Repackaging Attack:** Malware Authors catch and download popular application, dissembles them and enclose malicious payloads in them then reassemble them and upload or submit the new

Figure 4. Collusion attack

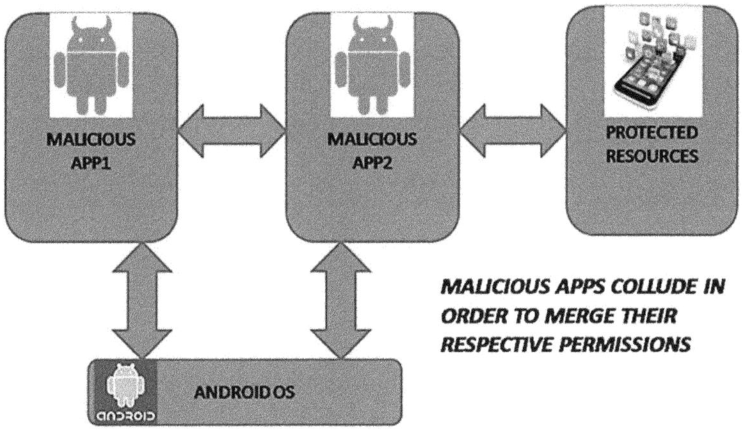

application on official android or alternative android market places. Users are made attracted to download or install these infected applications. AnserverBot Malware repackaged with a paid application "com.camelgames.mxmotor" is uploaded to the official android market. DroidKungFu uses com.google.search to disguise as the google search module and it follow up versions use com.google.update and pretend to be the official android update (Jiang, X., & Zhou, Yajin, 2013).

- **Update Attack:** These attacks are difficult to detect as they include malicious update component which update initial benign app at runtime only. Due to this, the static scanning fails to detect host application which is capturing the existing malware at run time by updating itself. Malwares families like BaseBridge, DroidKungFuUpdate, AnserverBot, Plenkton etc. belong to this category (Jiang, X., & Zhou, Yajin, 2013).
- **Drive by Download:** Android attackers not only directly exploit mobile browser's vulnerabilities but also they are essentially exciting users to download interesting and feature rich applications. Malware Families such as Zitmo, spitmo, Ggtracker etc. comes into this category.
- **Denial of Service (DoS) Attack:** Mobile devices have limited CPU, memory and battery. A malicious app overloads these already limited resources and restricts the users in executing their normal functions.

MALWARE ANALYSIS TECHNIQUES

To prevent from these attacks, malware analysis techniques have been proposed which are broadly classified into two categories namely static analysis and dynamic analysis. These are explained in detail below:

Static Analysis

Static analysis extracts software features by analyzing the Android APK, its Dalvik ByteCode and its source code without executing the app. In its simplest form, Static analysis involves identifying and querying cryptographic hashes like MD5, SHA-1and SHA256. Static analysis identify app as benign or malicious based on various static information like permissions, API calls, deployment of components, code-chunks (Tangil et al., 2014) . Some Static analysis based techniques read and parse app bytecode to create a control-flow graphs(execution paths in an app), which are further analyzed to identify malicious behavior. With these static features, signature database is constructed using training set which is used at the time of submission of new application. Signature based detection of apps is an example of static analysis and this is commonly used approach by antivirus products.

Static Analysis Tools

- **Androguard:** It is most widely used tool based on static analysis. It can be used for disassembling and decompiling Java source code from Dalvik bytecode. For detecting repackaged apps and known malware, Androguard can also compute a similarity value between any two android apps. This similarity can be used to classify apps as malicious. It can be used for parsing the app's manifest file and retrieving information from it. It is also used by many other dynamic analysis frameworks that are performing some form of static analysis prior to actual dynamic analysis and using this information in assisting dynamic analysis.
- **Dexter:** Dexter is a web based application which is designed for static analysis of Android applications. It has all features that are same as those of Androguard and APK inspector. Other than this, it also provides collaboration functionality to help in knowledge sharing among multiple researchers.
- **Dendroid:** It is a text mining based approach which automatically classifies malware samples and also analyze malware families based on the code chunks built from apps. Here, code chunk corresponds to a method associated with a class within the app.

Many tools based on static analysis for identifying malicious android apps are Mobile Sandbox, IDA Pro 6.1(and above), APKInspector, Dex2Jar, JAD, Dexdump, JD-Gui, Smali. Most of the tools are used for reverse engineering to decode apps to nearly get original form and then perform analysis based on various extracted static features.

Limitations of Static Analysis

For evading static analysis, many malwares incorporate obfuscation, encryption and compression techniques.. These malware also use native libraries written in C/C++ or other programming languages to hide system activities by calling functions outside the Dalvik/Java runtime library. Therefore, all the analyzers based on only Non-Native detection methods, fail to detect these malwares.

Dynamic Analysis

Dynamic Analysis monitors an application behavior in a controlled Environment called sandbox rather than inspecting the source code as in static analysis. It logs various operations performed by app like sending SMS messages, reading data from storage, and connecting to remote servers when it executes in sanboxeded environment and then analysis of these logs is done to identify malicious activity.

Dynamic Analysis Techniques

- **Droidbox:** Droidbox is an sandbox that uses combination of static analysis pre-check results, API monitoring and dynamic taint analysis. Activity leaks like cryptographic usage, net open/read/write, file read/write, SMS's, calls etc. are detected by tainting sensitive data and logging them in JSON files. Droidbox generates Behavior graphs between activities and timestamps along with JSON logs. It also logs relevant API function parameters and their return values for discovering a possible malware and report this malware for further analysis (Lantz, P., 2011). This is available as an offline tool.

On Droidbox, both benign and malicious samples are submitted. In this section, we are specifically presenting analysis of following samples:

- Two samples from ADRD malware family obtained from Malware GNOME Project (Malgenome).
- One samples from Pincer malware family obtained from Virustotal (Virustotal, 2015).
- One benign sample from google play store.

ADRD family performs following malicious actions: stealing users' personal sensitive information like IMEI, IMSI and sending them to remote servers, sending and deleting SMS messages, downloading unsolicited apps, and issuing HTTP search requests to increase websites search rankings on the search engine.

Pincer family malware forward SMS messages to its C&C server, and undertake functions based on commands it receives. The standard data sent to the C&C server includes phone number, device serial number, phone model, carrier, and OS version. However, The Pincer Family malware samples analyzed have the ability to identify that they are running in an Android Emulator, and therefore exhibiting normal behavior.

Each of these samples is analyzed using Android emulator version 4.1.1. Droidbox generates behavior graph between various activities performed by samples and timestamps as shown in Figure 5. These activities include Accessedfiles, apkName, cryptousage, dataleaks, dexclass, fdaccess, hashes, opennet, phonecalls, sendnet, sendsms, servicestart. These behavior graphs depict samples exhibit characteristics according to its pre-defined behavior in its family.

As shown in Figure 5(a) and 5(b), samples of ADRD malware family performs activities like file read, file write, sms and cryptographic usage, dexload, service start, opennet. The two samples for which analysis is presented are

`4de0d8997949265a4b5647bb9f9d42926bd88191.apk`

and

`5c70988fc9751f283b2f2b5733f9e7f2a54afa69.apk`

Respectively. Here, file read operation is required for accessing various system details and users private detail. Cryptographic activity may indicate that these values are encrypted before being transmission. The JSON logs corresponding to first sample shows that the encryption algorithm used is Data Encryption Standard (DES). The JSON file also logs the corresponding key used for encryption as shown in listing 1. Opennet activity indicated the presence of either transmission or reception of data/commands over some external IP address. Here also the logs shows destination IP address and destination Port address used for data transmission as shown in listing 2. To reduce the false positive (benign to be treated as malware, if that also includes activities like sms, cryptographic usage), more number of samples can be analyzed and a training set with sufficient samples can be used.

Sample

`7bd2f98aa1b28779f02f6883fa64d66dd1e228289de7d11a4c84c375014f4b00.apk`

Of Pincer family do not exhibit all of above activities as shown in Figure 6. This is because DroidBox runs it in emulator and as these samples detect its presence, they exhibit normal behavior. The file read log also shows the path of the file which is read. Therefore, by analyzing that path suspicious action can be guessed. For this pincer sample, file read is just read of cmdline not any system or shared file.

Figure 7 shows behavior of a benign app air.com.devoidgames.ScrollLockAndroid.apk, which is a game app requiring Adobe AIR . It can be noticed that it is not performing any suspicious activity.

From the logs of JSON files, feature set can be formed and the features of unknown android app can be compared with known malicious samples feature set and it can be predicted whether sample belong to malicious family or not. For the better accuracy, large number of malicious and benign samples can

Figure 5. Behavior graph for ADRD malware family

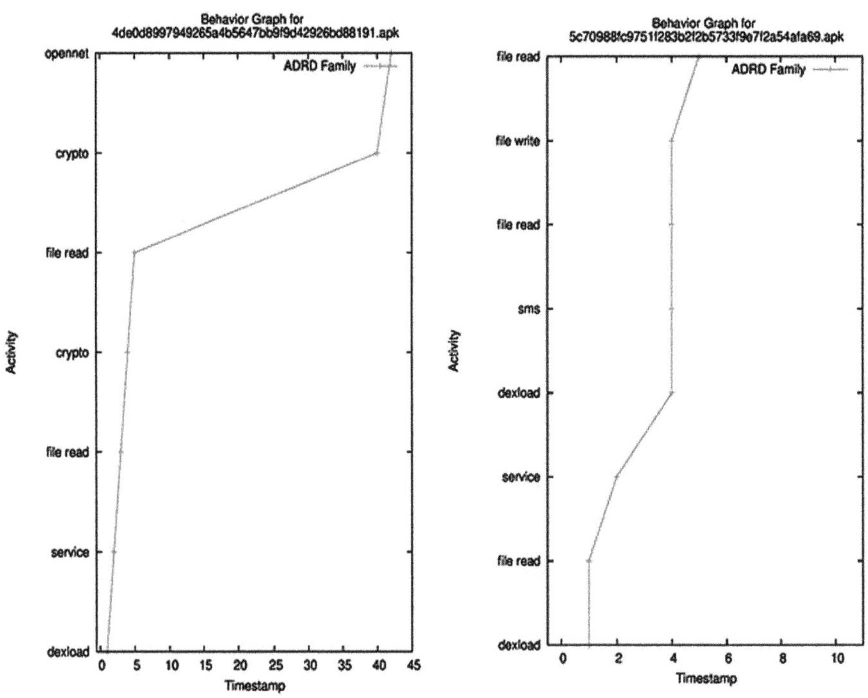

Listing 1. Log of cryptographic usage


```
"cryptousage": {
        "0.9532439708709717": {
            "algorithm": "DES",
            "key": "52, 56, 55, 51, 52, 49, 53, 52",
            "operation": "keyalgo",
            "type": "crypto"
        },
        "0.9859578609466553": {
            "algorithm": "DES/CBC/PKCS5Padding",
            "data": "357242043237511&310260000000000&1&1&1.1.1",
            "operation": "encryption",
            "type": "crypto"
```


Listing 2. Logs of Opennet activity


```
"opennet": {
        "1.073065996170044": {
            "desthost": "69.195.129.70",
            "destport": "80",
            "fd": "16"
        }
```


Figure 6. Behavior graph for Pincer family

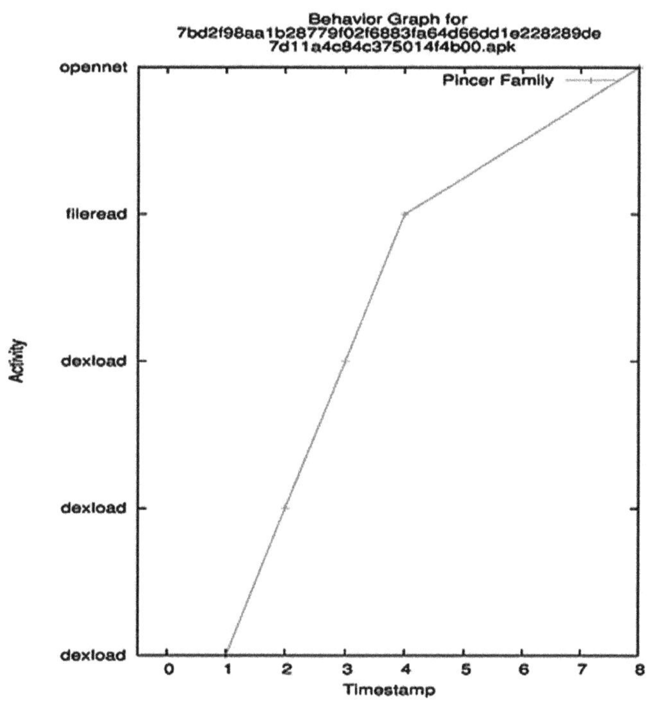

Figure 7. Behavior graph for benign sample

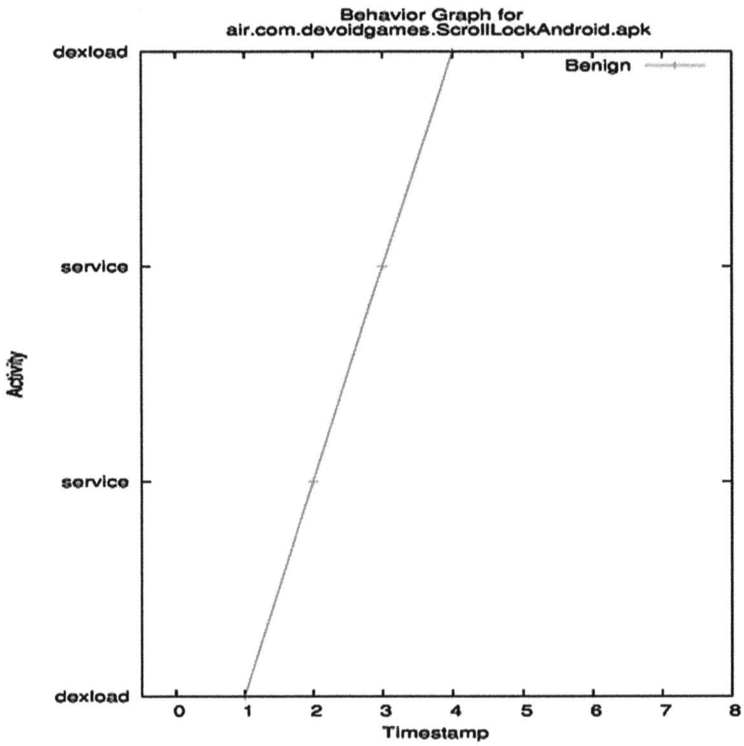

be analyzed to obtain the training dataset as input to machine learning techniques for automated classification of unknown samples.

- **DroidScope:** It is a Virtual Machine Introspection (VMI) based dynamic analysis Android framework. Unlike other dynamic analysis platforms, it stays out of the emulator and monitors the OS and Dalvik semantics. Hence, even the privilege escalation attacks on the Android kernel can be detected. It also makes the attackers task of disrupting analysis difficult. DroidScope is built upon QEMU emulator with a rich set of APIs to customize the malware analysis prototype. Android malware families DroidKungFu and DroidDream were analyzed and detected with this technique. However, DroidScope's effectiveness against other malware families remains to be tested (Faruki et al., 2015).
- **Mobile-Sandbox:** It is based on a novel combination of static and dynamic analysis and also applies machine learning techniques for classification. It is able to dynamically monitor malicious native API calls(outside Dalvik VM) along with malicious Java calls(within Dalvik VM) Has a publicly available web interface. It is able to analyze for android API level 13 and can be used for API level 17 in future and before this analysis is limited to API level 9 (Spreitzenbarth et al., 2014).

LIMITATIONS OF EXISTING SANDBOXES

Many sandboxes including DroidBox, DroidScope, TaintDroid etc. can be easily detected by malware using various evasion techniques (Vidas, T., & Christin, N., 2014, Jing, Y. et al., 2014, Petsas, T. et al., 2014) and thus can be evaded. Some sandboxes are not maintained now and some are not publicly available anymore. Some applications of the Obad family can evade DroidBox, Therefore, Mobile-Sandbox(which is based on Droidbox) will not be able to analyze malware of these families. In dynamic sandboxes many samples are in queue so new submitted sample will take too much time and this is very much undesired. SandDroid accepts samples for analysis but is not able to perform any analysis on the samples. Dynamic Analysis tools like SmartDroid, vetDroid are neither available in source code form nor having any web interface (Neuner et al., 2014).

CONCLUDING REMARKS

Lately mobile devices have become full-fledged computing platforms as they have started supporting operating systems, complex applications, Internet, data transfer and software development kits. This has also increased the necessity for human beings to become aware of various threats which are rapidly growing with this revolution. This chapter explores the various security issues and their analyses for mobile devices. Most of the market in mobile operating system is covered by Android. Because of this, the chapter was primarily focused on security issues related to android devices. By studying the intentions of various attacks, analysis methods like static analysis and dynamic analysis have been developed to detect malicious applications in Android Operating System. Researchers can use these analysis methods to provide better security solutions for mobile devices. Applying these solutions, users would probably feel more comfortable in storing private information in their mobile devices.

REFERENCES

Android Security Overview (ANDROID Security). (2014). Retrieved January 20, 2015, from http://source.android.com/devices/tech/security

CNN Global News View (CNN). (2014). Retrieved January 20, 2015, from http://money.cnn.com/2014/02/28/technology/mobile/mobile-apps-internet/

Faruki, P., Bharmal, A., Laxmi, V., Ganmoor, V., Gaur, M. S., Conti, M., & Rajarajan, M. (2015). Android Security: A Survey of Issues, Malware Penetration and Defenses. IEEE Communications Surveys and Tutorials.

IMAI. (n.d.) Retrived January 20, 2015, from http://www.artechhouse.com/uploads/public/documents/chapters/imai_520_CH03.pdf

Internet Usage. (2014). Retrieved January 20, 2015, from http://dazeinfo.com/2014/01/23/smartphone-users-growth-mobile-internet-2014-2017/

Jiang, X., & Zhou, Y. (2013). *Android Malware.* Springer.

Jing, Y., Zhao, Z., Ahn, G. J., & Hu, H. (2014). Morpheus: Automatically Generating Heuristics to Detect Android Emulators. *Proceedings of the 30th Annual Computer Security Applications Conference(ACSAC).* doi:10.1145/2664243.2664250

Lantz, P. (2011). *Droidbox - android application sandbox.* Retrieved January 20, 2015, from http://code.google.com/p/droidbox/

Malgenome. (n.d.). Retrieved 2014, from http://www.malgenomeproject.org/policy.html

Masud, M. Z., Sahib, S., Abdollah, M. F., Selamat, S. R., & Yusof, R. (2014). Android Malware Detection System Classification. Research Journal of Information Technology, 325-341.

Mobile/Tablet Operating System Market Share (MOBILE Share). (2015). Retrieved January 20, 2015, from https://www.netmarketshare.com/operating-system-market-share.aspx?qprid=8&qpcustomd=1

Neuner, S., Veen, V. V. D., Lindorfer, M., Huber, M., Merzdovnik, G., Mulazzani, M., & Weippl, E. (2014). Enter Sandbox: Android Sandbox Comparison. *Proceedings of the 3rd IEEE Mobile Security Technologies Workshop (MoST).*

Petsas, T., Voyatzis, G., Athanasopoulos, E., Polychronakis, M., & Ioannidis, S. (2014). Rage Against the Virtual Machine: Hindering Dynamic Analysis of Android Malware. *Proceedings of the Seventh European Workshop on System Security (EuroSec).* doi:10.1145/2592791.2592796

SE-Linux. Validating Security-Enhanced Linux. (2013). Retrieved January 20, 2015, from. http://source.android.com/devices/tech/security/se-linux.html

Sen, S. A., Clark, J., & Tapiador, J. (2010). Security threats in mobile Ad hoc networks. Security of Self-Organizing Networks: MANET, WSN, WMN, VANET. Taylor & Francis. doi:10.1201/EBK1439819197-9

Shih, D. H., Lin, B., Chiang, H. S., & Shih, M. H. (2008). Security aspects of mobile phone virus: A critical survey. *Industrial Management & Data Systems, 108*(4), 478–494. doi:10.1108/02635570810868344

Spreitzenbarth, M., Schreck, T., Echtler, F., Arp, D., & Hoffmann, J. (2014). Mobile-Sandbox: combining static and dynamic analysis with machine-learning techniques. International Journal of Information Security.

Tangil, G. S., Tapiador, J. E., Lopez, P., & Blasco, J. (2014). Dendroid: A Text Mining Approach to Analyzing and Classifying Code Structures in Android Malware Families. *Expert Systems with Applications, 41*(4), 1104–1117. doi:10.1016/j.eswa.2013.07.106

Types of Wireless Communication Technologies (WIRELESS Types). (n.d.). Retrieved January 20, 2015, from http://www.ehow.com/info_12157177_types-wireless-communication-technologies.html

Vidas, T., & Christin, N. (2014). Evading Android Runtime Analysis via Sandbox Detection. *Proceedings of the 9th ACM Symposium on Information, Computer and Communications Security (ASIACCS) Virustotal*. Retrieved January 20, 2015, from www.virustotal.com

Visual Networking Index (VNI). (n.d.). Retrieved January 20, 2015, from http://www.cisco.com/c/en/us/solutions/service-provider/visual-networking-index-vni/index.html#~mobilevni

Zhauniarovich, Y. (2014). *Android Security (and Not)*. Retrieved January 20, 2015, from Iwww.zhauniarovich.com/files/asani/asani.pdf

Chapter 4
Mobile Networks and Indian Agricultural Sector

Lokesh Jain
UIET, Panjab University, India

Harish Kumar
UIET, Panjab University, India

ABSTRACT

Information dissemination in agricultural sector for its growth using information and communication technology (ICT) as a tool is need of the hour. This can be achieved using information systems. ICT benefits are helpful in exchange and dissemination of information among farming stakeholders. By using the latest tool of mobile technology, farmers can get the current information related to their farming jobs around the clock and at any location, as the mobile network have touched every part/location of the India. Using the features of the mobile-phones like GPS etc. one can get the localized information. Only need is to structure the abundant information available across the various organizations. So, a mobile based agricultural information system framework 'mAgIDS' has been proposed employing the hybrid mobile application architecture approach. Client-server architecture using the location Application Programming Interface (API) has been proposed. Inference mechanism of the system has implemented on the basis of improved fuzzy rule promotion technique.

INTRODUCTION

Agriculture and its related functions are indisputably the largest source of livelihood to large number of Indian rural population. Agriculture sector accounts for approximately 14% of India's Gross Domestic Product (GDP) and 11% of its export. 68.8% of Indian population lives in rural areas. Income of 31.55% of this rural population depends upon agriculture and its related activities. Also, agricultural produce is a source of raw material for a large number of industries. Average rate of agricultural growth in India was 2.38% during 10th five year plan (2002-07) that rose to 3.6% per annum during 11th five year plan (2007-12). In 2012-13, the first revised estimate of agricultural growth has been declared at 1.42% per annum. In advance estimates for India-Macro Economic survey for 2013-14 (Annonymous, 2014), 4.64%

DOI: 10.4018/978-1-5225-2342-0.ch004

of agricultural growth has been expected to be contributed in Indian GDP. It is important to increase growth of agriculture production (Government of India, 2013) using technology to achieve targeted 8% GDP contribution during 12th five year plan (2012-17) and to meet escalating demands for food due to high population growth, increase in per capita income *etc.* To improve the quality and quantity of agricultural produce, a tremendous amount of efforts are being put by various organizations like Indian Council of Agricultural Research (ICAR), Central and State Agricultural Universities (SAUs) and their Regional Research Stations (RRSs), Krishi Vigyan Kendras (KVKs) along with other public and private organizations. It is very important for these organizations to disseminate appropriate, precise and latest information to farmers for timely action to improve the agricultural output. There is a greater need to lay emphasis on the dissemination of scientific information from laboratory and test fields to its actual stakeholders using latest technological tools. These institutions work dedicatedly for the welfare of agricultural community but their initiatives have been usually criticized for their limited coverage, sustainability and its effect on agricultural community. Therefore, it is required that agricultural inputs should be timely disseminated to its appropriate quarters using latest information technological tools provided relevant quarters are ready to adopt these tools required for information transfer. Improvement of agricultural and rural development is being focused through emerging field of E-Agriculture using improved information and communication processes. Print and electronic media such as news-papers, pamphlets, libraries, televisions and information centers (like Kissan Call Centers *etc.*) are playing very important role in providing information to the farmers. These information centers are basically providing information to farmers using the available resources to them. Improved data management for better decision-making can be achieved by the utilization of information with appropriate information and communication technology (ICT). Enhanced focus on research planning, monitoring and evaluation lead to better research and hence it can provide more benefits to farmers. For providing online information to distantly located stakeholders ICT can be used in breaking the blockades of these remote places, which is the major concern for information dissemination.

Information Systems

An Information system (IS) uses Information Technology (IT) to sustain various operations and management of people's activities. (Silver, Markus, & Beath, 1995) discusses two views on IS namely IS-centered view comprising of hardware, software, procedures, data and people, and managerial view that incorporates people, business processes *etc.* ISs are specifically designed to enable individuals to execute these tasks for which the individual understanding is not so proficient, such as, handling bulk amount of information, performing scientific calculations, and controlling many parallel processes. Thus, such systems are very helpful for providing the best and up to date information to associated stakeholders. IS acts as an interface between individuals, algorithmic processes, data and technology (Information System, 2011). It also refers to the manner in which individuals interact with technological tools in support of assembly of allied, planned actions or tasks that can fabricate a particular service or product or serve a specific goal for stakeholders. It can be envisioned as flowchart of a sequence of events to achieve a certain product or service. The activities of an IS are committed to process information by means of capturing, communicating, keeping, recovering, handling and exhibiting the information. As such, IS inter-relates with data and activity systems. These are a type of communication systems in which data is represented and processed to generate information which is useful to the end users. Human decision-making and action is supported by considering it as a semi-formal language. ISs are seen as having the ICT component to help in controlling the performance of business processes. When an individual realizes and appraises the processed information then it becomes knowledge (Kürşat DEMİRYÜREK, 2010). These are social systems whose performance is inclined to the aims, principles and philosophies of people, as well as the performance of the technology (Angell & Smithson, n.d.).

Information and Communication Technology

World Bank report (World Bank, 2011) mentions that private and public sector are involved in research to find efficient answers to address short and long term concerns and challenges in agriculture that includes means and methods for responding to the information needs of the farmers. ICT based solutions have incredible potential to improve agriculture in emerging countries like India. The costly, bulky and energy-consuming equipment which were once available to the very few to store and analyze agricultural and scientific data, are now easily available with the farming community. With the booming wireless, Internet and mobile communication industries and reducing cost of electronic gadgets, ICT has become a part of daily routine of poor smallholder farmers. Its ability to bring recharged momentum to agriculture appears more convincing in view of increasing investments in agricultural research, strong attention in the development & spread of ICT by the private and public sector and the expansion of organizations committed to the agricultural development programmes. As described in (Vu, 2011), emerging nations need a more strategic focus on promotion of ICT penetration as a vital source of development, this includes the up-gradation of the ICT infrastructure and reduction of costs of its use and focus to increase the vital effects of its penetration. This can be helpful in improvement of the availability of information and technology, nurturing the novelty and exchange, and augmenting the excellence of decision making.

Emergence of digital convergence, explosive growth of wireless communications and increasing penetration of broadband technologies all over the world have made the adoption of ICT necessary to explode socio-economic development (Armenta, Serrano, Cabrera, & Conte, 2012). Availability and utilization of Telecom Infrastructure is invariably linked to the living standards and economic development in emerging countries (Garbacz & Jr, 2007). Pricing for telecom services like mobile phones and value added services have emerged as a vital factor that drives the quick adoption of mobile phones in emerging nations like India, Sri Lanka, China *etc.* Information about the affordability of ICT services (i.e. fixed line phones, mobile phones and fixed and mobile broadband Internet subscriptions) as adjudged against average gross national income per capita is expressed in terms of ICT Price Basket has been devised by International Telecommunication Union (ITU) (International Telecommunication Union (ITU), 2011a). Recent statistics indicate that cost for Internet access and mobile phones have decreased considerably throughout the world. Specifically, the cost of ICT has significantly decreased in Asian countries in previous decade. Thus, access to ICT services has become quite reasonable in these nations. This situation has unlocked the gate for a range of mobile-based services and applications that can be implemented in the region.

Creation, assessment and manipulation of information have been enabled by ICT as a combination of IT and communication technology (Information and communication technology, 2011). It can play a considerable task in conserving the attributes of information as it comprises computer, communication and information. These are concerned for handling, trading and supervising data, information and knowledge. Benefits of ICT include:

- Preservation of properties of information during its processing, exchange and management;
- Capability to record text, diagrams, multimedia, procedure descriptions *etc.*, in digitized formats;
- Production of duplicates of requisite information at lower cost;
- Transfer knowledge and information instantaneously all over the world through networks;
- Processing of large information rapidly using predefined algorithms;
- Achieving greater interactivity in connecting, appraising, creating and sharing useful knowledge and information.

The encouraging effect of ICT is quite evident in service oriented sector, where the productivity has gone high. Business process outsourcing, financial sectors *etc.* are taking full benefit of the ICT revolution.

Agricultural Information Needs and ICT

The application of ICT in agriculture is important. E-Agriculture (Information and communication technology, 2011) is an emerging area that focuses on the escalation of rural and agricultural development through superior information and communication processes. It involves the conceptualization, strategy, expansion, appraisal and use of innovative ways to ICT in the rural area, with a major attention on agriculture. Various E-Agriculture initiatives include information dissemination system (IDS), weather projection, seed sowing using sensor, irrigation monitoring *etc.* It is expected that scope of E-Agriculture will further evolve, as the understanding in the agricultural area improves.

Information requirements of farmers have also increased as the agricultural sector has been affected by latest global technological advancements. The international market requirements and ecological issues along with the national food security must be addressed by the agricultural development. Indian agriculture being a part of the global agricultural system has to keep momentum with continuous and immensely changing technologies and have to develop new products and product varieties, expanding and harnessing ICT for development of ISs, new market tactics, new services and so on (P. Singh, 2004).

In past, pamphlets, posters, radio, television *etc.* were used for the dissemination of information to the farmers. Huge amount of time was wasted in reaching the information to the appropriate quarters. The need for providing the accurate agriculture related information to the concerned at appropriate time leads to application of ICT in agriculture.

Concepts of virtual community or e-community, E-publications of agricultural literature, various information networks, institutional repositories, social networks sites and information kiosks *etc.* are being employed for dissemination of agricultural information. Various ICT initiatives/tools like multimedia, web and mobile applications are being deployed for the dissemination of information among various stakeholders.

Need for Mobile Communication in Agriculture

Advancement and accessibility of ICT has been the ultimate communication revolution during the last few years. The decreasing cost of hardware, expansion of communication set-up and its availability at district and block level has opened up vast potential for agro-worker to reach the rural community in a more dedicated, accurate and specific manner. Understanding of hardware, software, ICT indicators and home/global networks can be a value addition to the knowledge base of extension functionaries. One such tool is mobile phone network. It is simple, economical and convenient to use. Illiterate people are also easily adopting the usage of the smart mobile phones as compared to the laptops/computers. Access to mobile networks is now available even in the remote regions. It is possible for everyone and everything to be connected and this connectivity has a huge impact on society. For community development, it is pronounced as the most transformative tool (Kirk et al., 2011). This connectivity not only offers people the ability to stay in touch with friends and family but it also provides access to finance, improved healthcare solution, supply chain efficiencies and increasingly automated mobility. ICT services are commercially successful and therefore, it is believed that these are more sustainable in long term and provide more benefits to society.

Countries like India have a rising mobile market, deep diffusion of mobiles and elementary ICT infrastructure. Mobile phone subscription has increased tremendously over the years and during 2005-10, there was 50.7% per annum rise in mobile phone subscription (UNESCAP, n.d.). In a report (International Telecommunication Union (ITU), 2011b), the number of mobiles subscription per 100 inhabitants was 61.4, where households having a computer were 6.1. Mobile phones is a reasonable tool for communication and learning than personal computers (The Economist, 2011). Mobile communication has become an important tool for voice and data transfer. Telecom Regulatory Authority of India (TRAI) has released data showing that there are 893.31 million mobile subscribers in India as on 31st January, 2014. About 40.75% of these subscribers belong to rural areas with a rural tele-density of 42.43 as against urban tele-density of 139.42 (TRAI, 2014). The mobile subscription base has been projected about 900 million in 2015-16 as compared to 90 million in 2005-06 (S. K. Singh, 2008). The base is already touching this projection in the year 2013-14. Mobile devices are now becoming the most suitable means to disseminate agriculture related information to farmers in rural areas with 1.21% monthly growth of rural mobile subscription. These devices have lot of benefits over computers/laptops such as low cost, easy interface and convenient usage by farmers and higher penetration in the emerging nations (Gichamba & Lukandu, 2012). Growth in the mobile subscriber base will have vital impact for the prospective plans of mobile service and infrastructure providers, handset manufactures and vendors. This projected growth may be due to Internet and customized applications for individual use on their mobile handset. Agriculture sector can also harness the advantage of this growth.

Mobile networks access is broadly available at district and block level as well as in remote areas. This has opened up wide opportunities for agricultural scientists and extension workers to work with agricultural community in more focused and personalized manner.

Information Dissemination Systems

Various initiatives have been taken by different organizations to cater to the needs of the agricultural stakeholders. These include development of Information dissemination systems like Experts System (ES) and Decision Support System (DSS) *etc.* Most of these have been focused for the use by the extension specialist and computer literate farmers. These systems have static database of information and knowledge, which is constantly being used to deliver the results. Agriculture is a complex process which has unpredictable and uncertain solutions. To get the inference for such solutions soft computing techniques are used. These techniques deal with inaccuracy, vagueness, fractional truth, and approximation to achieve practical, robust and low cost solution. As such it forms the basis of a considerable amount of machine learning techniques. Usage of these techniques in agricultural systems has given tremendous scope to provide the precise output.

Mobile Based Systems for Localization

Localized information needs of the farmers are very vital. Certain agricultural activities are prominent in some regions and may be irrelevant to the other nearby regions. Information processing and dissemination to the farmers on the basis of their geographical area is very crucial. Information provided in the native language is a boon for the farmers. This can generate interest among the farmers for information. Mobile systems /handsets with some additional features may help in getting the localized information. So, there is a need for the development of mobile based information dissemination systems for transferring the localized multi-lingual information from the knowledge repositories so that the information may be available round the clock.

Possible Gaps in Information Dissemination

Various ESs, DSSs for agricultural activities have been developed but most of these are standalone or web based systems or relevant to market related agricultural activities like commodity exchanges, web based information repositories *etc.*, These are confined to the knowledge base of domain expert and knowledge engineer.

Following are the possible Gaps for information dissemination in agricultural based activities:

Lack of Exploiting Soft Computing Techniques in Agriculture Decision Making and Information Dissemination

Agriculture itself is a vast field having complex activities in its domain. Understanding these complicated activities along with the techniques of soft computing, which are approximation to the accuracy of the decision and information dissemination, is a cumbersome job for an agricultural researcher. This area has been focused in this study.

Lack of Mobile Network Usage for Agricultural Information Dissemination

Various experts', decision support and fuzzy logic based systems for agricultural activities and processes have been developed. But most of these are standalone or web based system or they are related to agricultural marketing activities like commodity exchanges, AGMARKNET (www.agmarknet.nic.in), e-Choupal *etc.* IT knowledge is the basic requirement for a farmer to use these kinds of systems. It is felt that the mass understandable and economical technology needs to be exploited for dissemination of agricultural related information. The mobile phone network is one such popular technology that can be exploited easily for such information dissemination. Mobile phone networks are available in almost all the geographical areas of India and used by majority of farmers (Jain, Kumar, & Singla, 2011).

Localization of the Desired Agricultural Activity

Web based systems are not able to guess the exact location of the agricultural activity under review. Therefore it is required to use the telecommunication infrastructure to accurately define the location of agricultural activity under consideration. By getting this accurate information, the soft computing based information systems can give localized information for better decision making. It is possible to exploit the power of mobile technology to transfer this knowledge to the concerned at any location on 24x7x365 basis. The knowledge can be made available in the native language of the farmers. In this study, localization of desired agricultural activity has been achieved using the features of mobile phone like Global positioning System (GPS) for location detection, multilingual for local language of the farmers.

Information Dissemination in Agricultural Sector

Agricultural and its allied activities are the sources of income for most of the Indian rural population. Therefore it is very important to enhance their agricultural productivity, which is a complex and cumbersome activity. It requires a lot of practice and latest information about the latest tools, implements, procedures and knowledge about the various activities involved in practicing agriculture. Various methods have been used to transfer these latest information to the farmers that include, pamphlets, farmer fests, multimedia involving television and radio talks, public and private organizations (like ICAR's and SAU's research station, KVK's, ITC, TATA Consultancy Services), Internet based web and mobile applications etc.

ICAR is premier organization committed to developing and governing agriculture under Ministry of Agriculture, Government of India. It has a number of specialized research institutes for different crops, animals, soils, regions *etc.* Under its leadership, the agricultural education, research and extension programmes are also governed which involves the state agricultural and veterinary universities and other organizations. These institutes are committed to the transfer of knowledge and information available to them through research and other means to the farmers through different types of extension initiatives.

Role of ICT in Agriculture

The potential of ICT to contribute towards agriculture and rural development of a nation has been well recognized e.g. Establishment of Rural ICT centers in Iran has played an efficient role in rural development through improving e-governance in these areas (Alibaygi, Karamidehkordi, & Karamidehkordi, 2011). In order to make the most of this potential, nations have two issues: firstly, empowerment of small farmers with ICT services and assets, which can improve their productivity, incomes as well as which can secure their food security and livings, and secondly harnessing ICT commendably to compete in swiftly changing and complex global markets. These issues can be addressed by execution of a complex set of strategy, innovation, investment, and capacity-building processes *etc.*, in consultation with concerned stakeholders. This can inspire the growth of localized applicable, cheap, and viable ICT setup and facilities for the rural economy (World Bank, 2011).

The association of the usage and adoption of ICT by farmers practicing conventional extension systems as well as policy suggestions for bridging the evolving digital gap within the agricultural community have been described in (Anastasios, Koutsouris, & Konstadinos, 2010).

Nokia life tool is an example of agriculture based information systems on mobiles, which is a payment based information service intended for developing markets. It provides a vast variety of information service. It covers education, entertainment, healthcare and agriculture ("Nokia Life Tools - Wikipedia, the free encyclopedia," n.d.). mKrishi, an initiative of TATA Innovation Lab, is the one of the good mobile application involved in disseminating the agriculture related information to the farmers at the nominal cost. It associates farmers with a system that assists them to make sound decisions about agriculture which can drive higher profits and preserve the environment. With the help of Integrated Voice Response (IVR) technologies the software can also transport voice messages between farmer and the expert. aAQUA which stands for almost All Questions Answered, is an online question & answer and community discussion forum for generating and providing information significant to farmers (Ramamritham, 2004), (Ramamritham, Bahuman, Duttagupta, Bahuman, & Balasundaram, 2005). It allows members using a web browser on a computer to create, view and manage content in local languages. A large number of farmers have reported fairly affirmative changes in quality of yield, crop diversification due to Agropedia and aAQUA (Yadav, 2011) thus improving their income. Agropedia is a good resource of the agricultural information ("Agropedia," n.d.).

Use of Multimedia in Agriculture

Green revolution in India is possible because of the transfer of technology from the research fields to the farm field. In earlier years, the print media played vital role in transforming the agricultural productivity from extremely low to reasonably good point. Information dissemination using the CDs and DVDs have further strengthened agricultural output.

Multimedia, which includes all the customary forms of communication like – audio, video, photographs, graphics, animation and text, is an admirable tool for extension and communication. Each learning task is most effectively taught through the optimum usage of relevant media. Multimedia describes fusion of text, records, visuals, animation, optical storage, image processing and sound. Using the multimedia data types such as video, graphics, images or animation, any participant can introduce illustrative material to facilitate better communication. Incredible images and animations, audio, audio-visual clips and raw textual information can inform, educate, persuade and entertain the users. This can help in electrifying the public's mind. Thus, the process of transfer of technology, from the laboratories and research field, among the stakeholders must be strengthened and must reach the farmers at the earliest (Philip, Vennila, & Gomathy, n.d.).

Multimedia as transfer of technology can function in the field of agricultural extension. It can be used to illustrate the issues in an interactive, interesting and effective format. Photographs, pictures *etc.* of symptoms of diseases, insect pest damage and of insects can be made available using multimedia for their effective monitoring and management. Solutions of the problems can also be depicted using figures, photographs or pictures. Using a chat room, crop pest specialists can be consulted to know how to identify accurately and control the pests. Quality of extension material and process can be supplemented time to time by multimedia and virtual reality.

Now a day, work is being done to attain the second green revolution. This can be achieved by harnessing the power of ICT. Usage of various tools of ICT can be very helpful to agricultural community. The Digital Green (Gandhi, Veeraraghavan, Toyama, & Ramprasad, 2007) system propagates targeted agro-info to small and marginal land holders in India through digitized video. Concept of participatory learning is implemented in the system. Database of videos, which is generated by farmers and experts, is included in it. These videos are distributed to the farmers through a repository. Farmers are motivated and trained by these recorded experience. Development of Internet and online agricultural repositories has further added to enhance the agricultural output. The strengthening and broadening field of Internet over the mobile phones in developing nations can provide an exclusive interface for adoption of agricultural technologies using ICT.

Government policies and announcements can be made available to farmers and extension workers from time to time with the use of multimedia. The information on weather parameters, cropping systems availability of high yielding varieties, other related information on varieties, irrigation, fertilizer management, Integrated Pest management (IPM) *etc.* can be made available to farmers using multimedia and convince them easily. Multimedia offers immense potential today as well as for the future and it will help extension worker and scientist to integrate modelling, visualization and decision making processes associated with farming systems. This will definitely accelerate development in the field of agricultural extension and ultimately farmers of the country will be able to harvest the benefits at the desired level.

Web and Mobile Applications

ICT enables users to build, retrieve and handle the information. It is combination of IT and communications technology (Information and communication technology, 2011). In modern arena, it lays emphasis on the role of integrated communications and amalgamation of tele-communications, smart information managing arrangements and multimedia arrangements. Being an arrangement of three major technologies: Computer, Communication and Information Management, it is playing a vital role in preserving the characteristics of information. These tools are smeared for handling, trading and managing data, information and knowledge. With the help of various ICT tools, it is possible to produce exact copies of such information at considerably lesser cost, rapid transfer of information and knowledge globally through communication networks, develop standardized algorithms to synthesis quickly the vast quantities of information, achieve greater interactivity in producing, transferring, appraising, and sharing beneficial knowhow and information.

Mobile and Internet based communication are precise, fast and economical mode to transfer the information from research farms and laboratories for disseminating agriculture related information to farmers. It has also reached almost all the districts and administrative blocks of India (Chauhan, 2010). Farmers are eager and become concerned to get rapid, true and reliable information in globally changing agricultural scenario. Transfer of requisite and latest agricultural information to the farmers in scattered locations at diverse geographical situation in India is a very tough job. Still it can be said that development of technology and its dissemination to the farmers is a continuous process (Mehta, 2003). This has been tried to be accomplished by the various web portals (Raj, 2013).

Portable Chemical Analysis (PCA) using a mobile phone has been presented in (García et al., 2011). Built-in camera is used for clicking the image of a single-use colorimetric chemical sensor. A software application processes this image to obtain its characteristic H (hue) value related to analytic concentration. Using mobile phone the water and sand samples have been analysed using the reflectance measurement (Iqbal & Bjorklund, 2011). The screen of phone work as a light source and the detection is done by front view camera. PCA for the taxonomy of several compounds and their concentrations in water has been done using the reflected intensities of red, green, white and blue colours. For Arsenic detection in groundwater, discoloration caused, because of its presence in groundwater, was analysed with the help of model coloured sand samples. In (Sumriddetchkajorn, Chaitavon, & Intaravanne, 2013), it has been experimentally demonstrated that for analysis of chlorine concentration in water can be estimated with the help of embedded digital camera in a mobile device, which can be used to function as a colorimeter. The feasibility of using mobile phones cameras for DNA detection has been investigated in (Lee, Chou, Yeh, Chen, & Chen, 2011).

In (Confalonieri et al., 2013), a smartphone application has been discussed and it implements (App-L & App-G) methods for Leaf Area Index (LAI) estimation. LAI is an important parameter in environmental and agronomic studies. It can estimates the amount of radiation captured by the canopy and describes crop water requirements. It can be judged with the help of the developed application. This application is based on the usage of the sensors present in the most of the smart-phone and their processing powers. Gap fraction estimation at 57.5° (to obtain values that are almost free of leaf inclination) for luminance estimated above and below the canopy has been used in the first method (App-L). The gap fraction has been estimated via automatic processing of pictures obtained below the canopy using the second method (App-G). The data has been collected in scatter-seeded paddy farm-field in northern Italy. The performances of two methods implemented in application were assessed using this data. The results were matched with those of the LAI-2000 and AccuPAR ceptometer, by determining the methods' accuracy and linearity. The performances of App-G were similar to those shown by these instruments, whereas App-L attained the best true value, although it has resulted in lower precision, necessitating a great number of replications to provide trustworthy estimations. It has been shown that the proposed application can be considered as a substitute to commercially available instruments and useful in perspectives characterized by cost-effective resources or when the maximum portability is necessary.

A quick and mobile nitrogen management risk assessment tool, Nitrogen Index application for android smart-phones, has been developed that can be carried to the field and elsewhere (Delgado, Kowalski, & Tebbe, 2013). A desktop version of the Nitrogen Index, which can be used on desktop and laptop computer, has also been developed. Users can email the results from their Nitrogen Index application developed for smartphones/tablets in the field to their computer back at the office (or a farmer's or consultant's computer) using their portable device's Internet service. This portable technology can enable the use of small devices such as smart-phones to conduct an assessment on site and visit with a farmer at any given field site where a wireless connection is available. The Nitrogen Index is a quick tool, and its assessments of nitrogen loss risk are correlated with observed values. Nitrogen Index application is first of its kind in the world and it is assumed as the beginning of more advances to come in the field of portable agricultural technology. New advances in software (e.g., applications) and technology are bringing us to a new frontier of technology transfer, and portable technologies (e.g., smartphones, tablets) are making possible the development of '*smart* agriculture'.

Agricultural Information in Farmer's View

Mobile phone subscription is increasing day by day in rural areas of India. Farmers require the timely and precise information. To study the requirements of the farmer for information dissemination a survey had been conducted on a sample of farmers in Punjab, India(Jain, Kumar, & Singla, 2014). This survey included the usage of internet on mobile, computer for agricultural information needs. The information from the survey was compiled and it is found that 63% know Internet usage and only 38% of farmers surf the Internet to get agriculture-related information. Approximately 99% of farmers use mobile phones efficiently having multimedia features, etc., 95% of farmers opted for getting the information using SMS services etc.

Presentation of Agricultural Information

A lot of agricultural information is available and it is needed to be presented to the user so that he can best make the use of it to increase agricultural; productivity. Various information systems, expert systems, decision support systems have been developed in this direction.

Among ICTs for agriculture, Kissan Call Centre, an initiative of Government of India, e-Sagu (www.esagu.in), e-Arik (www.earik.in), India development gateway (www.indg.in) and e-aAqua (www.*aaqua.org*) are prominent examples (Saravanan, 2008), (Saravanan, 2010).

AgrIDS (Reddy & Ankaiah, n.d.) is a framework for an IT-based agricultural information dissemination system integrating IT and agriculture. The system have been developed using web technology and it aims to provide fresh and expert agricultural advice to needy farmers quickly and personalized manner for crop productivity improvement. It also aims to increase the cost-effectiveness of farmer by escalating the efficiency of agricultural inputs and decreasing the production cost. Some of the agricultural information systems have been provided in Table 1. Using Internet and communication networks, AgIS should capture various parameters like precise location, quality of soil, soil health, rainfall, agro-climatic conditions *etc.* of farm location to suggest best practices. Soft-computing techniques can process this set of parameters concurrently to deliver timely, precise and appropriate inferences. Various parameters like localized information along with generic information from the farmer are required to reach at a decision.

Farmers may not be comfortable or be trained to enter localized information into the computing systems. Automation can help to capture certain parameters. For example, longitude and latitude of the mobile device can be obtained using the communication networks automatically and this can be passed to the AgIS server, which in turn, can obtain the various other parameters like the type of soil, ground water level, weather conditions, prominent diseases *etc.* of the location. Soft-computing techniques, implemented at server level, process the generic and local information collected automatically using the AgIS. Processed information in the form of resultant expert like inference of agriculture practice can be delivered to the farmers in their native language, which can be helpful to them in decision making. It can enable them to improve their agricultural productivity and livelihood. However, there is need to involve agriculture experts, software developers and computer scientists to develop such a system in the local language of farmers. In some cases, language experts are also required to translate scientific knowledge

Table 1. Some agricultural information systems and their hyperlinks

Agricultural Information System	URL
aAQUA (almost All Question Answered)	http://www.aaqua.org
KISSANKERLA	http://www.kissankerla.net
Ministry of Agriculture: Farmer portal	http://farmer.gov.in
TNAU Agritech Portal	http://www.agritech.tnau.ac.in
DACNET	http://dacnet.nic.in
e-Krishi	http://www.ekrishi.org
InDG (India Development Gateway portal)	http://www.indg.in
RKMP (Rice Knowledge Management Portal)	http://www.rkmp.co.in
Agropedia	http://agopedia.iitk.ac.in
Mahindra's Kisan Mitra	http://www.mahindrakisanmitra.com
IFFCO Agri-Portal	http://www.iffco.nic.in
Agriwatch Portal	http://www.agriwatch.com
mKrishi, Tata Consultancy Services Limited	http://www.tcs.com/offerings/technology-products/mKRISHI/Pages/default.aspx

into the local language. All the diverse partners need to understand the practices of other professions, so that this kind of system should be in place.

The task of creating IS based on mobile application supporting native language of farmers to cater to the complex spectrum of Indian agriculture is a giant task. Complicated information may be compiled in a structured and user friendly AgIS available round the clock. Low cost mobile phone technology and its availability with most of the farmers can provide localized information to them. Content of AgIS in local languages can engender interest in farmers and others at grassroots level. Majority of farmers are not aware of such systems. Awareness can be generated by extension specialists at ADO, BDO, SAUs, and KVKs *etc.* Use of soft computing techniques to process information required in conjunction with mobile networks and technologies, for dissemination of complex agricultural information, is one of the ways in the future to deliver expert advice to farmers.

Mobile Based Agricultural Information System Framework

Creation of databases and applications to cater to the complex spectrum of Indian agricultural scenario is a challenging opportunity. Dissemination of agricfultural information to the farmers freely in their local language is a giant task which can be achieved using the knowledgebase of existing set-up of SAUs, agricultural research institutions, state agricultural departments *etc.* Skilled manpower available in these organizations can be useful in developing mobile applications. A structured AgIS can be formulated using this information stored in the database which can be made available round the clock freely using the electronic devices like mobile-phones, laptops, desktops *etc.* as shown in Figure 1. In the present conditions, penetration of Internet is increasing but the farming community is not well conversant with the use of Internet and computer. The commonly available mobile handset e.g. smart-phones are capable of installing these mobile based information systems for providing the localized information. Various backend techniques to process the information may be deployed for developing AgIs.

Harnessing the power of mobile based communication technology can provide the localized information which is very crucial at one location but may be irrelevant at another location. The Application Programme Interface (API), which is available in the mobile phone, can tell us the precise location and it may be used for fetching the various other parameters at that location such as soil type, soil fertility, agro-climatic conditions, ground level of water using the hybrid mobile application *etc.*

The following section discusses and proposes architecture to develop a mobile based dynamic and hybrid system, named as '*Mobile based Agricultural Information Dissemination System (mAgIDS)*'

Figure 1. Basic architecture for flow of agricultural information from/to AgIS

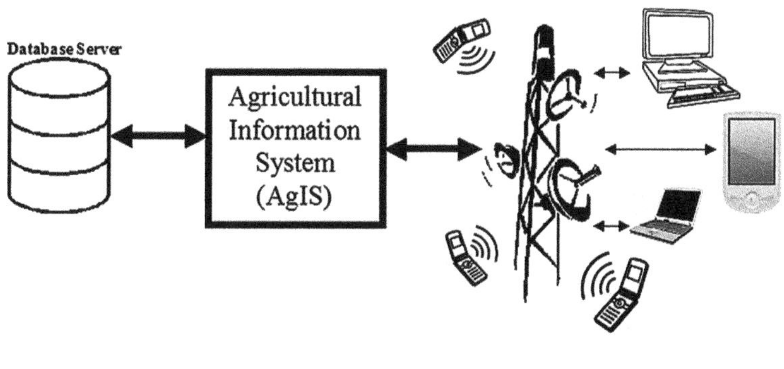

for various agricultural processes and activities based on the location of the mobile device being used and to propose the possible solutions as provided by the experts of agricultural process in the preferred language of the farmer.

Hybrid Mobile Application Architecture

Hybrid applications are those which install on the competing device but interact with the webserver for various data processes and requirements. The proposed architecture of the hybrid application is based on client-server methodology as shown in Figure 2. The application installed on an electronic device, for example, a mobile phone, is available to the web server through the Internet Cloud using either of the means like mobile operator provided Internet, Wi-Fi, hotspot *etc.* Web server interacts with database and application server for various internal processes and activities.

Database Server

The database server stores location maps, various agricultural activities and processes, their symptoms in the form of rule base, their weights as provided by agricultural experts to a symptom of an activity, user information and storage of final results of a selected location unit.

Application Server

Application server interacts with both web and database server for different activities. The web server sends a request to the application server with location parameters of the mobile device which in turns finds the exact smallest location unit name from the database using the predefined instructions employing

Figure 2. Hybrid mobile application architecture

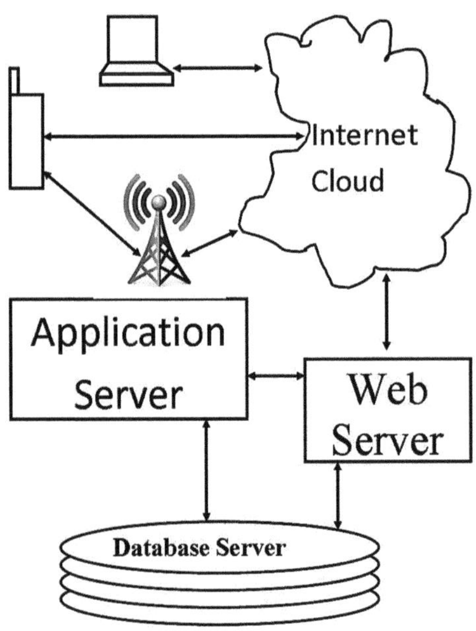

the location API (Jain, Kumar, & Singla, 2013). Then the web server sends other parameters related with agricultural activity under consideration to the application server for selecting symptoms of agricultural activity, which interacts with the database containing set of various activities, symptoms and rule-base of agricultural processes (Figure 3).

Localization of Information

Technological advances are opening up huge opportunities. With new technologies, unmatched developments, and booming economies, the face of the world of telecommunication and IT is changing rapidly. Integrating the locale into this change and adopting the technologies for maximizing the impact in rural areas is a challenging task (Nanda & Arunachalam, 2010). Mainly mobile technologies have created new channels to communicate with others through a well-located manner. The penetration of mobile applications in the rural market is increasing day-by-day due to reducing cost of smart phones, PDA, tablets *etc.* The mobile application developer can use the manufacture provided API of these phones/technologies to develop various applications for providing localized services to the farmers. For getting the location information at the micro-macro level (administrative block level), two techniques have been employed:

- Selection based techniques,
- Automatic location information.

There are 29 states and 7 union territories in India with overall 640 districts and 6451 administrative blocks (Census of India, 2011). Thus, the whole nation is divided into smaller units called administrative

Figure 3. Application server architecture

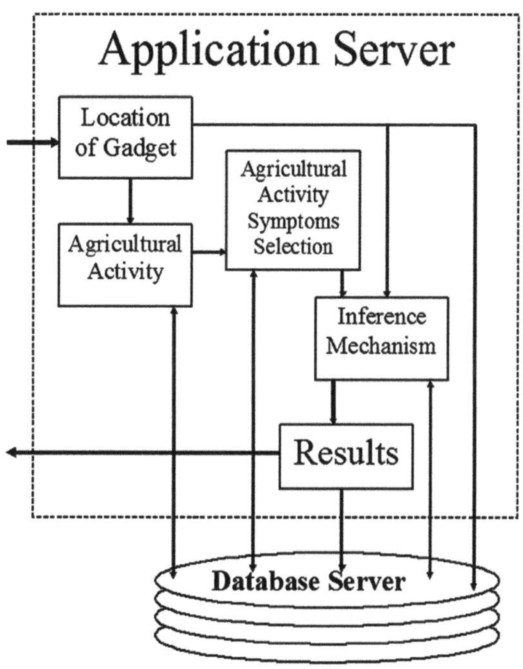

blocks. These blocks have diverse information related to geography, geology, administration, ecology *etc.* Thus localized information can be obtained if stored in the digital format. In the selection based techniques, a database is populated with the information of the state, district and blocks of the country, for example, in MNREGA (Mahatma Gandhi National Rural Employment Guarantee Act) database, desired information can be obtained by selecting the state, district and administrative blocks under observation.

In automatic location information technique, Global Positioning System (GPS) of a mobile device can pass information related to longitude and latitude to the server. Location API on server can process this information to get the geographical data of administrative blocks in which this device is located. Then based on this, geographic agricultural data can be extracted and disseminated to this device. The server Location API can access the Geographical Information System (GIS) maps, which is imported in the database server supporting spatial information. This information is used further for processing of the other parameters. Both techniques have been described in Figure 4.

Figure 5 describes the way an electronic device, whether GPS enabled or not, accesses the Location API and uses the other APIs to process the location information. This information is used in the ISs for getting localized knowledge. Storing the information about the local spoken language can further localize it. Based on this input, data can be disseminated in the localized format.

Inference Mechanism for an AgIS

Various techniques of Soft computing like fuzzy logic, artificial neural networks, Genetic algorithm, support vector machines, Bayesians approaches etc. are being used for making the inferences. These techniques give precise result which can be used for delivering the precise agricultural information.

Figure 4. Flow chart for implementation of location based search

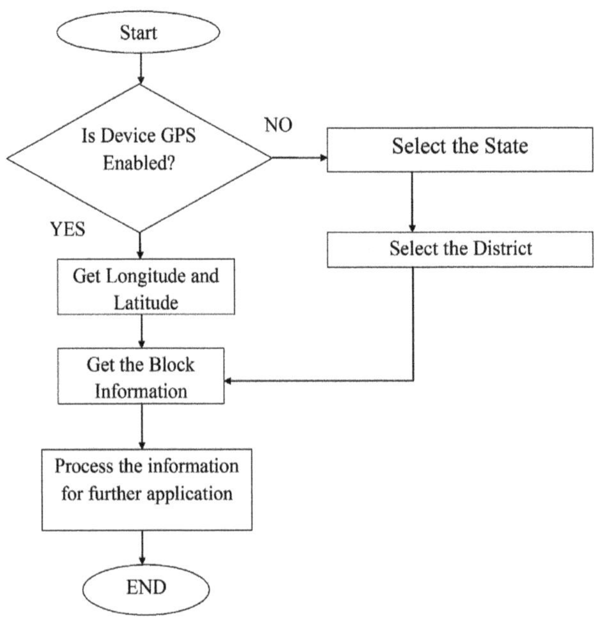

Figure 5. Client-server architecture using the location API

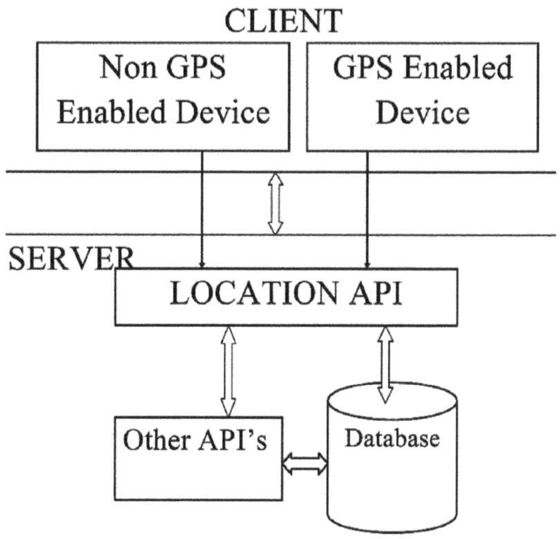

Mobile Based AgIS (mAgIDS)

Various initiatives have been taken to disseminate the agricultural information to farmers. An initiative name 'mAgIDS' has been taken to provide the agricultural information precisely and round the clock. This Android-based mobile phone application 'mAgIDS', using the improved dynamic fuzzy rule promotion (Jain, Kumar, Singla, Singh, & Singh, 2014) based technique to infer the advice to farmers, has been developed.

A dynamic fuzzy rule promotion approach (Kolhe, Kamal, Saini, & Gupta, 2011) for the promotion of confidence factor of a rule for every successful session for diagnosis of activity by using the specific rules have already been proposed in the literature. This technique has the limitation that error in the initial estimation of weights reduces linearly after every session the rule is being used. An improved approach has been proposed using square root of sum of squares of frequencies, which are spread around mean true value to reduce error around mean value. The improved approach decreases error in uncertainty of estimation of weight for rules after every successful session. This approach has been used in the 'mAgIDS' application.

The application uses Global Positioning System (GPS) for localized inference, and local language selected by the user. The required information is collected and transferred to the application server over the mobile networks. This information is processed incorporating the improved dynamic rule promotion technique. Application has been tested on the dataset of various symptoms (rules) for disease diagnosis (activity) of paddy crop. Chi-Square ($\chi 2$) test has been used to match the result of the mAgIDS system with the expert's inferences. It has been found that the results of the system and experts match with accuracy of 95.84% at 5% level of significance. Twenty two scientists working at various locations within Punjab state in India were contacted for validation of the system with fifteen relevant parameters. It has been found that the proposed system is statistically valid. The real time application of this system is in process and implementation of the system can provide immense and timely help to the farmers for taking appropriate decisions for their crop production.

CONCLUSION

Generic Information and knowledge furnished to farmers may not be relevant to them at their respective geographical locations. Dissemination of local level information through mobile networks can help them to improve their productivity and hence socioeconomic aspects. The experts can get the information about the location unit where a particular agricultural activity is more prevalent and it can help them to research that location unit. The information being requested can be processed for use in further sessions to get improvement in decision level and this can make the decisions dynamic based on previous experiences. Information dissemination in local language is one of the main requirements for proper transfer of agricultural information to the user and has to be explored.

REFERENCES

Agropedia. (2014, Feb 10). Retrieved Agropedia site from: http://agropedia.iitk.ac.in/

Alibaygi, A., Karamidehkordi, M., & Karamidehkordi, E. (2011). Effectiveness of rural ICT centers: A perspective from west of Iran. *Procedia Computer Science, 3*, 1184–1188. Retrieved from http://www.sciencedirect.com/science/article/pii/S1877050910005661

Anastasios, M., Koutsouris, A., & Konstadinos, M. (2010). Information and Communication Technologies as Agricultural Extension Tools: A Survey among Farmers in West Macedonia, Greece. *Journal of Agricultural Education and Extension, 16*(3), 249–263. doi:10.1080/1389224X.2010.489767

Angell, I. O., & Smithson, S. (1991). *Information Systems Management: Opportunity and Risk.* London: Palgrave Macmillan. doi:10.1007/978-1-349-21555-3

Annonymous. (2014). *Data Tables: Planning Commission, Government of India.* Planning Commission of India. Retrieved from http://planningcommission.nic.in/data/datatable/

Armenta, A., Serrano, A., Cabrera, M., & Conte, R. (2012). The new digital divide: The confluence of broadband penetration, sustainable development, technology adoption and community participation. *Information Technology for Development, 18*(4), 345–353. doi:10.1080/02681102.2011.625925

Chauhan, N. M. (2010). *Expectations of the Farmers from ICT in Agriculture.* Academic Press.

Confalonieri, R., Foi, M., Casa, R., Aquaro, S., Tona, E., Peterle, M., & Acutis, M. et al. (2013). Development of an app for estimating leaf area index using a smartphone. Trueness and precision determination and comparison with other indirect methods. *Computers and Electronics in Agriculture, 96*, 67–74. doi:10.1016/j.compag.2013.04.019

Delgado, J. A., Kowalski, K., & Tebbe, C. (2013). The first Nitrogen Index app for mobile devices: Using portable technology for smart agricultural management. *Computers and Electronics in Agriculture, 91*, 121–123. doi:10.1016/j.compag.2012.12.008

Demiryürek. (2010). Information systems and communication networks for agriculture and rural people. *Agricltural Economics – Czech, 56*(5), 209–214.

Gandhi, R., Veeraraghavan, R., Toyama, K., & Ramprasad, V. (2007). Digital Green: Participatory video for agricultural extension. In *Information and Communication Technologies and Development, 2007. ICTD 2007. International Conference on* (pp. 1–10). IEEE. doi:10.1109/ICTD.2007.4937388

Garbacz, C., & Jr, H. T. (2007). Demand for telecommunication services in developing countries. *Telecommunications Policy*, *31*(5), 276–289. doi:10.1016/j.telpol.2007.03.007

García, A., Erenas, M. M., Marinetto, E. D., Abad, C. A., de Orbe-Paya, I., Palma, A. J., & Capitán-Vallvey, L. F. (2011). Mobile phone platform as portable chemical analyzer. *Sensors and Actuators. B, Chemical*, *156*(1), 350–359. doi:10.1016/j.snb.2011.04.045

Gichamba, A., & Lukandu, I. (2012). A Model for designing M-Agriculture Applications for Dairy Farming. *The African Journal of Information Systems*, *4*(4). Retrieved from http://digitalcommons.kennesaw.edu/ajis/vol4/iss4/1

Government of India. (2013). State of Indian Agriculture. Ministry of Indian Agriculture, Department of Agriculture & Cooperation, New Delhi.

Information and Communication Technology. (2011). In *Wikipedia*. Retrieved June 15, 2011, from http://en.wikipedia.org/wiki/Information_and_communication_technologies

Information System. (2011). In *Wikipedia*. Retrieved June 15, 2011, from http://en.wikipedia.org/wiki/Information_system

International Telecommunication Union (ITU). (2011a, Jan 6). *ICT services getting more affordable worldwide*. Retrieved from International Telecommunication Union website: http://www.itu.int/net/pressoffice/press_releases/2011/15.aspx#.Ulf2LdJkPE0

International Telecommunication Union (ITU). (2011b, Nov 12). *World Telecommunication /ICT Indicators Database*. Retrieved from International Telecommunication Union website: http://www.itu.int/ITUD/ict/statistics/#.Ulf2LdJkPE0

Iqbal, Z., & Bjorklund, R. B. (2011). Colorimetric analysis of water and sand samples performed on a mobile phone. *Talanta*, *84*(4), 1118–1123. doi:10.1016/j.talanta.2011.03.016 PMID:21530787

Jain, L., Kumar, H., & Singla, R. K. (2011). A Review of Fuzzy Rule Promotion Techniques in Agriculture Information System. *International Journal of Computer Applications*, 55–60.

Jain, L., Kumar, H., & Singla, R. K. (2013). *Localization of Information Dissemination in Agriculture using Mobile Networks Changing Agricultural Scenario and Information Needs Use Multimedia in Agriculture Web and Mobile Applications Need for Mobile Phone based Communication in Agriculture Lack of*. Academic Press.

Jain, L., Kumar, H., & Singla, R. K. (2014). Assessing Mobile Technology Usage for Knowledge Dissemination among Farmers in Punjab. *Information Technology for Development*, 1–9. doi:10.1080/02681102.2013.874325

Jain, L., Kumar, H., Singla, R. K., Singh, P., & Singh, L. J. (2014). Error Reduction in Promoted Confidence Factor of a Rule Using Improved Fuzzy Rule Promotion Technique. *Cybernetics and Information Technologies*, *14*(1), 72–83. doi:10.2478/cait-2014-0006

Kirk, M., Steele, J., Delbe, C., Crow, L., Keeble, J., Fricke, C., … Bulloch, G. (2011). *Connected Agriculture: The role of mobile in driving efficiency and sustainability in the food and agriculture value chain*. Retrieved from http://www.accenture.com/SiteCollectionDocuments/PDF/Accenture-Connected-Agriculture.pdf

Kolhe, S., Kamal, R., Saini, H. S., & Gupta, G. K. (2011). A web-based intelligent disease-diagnosis system using a new fuzzy-logic based approach for drawing the inferences in crops. *Computers and Electronics in Agriculture*, *76*(1), 16–27. doi:10.1016/j.compag.2011.01.002

Lee, D., Chou, W. P., Yeh, S. H., Chen, P. J., & Chen, P. H. (2011). DNA detection using commercial mobile phones. *Biosensors & Bioelectronics*, *26*(11), 4349–4354. doi:10.1016/j.bios.2011.04.036 PMID:21621401

Mehta, P. (2003). Information Technology in Agriculture: Reaching the Unreached. *National Workshop on ICT for Agriculture and Rural Development*, Ahmedabad.

Nanda, S., & Arunachalam, S. (2010). Reaching the unreached (Vol. 7). Chennai: Jamsetji Tata National Virtual Academy (NVA), M S Swaminathan Research Foundation.

Nokia Life Tools. (n.d.). In *Wikipedia.* Retrieved February 10, 2014, from http://en.wikipedia.org/wiki/Nokia_Life_Tools

Philip, H., Vennila, M. A., & Gomathy, M. (2003). Multimedia: an effective communication technology for extension | OpenAgri: An Open Access Agriculture Research Repository. In B. Jirli (Ed.), International Conference on Communication for Development in the Information Age: Extending the Benefits of Technology for All., Varanasi, India. Varanasi, India: Department of Extension Education, Institute of Agricultural Sciences, Banaras Hindu University. Retrieved from http://agropedia.iitk.ac.in/openaccess/?q=content/multimedia-effective-communication-technology-extension

Raj, S. (2013). e-Agriculture Prototype for Knowledge Facilitation among Tribal Farmers of North-East India: Innovations, Impact and Lessons. *Journal of Agricultural Education and Extension*, *19*(2), 113–131. doi:10.1080/1389224X.2012.718247

Ramamritham, K. (2004). aAqua : A Database-backended Multilingual. *Multimedia Community Forum*, 1–4.

Ramamritham, K., Bahuman, A., Duttagupta, S., Bahuman, C., & Balasundaram, S. (2005). *Innovative ICT Tools for Information Provision in Agricultural Extension (December 2005)*. Retrieved from http://www.cse.iitb.ac.in/~krithi/papers/ ICTD2006.pdf

Reddy, P. K., & Ankaiah, R. (n.d.). A framework of information technology-based agriculture information dissemination system to improve crop productivity. *Current Science, 88*(12), 1905–1913. Retrieved from http://tejas.serc.iisc.ernet.in/~currsci/ jun252005/1905.pdf

Saravanan, R. (2008). *Agricultural extension: worldwide innovations*. New Delhi: New India Publishing Agency.

Saravanan, R. (2010). *Icts for@ Agricultural Extension: Global Experiments, Innovations and Expereience*. New Delhi: New India Publishing Agency.

Silver, M. S., Markus, M. L., & Beath, C. M. (1995). The Information Technology Interaction Model: A Foundation for the MBA Core Course. *Management Information Systems Quarterly*, *19*(3), 361–390. doi:10.2307/249600

Singh, P. (2004). Indian Agricultural Development in Changing Scenario Past, Present and Future. *Journal of the Indian Society of Agricultural Statistics*, *58*, 37–49. Retrieved from http://isas.org.in/jsp/volume/vol58/panjabsingh.pdf

Singh, S. K. (2008). The diffusion of mobile phones in India. *Telecommunications Policy*, *32*(9-10), 642–651. doi:10.1016/j.telpol.2008.07.005

Sumriddetchkajorn, S., Chaitavon, K., & Intaravanne, Y. (2013). Mobile device-based self-referencing colorimeter for monitoring chlorine concentration in water. *Sensors and Actuators. B, Chemical*, *182*, 592–597. doi:10.1016/j.snb.2013.03.080

The Economist. (2011). Beyond the PC. *The Economist*. Retrieved November 12, 2012, from http://www.economist.com/node/21531109

TRAI. (2014). *Telecom Subscription Data as on 31st January, 2014*. Retrieved from http://trai.gov.in/Content/PressDetails/2135_0.aspx

UNESCAP. (2011). *Statistical Yearbook for Asia and the Pacific 2011*. Retrieved from http://www.unescap.org/stat/data/syb2011/

Vu, K. M. (2011). ICT as a Source of Economic Growth in the Information Age: Empirical Evidence from the 19962005 Period. *Telecommunications Policy*, *35*(4), 357–372. doi:10.1016/j.telpol.2011.02.008

World Bank. (2011). *ICT in Agriculture: Connecting Smallholders to Knowledge, Networks, and Institutions*. World Bank.

Yadav, K. (2011, April 4). *Impact Assessment of ICT – enabled Knowledge Sharing Agri-portals in Uttarakhand*. Govind Ballabh Pant University of Agriculture & Technology, Pant Nagar, Uttrakhand. Retrieved from http://oar.icrisat.org/id/eprint/5845

Chapter 5
Wireless Networks:
Greedy Routing in Vehicular Area Networks

Monika Rani
UIET, Panjab University, India

Harish Kumar
UIET, Panjab University, India

ABSTRACT

VANET is a type of Mobile Ad-hoc network (MANET) that provides wireless communication between close by moving vehicles as well as between vehicles and road side fixed equipment. It requires the vehicles to be equipped with sensors and wireless devices and fixed roadside wireless infrastructure. There are number of research issues like Highly Dynamic Topology, Signal fading, Limited Bandwidth, Intermittent Connectivity, Mobility Modelling, Propagation Model, Routing which needs attention. The scope of this chapter is focused on various advances in Greedy Forwarding routing techniques. Various VANET routing protocols are reviewed and compared in this chapter.

1. INTRODUCTION

The automotive industry has been the fastest growing industry in economic sector that employs millions of employees throughout the world. Over decades of years, more and more efforts are being paid to road infrastructure and automotive technology. The dynamic evolution in electronics has transformed the mechanical machines into intelligent shipper. The pace with which transportation industry is moving ahead has a great impact in the growth of our society. But on the darker side, increasing number of vehicles has also increased traffic and the number of road accidents. This is considered as the biggest issue being faced in metro cities. To avoid these accidents, automotive industry is focused towards creating and improving safety systems inside the vehicles e.g. security belt, airbags. However, these systems can decrease the consequences of accidents but not the number of accidents. The solution is to practice proactive approaches for accident prevention e.g. to increase the driver's knowledge about traffic density and surrounding environment. It is common to find features like Adaptive Cruise control in today's cars

DOI: 10.4018/978-1-5225-2342-0.ch005

that use forward looking radar to detect the speed and distance of the vehicle moving ahead to maintain safe following distance. But these sensors have a limited action area.

The development in radio frequency technologies and above issues gave rise to the development of new type of wireless communication especially for vehicles that is Vehicular Ad-hoc Networks (VANET). VANET is a type of Mobile Ad-hoc network (MANET) that provides wireless communication between close by moving vehicles as well as between vehicles and road side fixed equipments. It requires the vehicles to be equipped with sensors and wireless devices and fixed roadside wireless infrastructure. It is dynamic wireless sensor based network in which vehicles coming in signal range join a network and going out of signal range are dropped from network. It employs IEEE 802.11p technique i.e. a Dedicated Short Range Communication (DSRC) that enables various vehicles to communicate with other vehicles on the road as well as with the road side preset equipment. DSRC has been allocated 5.9 GHz spectrum band and 75 MHZ of bandwidth with a range of 1000m.

VANET is different from MANET in various ways such as architectures, characteristics, and applications. In MANET, nodes are free to move in any random direction with any speed. They require no infrastructure to support wireless communication. Nodes are restricted by limited battery power. However, vehicles in VANET tend to move in an organized manner restricted by roads, traffic and its regulations. Due to restricted movement, VANET has to be supported by some fixed roadside infrastructure to provide network services. As wireless devices in vehicles are connected to powerful batteries in vehicles, VANETs are not limited/constrained by low battery power. VANET allows vehicles to share and disseminate various information regarding route maps & traffic information like jams, accident prevention, safety measures and precautions thereby reducing traffic jams/ accidents and improving efficiency and safety of transportation. Its application comprise of driver assistance including warning about secure distance, brake application and vehicle collision, road and local weather conditions. It includes local information as next free parking space, fuel prices and services at approaching service station or sight-seeing information for tourists. It can be used for infotainment such as the Internet access, chatting and interactive games between vehicles moving close to each other.

The scope of this chapter is focused on various advances in Greedy Forwarding Technique. The chapter comprises of four sections. Section-I gives introduction of VANET and its real time applications. Various VANET Issues have been discussed in Section –II. Section-III portrays Advanced VANET routing protocols based on Greedy Forwarding. Section-IV provides a conclusion and future work.

2. VEHICULAR AD HOC NETWORKS (VANET) ISSUES

VANETs are used for both safety and non-safety applications. Various aspects having a significant impact on VANET operation are needed to be considered for deployment of VANET for various applications. The key challenges from practical perspective are as follows (Ranjan & Ahirwar, 2011)

1. **Highly Dynamic Topology:** It is an extremely dynamic network in which vehicles coming in signal range join a network and vehicles which fall out of signal range are dropped out of the network. Network is frequently disconnected due to changing node density. The topology formed by VANETs is rapidly changing as vehicles are moving at high speed in different directions. On highways, for vehicles moving at a speed of 60-70 kmph and radio range between two vehicles as 125 m, then the link between the two vehicles would last at most 10 sec.

2. **Signal Fading:** The antennas installed in vehicles are relatively at low height. Thus, line of sight (LOS) can easily be blocked by obstacles either static (e.g., buildings, hills, trees) or mobile (other vehicles on the road) that are distributed along roads particularly in the cities. These obstacles cause shadowing, multipath, fading effects, significant signal attenuation and packet loss. The free space radio waves are obstructed or get faded due to these obstacles in the way of transmission which decreases the efficiency of VANETs (Meireles, Boban, Steenkiste, Tonguz & Barros, 2010).
3. **Limited Bandwidth:** There is no central coordinator to control communication between nodes, manage fair usage of bandwidth and conflict management. In high density traffic environment, channel congestion can occur due to limited range of bandwidth frequency (10-20 MHz for VANET applications). Thus available bandwidth has to be utilized very efficiently. Unavailability of bandwidth compels vehicles desiring to communicate to wait for getting a turn to transmit which results in increase of waiting time and latency.
4. **Intermittent Connectivity:** Due to high mobility and dynamic topology, networks get frequently fragmented which leads to short connectivity and throughput degradation. The link between two communicating vehicles gets disappeared very quickly. During non-peak rush hours, node density is very low which results in disconnection of nodes. This frequent disconnection has to be recognized and an alternate link should be provided very quickly to ensure uninterrupted communication.
5. **Mobility Modeling:** Mobility of Vehicles is restricted to the underlying road patterns, traffic rules, traffic lights, speed limits, traffic conditions and driving behavior of drivers. Depicting such mobility patterns can be possible through traces obtained from the traffic patterns itself.
6. **Propagation Model:** VANET has to consider different scenarios like city, rural and highway. On highways, propagation model is approximately free space. But interference from reflected signal from walls along the road or vehicles on road and obstruction of signal due to various vehicles between communicating vehicles has to be considered. However, in cities, VANET propagation model is not a free space but has complex surroundings with vehicles, trees and buildings distributed along/over roads. Similarly, in rural/forest area, signal reflection and interference occurs due to topographic entities like hills, forests, fields etc. In all these scenarios, signals from other vehicles and Road Access points also interfere. Thus Propagation model for VANETs has to analyze effect of these obstacles in the path of transmission as well as potential interference of wireless communication from other vehicles or road access points (Da Cunha, Vilas, Viana & Loureiro, 2014)
7. **Unlimited Battery Power and Storage:** The nodes in VANET are not restricted to battery power and storage as in other sensor networks. Nodes have sufficient energy and computing power and thus optimizing computation and power usage is not so important.
8. **On-Board Sensors:** Nodes are assumed to be equipped with sensing units to provide location, speed and direction information used for making routing decisions. Location and direction information can be provided by GPS unit from on –board Navigation system. Speedometer installed on vehicles provides speed information about vehicle. This information is used by routing protocols for making quick and efficient routes for vehicular communication.
9. **Redundant Data Traffic:** Multiple vehicles and road sensing units may generate redundant data traffic that needs to be verified by routing protocols for efficient bandwidth utilization and accurate information provision. Redundant data traffic may also lead to medium contention and collisions. To remove redundancy, data aggregation techniques are applied to information being broadcasted. But these techniques increase delay in communication and thus have to be considered for a tradeoff between delay and communication overhead. Also, these techniques cannot be applied in highly

mobile vehicular environments. Yu, Xu, and Guo (2012) introduces a Catch-up scheme that tries to reduce delay while also reducing communication overhead.

10. **Small Effective Diameter:** Due to small range of radio propagation, the effective network diameter of VANETs is also small. This results in poor connectivity between nodes. Due to small effective diameter, complete global topology of the network cannot be maintained. Lack of global topology information restricts application of many existing algorithms to VANETs.
11. **Security and Privacy:** Providing secure communication between nodes is a challenging issue in VANETs. As nodes frequently join and leave a VANET, the nodes cannot be trusted for secure communication. VANET is vulnerable to various security attacks that can inject malicious packets by compromising the node, disrupt the traffic, prevent from network formation or even adverse can saturate the network with a large amount of broadcast messages. Nodes may be forced to store incorrect routes thereby upsetting network topology or driver's private information can be stolen. Various security Attacks in VANET include Selective forwarding, Identity spoofing, Denial of Service attack, Black Hole attack, spamming, Sybil attack, GPS spoofing etc. (Ganesh & Ranjani S, 2013).
12. **Highly Delay-Sensitive Applications:** Safety message broadcasting is considered as highly delay sensitive application of VANETs. In an automatic highway system, on brake application, an immediate message must be transmitted to following vehicles to avoid collisions. In these applications, minimum delay is more crucial than maximum data rate. A delay in seconds can make the message and the VANET services meaningless and the consequence will be very dreadful. A reliable end-to-end packet transmission with minimal latency should be provided for highly delay-sensitive life-critical emergency messages (Cheng, Sha & Zhuang, 2011).

3. ROUTING IN VANETS

To enhance communication among vehicles, a variety of messages are sent from vehicles to vehicles or from road side units to vehicles. Routing decision in VANET is a challenging task due to high mobility, highly dynamic topology, limited network life time and restricted mobility of nodes. Frequently disconnected network caused by highly dynamic topology effect the performance of data transmissions. While transmitting, security and privacy of driver's information should not be compromised. The message routing should ensure network availability and receipt of trustworthy information. A number of protocols are used for making routing decisions for wireless networks. However due to complexity of VANETs, all these protocols are not applicable due to one or more limitations. Routing protocols can be classified in various ways (Ferreiro-Lage, Gestoso, Rubinos, & Agelet, 2009):

- **According to Range:** Unicast, multicast, geocast, broadcast.
- **According to Geometrical Features:** Topology based, position based.
- **According to Route Discovery Technique:** Proactive, reactive, hybrid
- **Based on Algorithm Used:** Link state, vector distance, greedy forwarding, minimum distance etc.

Unicast routing is used for sending information from one source to one receiver only. Multicast routing sends the same information from one source to a previously created group of nodes by multi-hop communication. Geocast routing is used to deliver a geocast packet to a particular geographic region. Broadcast routing is utilized for sending information from one source vehicle to all other vehicles in the network.

Topology based routing protocols use link information that exists in the network for forwarding packets. This link information of whole network can be maintained in advance (proactive approach) or only active route link information can be collected on-demand (reactive approach). A hybrid approach maintains link information of nearby neighbors in advance (proactively) while that of far away neighbors on-demand (reactively). Topological routing, that is highly used in MANETs, suffers from route instability and route maintenance overhead in highly mobile VANET. Its performance degrades on increase in network density thus is not scalable. Thus VANET employs Position Based Routing which assumes that each node has knowledge about its physical location by some other location sensing device/service. Position Based Routing uses geographical position information of participating nodes in a network to make routing decisions. Beaconing is used to collect information about neighbors of a node. Packets are forwarded hop by hop using: Restricted directional flooding, Geodesic packet forwarding or Greedy forwarding. In this chapter, basic greedy forwarding and its recent enhancements have been discussed.

4. ROUTING PROTOCOLS IN VANETS

Greedy Forwarding

It is based on a 'greedy' heuristic: "Each message holder forwards the message across a connection that brings it as close as possible to the target in lattice distance" (Kleinberg, 2000). Greedy Forwarding tracks consecutively closer geographic nodes, until the destination is reached. It believes that locally best selection of next forwarding hop is a neighbor geographically closest to destination. The neighbor position is updated by periodic beaconing (Pham, Young, & Won, 2006). The source node forwards a packet to the neighbor closest to the destination. On receiving packet, the neighbor node forwards the packet to next node that is closer to destination than itself. If there is no neighbor closer than it, packet is simply dropped. This stage is known as local maximum. As it relies on immediate next neighbor, state information required is insignificant. As the network density increases, Greedy forwarding approximates to shortest path. Its limiting issue is local maximum which decreases its throughput and overhead due to periodic beaconing. In VANETs, as nodes are highly mobile, greedy forwarding results in more packet losses and frequent retransmissions.

Enhanced Protocols Based on Greedy Technique

1. **Greedy Perimeter Stateless Routing (GPSR) (Karp & Kung, 2000):** It is presumed that all nodes know geographical location of each other and destination node. Data packet contains the identification ID of its destination and its geographical position in two four byte float numbers. Location information is recorded and maintained in a neighborhood table by all nodes. This information is used to route data packet from source to destination through shortest path. Current geographical

Figure 1.

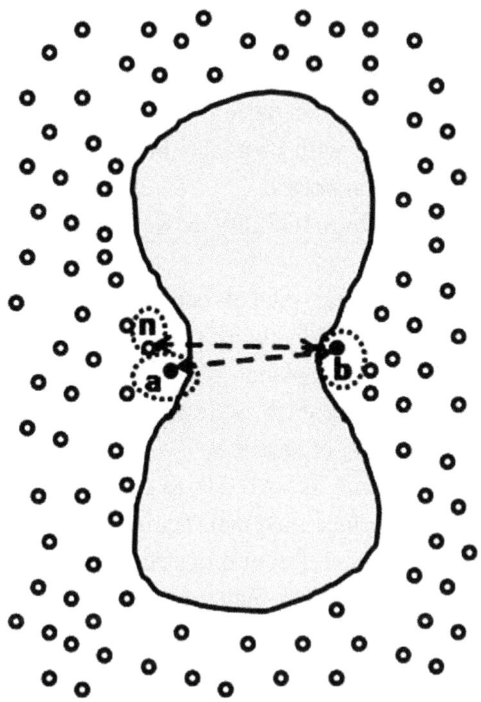

co-ordinates of node are updated to its neighbor nodes by using periodic beaconing. Node position information is piggybacked to forwarded data packets to reduce overhead due to periodic beaconing. Information of node's whose beacons are not received is removed from the table. It constructs a planar graph (a graph in which no two edges cross) to make routing topology and then forward packets. A set of nodes with radios having a specific open space range can be considered as a planar graph. A Graph whose edges are stated by a threshold distance between vertices is termed as a unit graph. GPSR uses Relative Neighborhood Graph (RNG) and Gabriel Graph (GG) algorithms that yield a connected planar graph as long as the unit graph assumption is obeyed by connectivity between two nodes. Two methods for forwarding packets are used: 1) Greedy forwarding for sending packets greedily to those intermediate nodes that are constantly getting closer to target. 2) Perimeter forwarding in case an intermediate node cannot find a neighbor that is closer to target than itself (local maximum). Algorithm recovers local maximum by routing around perimeter of the region by applying right hand rule to select next hop, until it reaches a node that is in the direction/closer to target. From here, it resumes greedy forwarding. Advantages for VANETs:

a. **Scalable:** It is a stateless protocol that keeps state of only first hop neighbor which makes it more scalable than shortest-path and ad-hoc routing protocols.
b. **Bandwidth Efficient:** It consumes significantly less bandwidth and energy than protocols which maintain and distribute state periodically throughout network region.
c. **Fast:** It can find new routes dynamically in minimal time by using its local topology information.

Disadvantages for VANETs:

a. Due to high mobility of VANETs, node's neighborhood table contains mostly stale information about neighbor node's position. Despite the fact that the destination node is also moving, the information about current position of destination is never updated.
b. It works best in a free open space scenario than in city scenarios with buildings and trees in way of traffic. This is because with obstacles in the path, it becomes difficult to establish direct communications between nodes.
c. It incurs link failure due to the high mobility network, frequent topology changes and network fragmentation.
d. In perimeter forwarding, sometimes packets can be travel in wrong direction which may result in high packet loss/increased delay/network partitions.
e. Delay is caused by construction of planar graph to make routing topology.
f. Maintenance of planar graphs at each node introduces a significant overhead and makes recovery mode state-full instead of stateless.
g. Unit graph hypothesis is not true in VANETs as obstacles and restricted mobility cause signal fading. Also, planar graph face traversal requires strictly identical radio ranges for graph constructions. But in VANETs, reflection/obstruction of signal from obstacles and signal interference modify radio ranges. Thus planar graphs are usually hard to construct in VANETs (Tian, Han, & Rothermel, 2003).
h. Routing loops may occur where the graph has crossing edges (non-planar graph).
i. It also incurs overhead of periodic beaconing and beacon loss due to high mobility/ channel destruction. When beacons get lost due to transmission error, neighbor information of corresponding beacons is deleted from neighborhood table by nodes. Thus some nodes may become unaware of nearby neighbors that results in decreased throughput.
j. It does not take into account mobility pattern of vehicles restricted through underlying roads and traffic signals that can result into wrong decision of greedy forwarding and later on switching to recovery mode (Tian, Han, & Rothermel, 2003).
k. It does not take into account velocity and direction of vehicle when selecting next hop neighbor closer to the destination.

2. **Spatially Aware Packet Routing (SAR) (2003):** This protocol uses spatial awareness to remove some shortcomings of existing GPSR. It defines topology hole as a state when a data packet reaches a node that does not have any neighbor closer to the destination than itself. Transient topology holes are caused due to node movement and subsist for a short time only. While permanent topology holes are caused by roads. SAR recommends that these permanent topology holes can be known in advance with knowledge about spatial environment (Tian, Han, & Rothermel, 2003) shows that the greedy forwarding failures and forwarding of packets in wrong direction as in perimeter forwarding can be avoided by choosing a more suitable neighbor with spatial information. It uses a proactive solution than a passive recovery solution. It makes use of digital road maps provided by Geographic Information System (GIS). Road topology information is extracted through a parser in Geographic Data Files (GDF) format. A spatial graph is constructed with significant places at vertices and interconnections between places as edges. Vehicles are considered as moving along

edges from one vertex to another in the graph. This spatial model can be represented at three layers of granularity: 1) Layer1 represents only main roads and junctions; 2) Layer 2 represents the whole street-level topology with small roads; 3) Layer 3 represents extra semantic information, like speed limit, number of lanes, etc.

SAR utilizes GSR (Geographic Source Routes) based forwarding. A node maps itself, neighbor positions and destination position on a graph-spatial model and computes a shortest path to destination. It embeds some intermediate nodes along this path into packets forwarded. These intermediate nodes are called Geographic Source Routes (GSR). A neighbor with shortest path along GSR to destination on this graph model is chosen as next forwarding hop instead of geographically nearest node as in GPSR. Whenever an intermediate node is reached in GSR, it is removed from GSR. If a forwarding node can't find a suitable neighbor along this GSR, packet forwarding is suspended for some time or greedy forwarding is started from this node, or, GSR is recalculated to find a new shortest path. Advantages include:

a. It improves recovery strategy of GPSR to avoid local maximum.
b. It includes spatial information which leads to better routing in case of city scenarios where more permanent topology holes exist.

Disadvantages:

a. A forwarding vehicle cannot always find a suitable neighbor on the GSR which leads to GSR to be recomputed and increases delay (Bernsen & Maniyannan, 2009).
b. Suspending a packet for a predefined time may lead to high delay due to obsolete neighbor information. This may also lead to packet loss if no neighbor with progress to next vertex in GSR is found in this time.
c. Intermediate nodes in GSR may become obsolete due to high mobility of VANETs.

3. **Anchor Based Street and Traffic Aware Routing (A-STAR):** Seet, Liu, Lee, Foh, Wong, and Lee (2004) incorporates street and traffic awareness by using statistically rated and dynamically rated maps while forwarding packets through a sequence of anchors (junctions) greedily. It hypothesized that city buses travel on major roads that have more likely heavy traffic density. Street maps are generally static but traffic density at various roads is different at different times of the day. The map which uses preconfigured information about streets and traffic is called a statistically rated map. The map which can reconfigure itself with dynamic traffic density is called dynamically rated map. The statistically rated maps are used to ensure high connectivity by mapping routes of city buses. The dynamically rated maps are used to incorporate changing traffic density to select optimum anchors in route path.

A-STAR uses information about street map and traffic density at roads to select sequence of anchors through which a packet must travel from source to destination. An anchor path is computed using Dijkstra's least weight path algorithm. When a local maximum is reached, A-Star computes a new anchor path from the local maximum. The path at which local maximum occurred is marked as "out of service" for some duration temporarily so that other packets should not traverse through the same void area. This information is piggybacked to packets and distributed in the network. After receiving these packets, nodes update their local map by marking this path as "out of service" and don't use it in anchor path computation during out of service duration. "Out of service" streets resume operational after the time out duration. Advantages include:

a. Path selection mechanism of A-STAR provides high connectivity.
b. It has approximately 40% higher packet delivery ratio and low end-to-end delay than GPSR when applied with local recovery.
c. It employs more efficient local recovery strategy to recover local maxima.
d. It is suitable for VANETs in both city and highway environment as it incorporates street and traffic information.
e. It ensures finding an end-to-end connection even in low traffic density.

A major disadvantage is that it has end-to-end delay lesser than GPSR but still it is not good for delay sensitive applications.

4. **Greedy Perimeter Coordinator Routing (GPCR) (Lochert, Mauve, Füßler & Hartenstein, 2005):** GPCR makes use of natural planar graphs formed by streets and junctions without using any global or external information such as a static street map as used by A-STAR. It applies to methods of forwarding methods: 1) Restricted Greedy forwarding and 2) Street and junction based repair strategy that does not require graph planarization algorithm of GPSR. In restricted greedy forwarding, packets are traversed along the streets only due to obstruction of radio signals from road side buildings and trees. Actual routing decision is required to be taken while approaching some junction. Nodes that are positioned in the region of junction are called coordinator. Thus packet will be forwarded to a node that is on the junction (coordinator node) rather than to a node that is beyond the junction (non-coordinator node). This avoids getting trapped in local maximum by selecting better node at junction.

The coordinator node broadcasts its designation and location to other nodes. When a node receives a packet from its predecessor node, it checks whether it is a coordinator node or not. If it is a non-coordinator node along the street, it checks if some eligible coordinator node is in its range. As long as no coordinator node is in range, a node with largest distance to the source is chosen as the next hop. When some coordinator nodes come in range, a random coordinator is chosen as the next hop. At this coordinator node, decision is required to be made about the street that the packet should follow. The neighboring node with the largest progress to the destination is chosen greedily as the next hop.

However GPCR is not free from the problem of local maximum. As a result, repair strategy has to be applied to overcome local maximum which consists of two parts: 1) at each junction, right hand rule is applied at the coordinator node to choose next street to follow until a coordinator node is found whose distance to destination is less than the node from where the repair strategy was started. 2) In between junctions greedy routing is followed. To find nodes at junction, periodic beaconing is used to update current locations. The node x which has two other nodes y, z as neighbors where y and z are not neighbors to each other, then node x is taken as a node at junction. y and z are assumed to be separated by some obstacle. Simulations show that GPCR has a better average delivery rate and less hop count than GPSR. Disadvantages include:

a. Selection of junction node is very critical and missing a junction node will risk the successful delivery of packets.
b. Junction nodes are periodically advertised which cause overhead to the protocol.

   c. Forwarding packets to a node at junction rather than to a maximum distance covering node (greedy forwarding) increases delay. Also forwarding packets to a node at junction is not always necessary as all junctions are not critical.
   d. Poor connectivity in low traffic density.
5. **Spatial and Traffic Aware Routing (STAR) (Giudici & Pagani, 2005):** It takes into account street topology and traffic density along the streets to make effective routing decisions. To prevent local maximum, it avoids streets with low or no traffic density. Such streets are considered as "broken links" in the street topology. It translates digital road maps into graphs with crossings as vertices and streets between crossings as edges. It works in two layers with lower layer collecting information about local network topology and a higher layer computing on-demand partial path to destination on this topology. Dijkstra's algorithm is applied to the graph in order to compute the shortest path. Packets are forwarded on some anchor points on the computed path. Periodic beaconing is used to monitor and update neighbor node positions and vehicular traffic conditions. A node maintains information about neighbor nodes and traffic conditions in neighbors-table and traffic-table respectively. The neighbors-table and two vectors: presence vector and persistence vector are used to determine density of traffic on streets. The presence vector counts the number of neighboring nodes in each prime direction. The persistence vector determines the traffic condition that persisted for a time. Whenever a value in presence vector is found below some low value, persistence vector is updated for traffic condition. If traffic is found to significantly sparse or dense in persistence vector, then this information is updated in traffic-table. In case of routing failure/local maximum, STAR computes a new route from current node using current traffic information. As it computes partial route to destination, the route is better optimized to destination movement and local traffic information. The packet header length is also fixed due to partial route. Disadvantages include
   a. Periodic beaconing results in wasted bandwidth.
   b. Size of Beacon messages is approximately six times size of beacons than in GPSR and SAR.
   c. When a vehicle moves at a high speed on a highway in one direction for a large time, the vehicle has less time to exchange large beacon messages thus resulting in inaccurate presence vector (Bernsen & Manivannan, 2009).
   d. It suffers from high collision probability which degrades its performance.
6. **Greedy Traffic Aware Routing Protocol (Improved) GyTAR (Jerbi, Senouci,, Rasheed, & Ghamri-Doudane, 2009):** It is a junction based routing protocol that takes into account vehicle speed and direction information in addition to real time traffic density and street map. Real time traffic density between two junctions is calculated using the Infrastructure-Free Traffic Information System (IFTIS). The protocol works in two modules: 1) Junction Selection and 2) packet forwarding. In junction selection module, a node examines the position of its neighbor junctions using digital street map. Then it assigns a score to each candidate junction which is a combination of traffic density between node and the junction; and curvemetric distance i.e. progress towards the destination in terms of distance. The junction with the highest score is selected as the next anchor to forward the packet to destination. Thus the selected junction is one which has highest traffic on it and is nearest to the destination. The next junction to forward packet is selected again at the junction when the packet arrives on it in the same fashion. Thus junction selection process is partial and dynamic and can adapt to the real time traffic variations.

While forwarding packet between node and junction, improved greedy forwarding is used. Each node maintains a neighbor table containing position, velocity and direction of its each neighbor. Periodic hello messages are used to update information in this table. Before forwarding a packet to next neighbor, the node predicts the position of each neighbor with the help of information in neighbor table and then selects a next hop neighbor which is closest to destination. When a packet gets struck in local maximum, carry and forward strategy for recovery is used. Advantages include:

a. It uses network resources efficiently by limiting control message overhead.
b. It has improved connectivity thus reduced end-to-end delay and low packet loss.
c. Unlike A-star, it selects junctions dynamically and can adapt to varying traffic conditions on road and destination movement.
d. The size of packet header is fixed due to partial path selection.

Disadvantages:

a. If wrong junction is selected, the packet can get struck to local maximum.
b. In junction selection process, it doesn't consider direction of vehicles which may cause selection of high density junction in opposite direction of destination. This can result in increased delay and low packet delivery ratio (Bilal, Madani & Khan, 2011).
c. It also doesn't consider variations in road lengths which may affect average traffic density values/unit distance Boba, Nor, & Nagar, 2014).

7. **Reliable Inter-Vehicular Routing (RIVER) (Bernsen & Manivannan, 2012):** It is a position based protocol that employs street awareness and periodic beaconing to greedily forward packets. It uses optimized greedy strategy and sends messages on reliable routes. Street awareness is gained through static preloaded data. It takes advantage of an undirected graph of street layout with vertices of graph representing junctions at which streets intersect and graph edges as street segments between these junctions. The reliability of routes is studied with traffic monitoring by sending probe messages along streets (active monitoring) and observing messages routed between adjacent junctions (passive monitoring). The Active monitoring determines connectivity of streets from source to destination. The passive monitoring determines connectivity of junctions through a graph edge i.e. if a packet is received from one junction to adjacent junction through an edge then the edge has sufficient number of nodes for the packets to traverse and cross the edge. The reliability information is propagated through network by piggybacking it to routing messages, probes and beacons. Using reliability information and street awareness, each node assigns a weight to every known edge in its street graph. More the assigned weight to an edge, lesser is its reliability.

When a sender node has to send some information to a destination node, it calculates its distance to destination. If this distance is small, it greedily forwards packets to destination. Otherwise, it uses Dijkstra's least weight path algorithm to find a least-weight path based on the sum of the weights of edges on the path. Thus calculated path is the most reliable routing path to destination and is free from routing loops. The geographic location of junctions in the computed path form anchor points on the route. This path with anchors is written into the packet header. Between anchors, packets are forwarded greedily by finding next hop in the neighbor table. As the packet arrives at each hop, edge weights are updated in the weighted route of the packet. If a node cannot find suitable next hop then the corresponding edge is marked as disconnected edge by assigning it maximum weight possible. The Dijkstra's least weight path algorithm is used again. If the algorithm finds a path whose mean weight is less than previous route's mean weight by a significant value then this route is overwritten in packet header. If weight of next edge in recomputed path is equal to disconnected edge, the packet is dropped. However this recomputed path may have routing loops. To reduce routing loops, last-known weight for each edge is maintained and updated in each packet header. The packet header also maintains a known-edge list for adjacent edges traversed during the packet's lifetime. Advantages include:

a. Its packet delivery throughput is far better than GPSR and A-STAR.
b. It disseminates reliability information by piggybacking it to packets on transit. It does not require broadcasting, network flooding and thus avoids network congestion.
c. It handles network gaps by traversing only the highly reliable paths.

Disadvantages:

a. It incurs more delay and traverses more no. of hops than GPSR and A-STAR.
b. As this protocol uses streets and junctions to find paths to destination, it cannot be used in highways.
c. Even with high throughput, some packets are still dropped due to routing loops and other inescapable issues.

See Table 1 for a comparison of vanet routing protocols .

5. CONCLUSION

Vehicular Ad-hoc Networks (VANET) is one of the recent developments in the transport sector. By integrating the wireless communication technology with the vehicle to vehicle communication a lot of new applications can be developed. This chapter describes various issues which need the attention of researchers related to VANET particularly stressing on the routing issue. These issues include Highly Dynamic Topology, Signal fading, Limited Bandwidth, Intermittent Connectivity, Mobility Modeling, Propagation Model, Redundant Data Traffic, Small effective diameter, Security & privacy and Highly Delay-sensitive applications. Routing is one of major activities in the whole vehicle to vehicle communication. Various VANET routing algorithms which include Greedy, GPSR, SAR, A-STAR, GPCR, STAR, GyTAR and RIVER have been summarized and compared in this study. This Comparison is based on the parameters like Forwarding strategy, Recovery strategy, Traffic Aware, Realistic Traffic Flow, Location service required, Map Required, Local maximum, Delivery, Delay, Privacy protection, Load balancing, Path maintenance, Mobility Model, Propagation Model and Scalability.

Table 1. Comparison vanet routing protocols

	Greedy	GPSR	SAR	A-STAR	GPCR	STAR	GyTAR	RIVER
Forwarding strategy	Greedy	Greedy	Greedy	Anchor based greedy	Restricted Greedy	Greedy	Improved Greedy	Least weighted edge based Optimized Greedy
Recovery strategy	No	Perimeter forwarding	Forwarding suspended, greedy forwarding, recomputed GSR	Recomputed anchor path	Carry and forward	Compute new route	Carry and forward	Recomputed least weight path
Traffic Aware	No	No	No	Yes	No	Yes	Yes	No
Realistic Traffic Flow	No	No	-	-	No	Yes	Yes	Yes
Location service required	-	Yes	Yes	Yes	Yes	Yes	Yes	Yes
Map Required	No	No	Yes	Yes	No	Yes	Yes	Yes
Local maximum	Packet dropped	Occur and Recovered	Occur and recovered	Occur and recovered	Occur and recovered	Doesn't occur	Occur and recovered	Occur and recovered
Delivery	Not guranteed	Best Effort	Best Effort	Best Effort	Best Effort	-	High	Best Effort
Delay	High	More	More	Less	Less	-	Less	High
Privacy protection	No	No	No	No	No	No	No	No
Load balancing	No	No	No	No	No	No	No	No
Path maintenance	No	-	Yes	-	Passive	Partial	Partial	Active
Mobility Model	-	Multiagent Traffic Simulator	Random Way Point	M-Grid Mobility	VanetMobisim	City model with Manhattan street topology	Free Way	Manhattan city model
Propagation Model	-	Probabilistic shadowing	Road Blocking	Road Blocking	Road Blocking	-	Free Space	Two-ray ground
Scalability	Good	-	Good	Good	Good	-	Good	-

Bernsen & Manivannan, 2012; Indra & Murali, 2014.

REFERENCES

Bernsen, J., & Manivannan, D. (2009). Unicast routing protocols for vehicular ad hoc networks: A critical comparison and classification. *Journal of Pervasive and Mobile Computing*, *5*(1), 1–18. doi:10.1016/j.pmcj.2008.09.001

Bernsen, J., & Manivannan, D. (2012). RIVER: A reliable inter-vehicular routing protocol for vehicular adhoc networks. *Journal of Computer Networks*, *56*(17), 3795–3807. doi:10.1016/j.comnet.2012.08.017

Bilal, S., Madani, S. A., & Khan, I. (2011). Enhanced junction selection mechanism for routing protocol in VANETs. *International Arab Journal of Information Technology*, *8*(4), 422–429.

Boba, M. S., Nor, S. M., & Nagar, S. A. (2014). A survey of unicast routing protocols based greedy Forwarding strategies for vehicular Ad–hoc networks in urban scenario. *Journal of Theoretical and Applied Information Technology*, *62*(1).

Cheng, H. T., Shan, H., & Zhuang, W. (2011). Infotainment and road safety service support in vehicular networking: From a communication perspective. *Journal of Mechanical Systems and Signal Processing*, *25*(6), 2020–2038. doi:10.1016/j.ymssp.2010.11.009

Da Cunha, F. D., Vilas, L. A., Viana, A. C., & Loureiro, A. A. F. (2014). *Data communication in VANETs: A Survey, Challenges and Applications*. Research Report.

Ferreiro-Lage, J. A., Gestoso, C. P., Rubinos, O., & Agelet, F. A. (2009). Analysis of Unicast Routing Protocols for VANETs. *5th International Conference on Networking and Services (ICNS '09)*, 518-521. doi:10.1109/ICNS.2009.96

Ganesh, S., & Ranjani, S. N. (2013). Security Threats on Vehicular Ad Hoc Networks (VANET): A Review Paper. *International Journal of Electronics Communication and Computer Engineering, 4*(6).

Giudici, F., & Pagani, E. (2005). Spatial and traffic-aware routing (STAR) for vehicular systems. Lecture Notes in Computer Science: Vol. 3726. High Performance Computing and Communications (pp. 77-86). Berlin: Springer. doi:10.1007/11557654_11

Indra, V. A., & Murali, R. (2014). Routing Protocols for Vehicular Adhoc Networks (VANETs): A Review. *Journal of Emerging Trends in Computing and Information Sciences*, *5*(1).

Jerbi, M., Senouci, S.-M., Rasheed, T., & Ghamri-Doudane, Y. (2009). Towards efficient geographic routing in urban vehicular networks. *IEEE Transactions on Vehicular Technology*, *58*(9), 5048–5059. doi:10.1109/TVT.2009.2024341

Karp, B., & Kung, H. T. (2000). GPSR: Greedy Perimeter Stateless Routing for Wireless Networks. *Proceedings of 6th annual International Conference on Mobile Computing and Networking*, 243-254. doi:10.1145/345910.345953

Kleinberg, J. (2000). Navigation in a small world. *Nature*, *406*(6798), 845. doi:10.1038/35022643 PMID:10972276

Lochert C., Mauve, M., Füßler, H., & Hartenstein, H. (2005). Geographic routing in city scenarios. *ACM SIGMOBILE Mobile Computing and Communications Review, 9*(1), 69-72.

Meireles, R., Boban, M., Steenkiste, P., Tonguz, O. K., & Barros, J. A. (2010). Experimental study on the impact of vehicular obstructions in VANETs. *IEEE Vehicular Networking Conference (VNC)*, 338–345. doi:10.1109/VNC.2010.5698233

Pham, N. N., Youn, J., & Won, C. (2006). A Comparison of Wireless Sensor Network Routing Protocols on an Experimental Testbed. *IEEE International Conference on Sensor Networks, Ubiquitous, and Trustworthy Computing (SUTC'06)*, *2*, 276–281. doi:10.1109/SUTC.2006.1

Ranjan, P., & Ahirwar, K. K. (2011). Comparative Study of VANET and MANET Routing Protocols. *Proceedings of International Conference on Advanced Computing and Communication Technologies (ACCT 2011).*

Seet, B. C., Liu, G., Lee, B. S., Foh, C. H., Wong, K. J., & Lee, K. K. (2004). A-star: A mobile ad hoc routing strategy for metropolis vehicular communications. *Proceeding NETWORKING*, *3042*, 989-999. doi:10.1007/978-3-540-24693-0_81

Tian, J., Han, L., & Rothermel, K. (2003). Spatially aware packet routing for mobile ad hoc inter-vehicle radio networks. *Proceedings IEEE Intelligent Transportation Systems*, *2*, 1546–1551.

Yu, B., Xu, C.-Z., & Guo, M. (2012). Adaptive Forwarding Delay Control for VANET Data Aggregation. *IEEE Transactions on Parallel and Distributed Systems*, *23*(1).

Chapter 6
Direction-Aware Routing Protocol for Delay-Tolerant Network:
Architecture, Design, and Implementation

Ramchandra S. Mangrulkar
Bapurao Deshmukh College of Engineering, India

Mohammad Atique
PG Department of Computer Science and Engineering, India

ABSTRACT

The main applications of delay tolerant network lie in space research, military, tactical and other security sensitive operations. In these and other applications, secure and energy efficient routing is an important issue in homogeneous and heterogeneous network. This chapter contributes a research work that demonstrates the design, simulation, testing and evaluation of secure, energy efficient and environment independent routing protocol for delay tolerant network. The proposed protocol not only proposed secure approach but also energy efficient approach in homogeneous and heterogeneous environment. The proposed approach utilize sectorization technique to achieve its goals of security, energy efficiency and environment independence. Described in this chapter is a detailed architecture, design, simulation, testing and evaluation of the proposed protocol. Having evaluated these protocols in various simulators, it is concluded that the proposed routing protocol utilized minimum energy and bandwidth and also provided secure transmission.

INTRODUCTION

As the Mobile Operating Systems evaluation is variant and competent for vendors, such as iPhone OS, Android, and Windows Phone OS, nature of mobile phones got changed from simple voice communication means into powerful devices, providing service on demand such as transfer the videos or music in a peer-to-peer (P2P) way through the short range communication technologies (Blue- tooth, WiFi,

DOI: 10.4018/978-1-5225-2342-0.ch006

etc.) in them. Limited communication range or mobility nature makes hard to maintain the continuous end-to-end path between them as far as classic Internet P2P applications are considered. Delay tolerant network (DTN) is opportunistic network where every node searches best opportunity to deliver the message called bundle to the destination. Mobile ad hoc network supposed to be the origin of DTN. In traditional MaNet, source node requires end-to-end connectivity to transfer messages. This almost produces the high possibility of data transmission even in a short span of time with minimum delay. In DTN, end-to-end connectivity is never ensured between source and destination which becomes main cause of increasing delay. The source node is always searching the best opportunity to route messages to the destination through some sympathetic and trustworthy relay nodes. Support and trust are wished as a source has to believe and work in cooperation on intermediate nodes. Therefore, DTN transmission is also called as Opportunistic Cooperative Transmission. DTN uses store and forward switching technique of intended messaged by introducing a new protocol layer called the *Bundle Layer* on top of the transport layer. The bundle layer is responsible for storing and forwarding entire bundles between a source and destination. Nodes wish to send data must have the bundle layer support on top of the TCP layer. The transport layer protocol, either TCP or UDP below the bundle layer is selected based upon the reachability and accessibility properties estimated about each region in and around the intended node in the network.

DTN is concerned with highly challenged network in which application layer sessions commonly lack contemporaneous end-to-end connectivity. In networking context, DTN is also called opportunistic network where every node searches best opportunity to deliver the message called bundle to the destination. The design of routing protocol for DTN is fundamentally determined by how efficiently the packets are being transferred from source to destination using available network resources.

In traditional mobile adhoc network(MaNet), source node requires end-to-end connectivity to transfer messages. This almost produces the high possibility of data transmission even in a short span of time resulting in minimizing delay. In DTN, end-to-end connectivity is never ensured between source and destination which increase the delay. Therefore, a source node is always searching the best opportunity to route messages to the destination through some sympathetic and trustworthy relay nodes. To achieve support and trust, the source has to believe and work in liaison along with intermediate nodes.

A considerable amount of research has been reported in providing shortest approach to deliver the packets to the destination in DTN. Early work in DTN focused on real-time aspects to increase the contract schedule. Researchers have attempted to implement ferry node approach to increase the contacts between various nodes along the trajectory. This eventually increases the amount of traffic and also increases energy consumption. Nevertheless, it remains an open problem as how to decrease energy and bandwidth consumption in DTN as desired.

This chapter is concerned with the decision making required to define the sequence of processes in finding an optimum trajectory in DTN; the principal objective is to develop a novel routing protocol for DTN. The issue with which authors deal range from the node selection on an optimum trajectory to complexity and time constraints of the DTN in both homogeneous and heterogeneous environments.

Although a lot of research efforts are directed towards the advancement of optimizing the packet delivery ratio, many remains are to be investigated. Some of the most challenging issues faced by the current research on DNT with respect to routing protocols are security, energy efficiency and environment independence. This research is motivated by an attempt to address these issues of security, energy efficiency and environment independent routing in DTN.

In general, the problem of routing in DTN, while it is touched upon by the routing literature in the form of "due date" problem, has not been as widely studied as the secured and energy efficient routing problem.

Normal routing problems, minimizing resources are almost always shown to be attractive in either homogeneous or heterogeneous environment. This chapter contributes a research work that attempts to address the security issues, energy efficiency and environment independence with respect to DTN. The approach adopted applies not only to optimize trajectory, but also to minimize energy consumptions and increasing security. The proposed approach utilizes sectorization technique to achieve its goals of security, energy efficiency and environment independence by

1. Destination aware routing is applied to locate the particular sector for effective flooding.
2. Optimal trajectory is determined to locate cooperative neighbor nodes.
3. A secured and energy efficient approach is used for routing of messages.

This mechanism also has the capability to provide resource sharing across both homogeneous and heterogeneous environments.

BACKGROUND

Delay Tolerant Network (DTN) is evolved from the Mobile Adhoc Network, Manet. It is sparse and intermittently connected mobile adhoc network where reliable communication and end-to-end connectivity is not available for message transmission. In DTN, the main principle for routing message is "Store and Forward". Each node in DTN stores incoming message in the buffer and then delivered it to the other desirable node towards destination whenever contact is initiated. The Intermediate node copies incoming message and work as a relay node in order to increase the probability of message delivery. DTN is called opportunity network because the intermediate node is always searches opportunity to relay message from source to destination.

PROPERTIES OF DELAY TOLERANT NETWORK

Basic properties of Delay Tolerant Network is store and forward. Some more are listed in this section.

High Latency and Low Data Rate

High latency is referred as the time duration required by the application to transmit data. Data rate is the transmission rate of data. DTN supports the application which has high latency and Low Data rate. This is because in DTN, the nodes come in contact for very less duration. The transmit rate may be considerably low and largely asymmetric with the long latency of data delivery.

Frequent Connection Breakup

In DTN, all nodes posses the mobility nature which causes frequent connection breakup due to a network partition. This increase the latency.

Queuing Delay

The end-to-end latency of data is often dominated by queuing delay. In DTN, frequent connection breakup occurs which results in queuing delay. Queuing delay is extremely large.

Limited Longevity

This refers to the lifetime of the end nodes in DTN which is sometimes called survival time, which is less than the data latency.

Limited Resources

Generally, nodes in DTN have limited resources such as battery and memory. Due to limited resources the intermediate nodes, which stores data temporarily to forward to next node must have enough power to survive for completing data transmission. Also, this intermediate node may have its memory storage full which doesn't allow this node to accept any more data packets which limit it to participate for data Transmission in DTN network.

ROUTING ISSUES IN DELAY TOLERANT NETWORK

Designing a foolproof Routing protocol, capable of working in homogeneous and heterogeneous environment is an important issue. Following constraints shall be considered during designing.

Routing Objective

The ultimate goal of routing in DTN is to increase the delivery of message probability by minimizing resource utilization such as memory space, battery and network bandwidth. The delay minimization is essential routing objective as DTN is expected to be tolerant of delay.

Resource Allocation

The routing protocol must allocate resources which are limited such as battery and memory space. The minimum resource allocation policies are strictly followed during resource utilization.

Effective Buffer Space Utilization

In DTN, the buffer is used to store incoming messages during "contact" period. The buffer must store all the messages and rearrange them depending upon their priorities. Also the fairness policies must be adopted for pending messages provide equal treatment to all the nodes in DTN.

Energy Constraints

Usually, due to frequent mobility in DTN, uninterrupted power supply can't be provided to DTN nodes. However a lot of energy is consumed during message transmission and message reception. This results in a DTN node to die out early. Therefore, the energy efficiency issue is very important while designing routing protocol.

ROUTING PROTOCOLS IN DELAY TOLERANT NETWORK

Routing in Delay Tolerant Network in a broad sense, categorize into Flooding and Forwarding protocols. Flooding protocols are based on the principal of replicating messages to enough nodes in order to deliver to the targeted destination. Forwarding protocols make use of predicted and heuristic knowledge about network to select the best path to the destination.

Flooding Protocols

In this category, the multiple copies of received message will be created and delivered to relay nodes. Relay nodes store the messages until they come in contact with the destination node. During "contact" phase, the messages are delivered with the relay nodes. Many of the strategies were workout by the researchers. Earlier strategies were studied in the context of MaNet. Message replication strategy is used sometimes to increase Message Delivery Probability. This category of protocols does not require any prior global or local knowledge of networking.

Flooding Based Routing Strategies

The state of the art flooding routing strategies is summarized as follows.

Direct Delivery Routing

Direct Delivery Routing (DDR) is the simplest strategy to transmit the data from source to destination where the messages are transmitted immediately the source comes in contact with destination directly. This constraint creates more possibility to deliver a message when two ends are one hop apart or immediate neighbor of each other. Simple as no relayed messages. Advantages as very less resources are required.

First Contact Routing

In this strategy source node along with the nodes which at the First instance come in contact with the source node work in a cooperative manner to increase message delivery probability to successfully deliver the message to the destination. This approach increases the message delivery probability. Also, it increases the bandwidth and storage consumption. This strategy has same fundamental advantages and limitations as direct transmission.

Epidemic Routing

Epidemic routing is an early sparse network routing protocol proposed for Delay Tolerant Network. The main assumption is unlimited storage space and bandwidth. Every node stores all the messages transmitted during the *contact* phase. The concept of database replication is used. A relay node can exchange the entire message during the *contact* phase. Each node argues list of messages called *summary vector* which is exchanged first. Later on only absent messages are exchanged with each other.

Prioritized Epidemic Routing

The main idea in prioritized epidemic routing is to impose a partial ordering of message called bundles. The priority functions of transmission and deletion of messaged are used based on estimated priority. The parameters used to evaluate priorities are current cost to destination, current cost from source, the expiry time and the generation time etc.

Spray and Wait Routing

Spray and wait approach consists of two-phase, spray phase and wait phase. During the spray phase, relay node will flood (spray) message to L no. of nodes during contact. This L is initialized by the source node. If destination meets, message transmission is successfully terminated. If not, wait phase is started. Parameter L which is a function of node density and their distribution and its mobility profile is used to control flooding.

MaxProp Protocol

In MaxProp protocol, city environment are considered where nodes are city buses which have high probability to meet again . The cost estimation technique is used to select a path to the destination. MaxProp divides the buffer into two phases. First phase order stored messages based on hop count information of each message from low to high. The second phase is to order messages by cost from high to low. The buffer is used from both the end. The front end of buffer is used by a first phase, whereas rear end is used by the second phase.

Spray and Focus Protocol

Spray and Focus protocol is the modification of spray and wait protocol. The nodes are assumed to move in a small area. Spray phase is similar to the spray and wait protocol. The difference lies in the second phase. In focus phase, single copy of the message is used to focus limited relay node to route the message to the destination.

Opportunistic Routing with Window-Aware Replication Protocol (ORWAR)

In ORWAR, the main goal is to reduce resource utilization and partial and dropped messages during transmission. Partial messages arise when a node moves out of transmission range during contact. ORWAR estimate number of bytes that can be transferred during each contact time. However, to deploy ORWAR, every node needs to be able to measure its speed and direction, which is possible when they are equipped with GPS like devices.

CEECR Protocol

The CEECR routing protocol is based on erasure based coding. This technique divides the original message into subsets called blocks. These blocks are individually passed on. Decoding technique is used to get the original message at the destination by receiving only subset of the original message or fewer blocks.

CCQR Protocol

CCQR routing protocol restricts the message flooding within only high quality nodes in DTN. The nodal cost is used as parameter which refers to the total number of messages received by a given node during the entire routing process. Homogeneous and heterogeneous cost model is used to estimate nodal cost. These measures are used to identify the best quality nodes along the path to the destination.

ASMART Protocol

ASMART protocol is modified version of the selective controlled flooding scheme which utilizes travel companions of the destinations which improves the delivery opportunities, by controlling flooding overhead. Companion nodes are organized to form an ANYCAST group. Nodes broadcast group identities and hop distance to build routing tables periodically.

FORWARDING PROTOCOLS

Forwarding protocols make use of the network topology, predictive and heuristic knowledge to find an optimistic route to destination. Optimistic route is selected depending upon parameters viz. probability of message delivery, Message transit time. These parameters are evaluated on the basis of heuristic information. This approach does not replicate the message.

Location Based Routing Protocol

In this, Nodes are assigned with the coordinates. Coordinates may physical such as GPS coordinate. Distance formula is used to estimate the cost of delivering messages from one node to another. Best path

Table 1. Comparison of flooding routing strategies

Sr. No.	Name of Routing Protocol	Message Storing Nodes	No. of Hops	No. of Copies of Message	Resources Required Battery Power/ Storage Buffer	Routing Table/Vector	Message delivery ratio	Latency
1	Direct Transmission	Source and Dest	1	No Copy	Source and Dest (Less Resources)	No	Min	Long
2	Two-Hop Relay	Source, Dest and One Relay	2	Single Copy	Source, Dest and 1 Relay	No	Low	Long
3	Tree Based Routing	Source, Dest and Log(N/2) Relay	Log(N/2)	Log(N/2)	Source, Dest and Log(N/2) Relay	No	Low	Long
4	Epidemic	All the nodes encountered	Many	Proportional to the no. of encountered relays	Proportional to the no. of encountered relays	Message Vector	Normal	Normal
5	Prioritize epidemic routing	Only those relays satisfying evolution criteria	Equal to no. of relays satisfying evolution criteria	Only those relays satisfying evolution criteria	Equal to no. of relays satisfying evolution criteria	Message Priority Vector	Good	Normal
6	Probabilistic Routing	Only those relays estimating high Probability	Equal to no. relays estimating high Probability	Only those relays estimating high Probability	Equal to no. relays estimating high Probability	Delivery Probability Vector	Good	Normal
7	Spray and Wait	No. of nodes specified by source, 'L'	L, if dest encountered, otherwise > L	L, if dest encountered, otherwise > L	L, if dest encountered, otherwise > L	No	Good	Normal
8	Spray and Focus	No. of nodes specified by source, 'L'	L, if dest encountered, otherwise > L	L, if dest encountered, otherwise > L	L, if dest encountered, otherwise > L	No	Good	Normal
9	MaxProp	No. of nodes encountered during contact 'P'	P, if dest encountered, otherwise > P	P, if dest encountered, otherwise > P	P, if dest encountered, otherwise > P	No	Good	Normal
10	ORWAR	Direction and byte predicted nodes	Direction and byte predicted nodes along path to dest.	Direction and byte predicted nodes along path to dest.	Direction and byte predicted nodes along path to dest.	No	Good	Normal
11	CEECR	nodes selected for Transmission of Encoded message	Can't predicted	Multiple copies	nodes selected for Transmission of Encoded message	No	Good	Normal
12	CCQR	Quality nodes satisfying criteria	Quality nodes satisfying criteria	Quality nodes on selected path	Quality nodes on selected path	Quality of Nodes	Good	Normal
13	ASMART	Travel companion nodes in Any cast group	Travel companion nodes in Any cast group	Travel companion nodes along the path	Travel companion nodes along the path	Travel compatibility	Good	Normal

estimation is based on the coordinates of source, destination and intermediate node. Finally the message is transferred through the best estimated path.

Gradient Routing Protocol

In this approach weights are assigned to the nodes in the network. The relay nodes contact another node that has a better metric for the message destination and pass on the message. This is called gradient strategy because it follows a gradient of improving utility function values towards the destination.

Label Based Forwarding Protocol

In this strategy first task is to create groups and label them. Particular groups are chosen as next node or next groups to send messages to the destination. This results in large improvements as the labeled nodes work in groups. The group based routing efficiently utilizes bandwidth and consumption.

Bubble Rap Forwarding Protocol

The Bubble Rap technique focus on two issues: community and centrality. In the bubble Rap strategy, if a node has a message to the destination node, this node bubble the message up in the hierarchical rankings tree using the global ranking, until it reaches a node which is in the same community as the destination node. Then local ranking is used to further bubble the message until the message is reached to the destination or expires.

NECTAR Protocol

NECTAR is based on the contact history in order to create a neighborhood index and then discover the most appropriated routes. The protocol uses heuristic knowledge based on mobility. The main consideration is the nodes meet recently have a large probability of meeting again. NECTAR allows sending messages by a relay in a controlled manner to limit the traffic in the network. This protocol significantly improves the number of delivered messages, consuming fewer resources, in limited resource networks.

CREST Protocol

CREST protocol based only on a limited knowledge of the contact behavior of the nodes in the network. The protocol uses *conditional residual time* as the parameter which accurately estimates the time remaining for a pair of nodes to meet using only the local knowledge of their past contacts. CREAST has a lower end-to-end delay.

CSPR Protocol

CSPR protocol based on human mobility traces and some findings of previous work. The protocol uses *conditional intermeeting time* as the parameters which compute the average intermeeting time between two nodes relative to a meeting with the third node using local knowledge of the past contacts. This *conditional intermeeting time* is used to estimate best path towards a destination in the shortest path routing algorithm.

QoNSW Protocol

QoNSW protocol uses the Quality of Node (QoN). QoN refers to the activity of node, or the number a node meets other various nodes within a given interval of time. Node meeting many more nodes in the network for giving interval has greater QoN. This measure is used to find the best suitable relay node to route messages.

SCR Protocol

Single Copy, Replication Routing (SCR) protocol, which duplicates the message to the node selected as best relay nodes. The protocol uses *Time Interval* as parameter to select the node with the least time interval among neighbors as the next hop. SCR allows to start message transmission in Critical Transmission Rang (CTR) and stopping transferring out of communication range, which reduces the overhead by decreasing message delivery failure during contact.

CRHC Protocol

CRHC protocol based on Hierarchical forwarding and cluster control mechanism. This protocol control routing in the same cluster efficiently using Cluster Head and reduces message transmission by allowing the transmission permission to Cluster Head. This protocol has higher delivery rate and lower delay.

PROCCS Protocol

PROCCS protocol is based on repetitive contact patterns and their time sequencing to improve routing. Greedy path probabilistic approach is used to calculate probability of the path and then the best path is selected as the path for routing message. Contact graph is used for particular time issued by PROCCS to calculate best path.

DESIGN AND IMPLEMENTATION OF DIRECTION AWARE ROUTING PROTOCOL FOR DELAY TOLERANT NETWORK

Before proposing a new routing protocol, study of the layer and node communication is an important issue. Traditional TCP/IP protocol suit needs to be modified to configure a node in DTN.

Motivation

DTN are mostly vulnerable to many security threats and also gave birth to new security challenges. In particular, the use of open networks to transmit data offers unprecedented opportunities for security at-

Table 2. Comparison of forwarding routing strategies

Sr. No.	Name of Protocol	Parameter Used to Estimate Best Path	Message Storing Nodes	No. of Hops	Resources Required Battery Power/ Storage Buffer of	Message Delivery Ratio	Latency
1	Location Based Routing	Distance between neighboring nodes	Nodes along the minimum path to destination	Nodes along the minimum path to destination	Nodes along the minimum path to destination	Good	Normal
2	Gradient Routing	Suitability to deliver message	Nodes on gradient path to destination	Nodes on gradient path to destination	Nodes on gradient path to destination	Good	Normal
3	Label Based Routing	Nodes with same label	Nodes on path to destination with same labels	Nodes on path to destination with same labels	Nodes on path to destination with same labels	Good	Normal
4	Bubble Rap Routing	Nodes in same Community	Nodes on path to destination with same community as well as centrality	Nodes on path to destination with same community as well as centrality	Nodes on path to destination with same community as well as centrality	Good	Normal
5	NECTAR	Previous Contact History	Only those relays which meets frequently in past.	Only those relays which meets frequently in past.	Only those relays which meets frequently in past.	Good	Normal
6	CREST	Conditional residual time	Only those relays which estimated to meet again during schedule	Only those relays which estimated to meet again during schedule	Only those relays which estimated to meet again during schedule	Better	Normal
7	CSPR	Conditional intermeeting time	Estimated nodes only	Nodes on Estimated path only	Nodes on Estimated path only	Better	Normal
8	QoNSW	Quality of Node	High quality Nodes	High quality Nodes along the path	High quality Nodes along the path	Good	Normal
9	SCR	Time Interval	Least time interval nodes	Least time interval nodes along the path	Least time interval nodes along the path	Good	Normal
10	CRHC	------	Cluster heads	Cluster heads along the path and bounSEDARy nodes on cluster	Cluster heads along the path and bounSEDARy nodes on cluster	High	Less
11	PROCCS	Contact pattern and time sequencing	High greedy probability value nodes	High greedy probability value nodes along the path	High greedy probability value nodes along the path	Good	Normal

tacks, and allows attackers to compromise information integrity, authenticity, user privacy and system performance. For example, in DTN, malicious nodes called routers can arbitrarily insert false information into the incoming messages. If innocent routers further continue this transmission of falsified messages, the attacks may generate large amounts of unwanted traffic to the network. This extra traffic may pose a serious threat on the operation of DTN. Further, unauthorized access and utilization of DTN resources are another serious concern in terms of DTN security. DTN represent a new network concept and therefore introduce new unique security characteristics and challenges. These unique characteristics make the existing security protocols designed for the conventional adhoc networks unsuitable for DTN. Therefore, a series of new security protocols should be carefully designed to address various node misbehavior or attacks.

Chennn-Jung Huang et al. are the first researchers to address the issue regarding routing in heterogeneous network. They propose prediction based DTN routing protocol to deliver the message in the heterogeneous network with the help of Edge servers. Location based user log is used to deliver the message to the exact destination by using predicted node's destination. This is useful to find the best Edge server towards the destination. The message successfully stored in the nearest edge server towards destination. As soon as destination comes in contact with the Edge server, the message is successfully delivered to the destination. The Fuzzy controller is used to select the best Edge server near to destination. This technique is good where all the DTN nodes follow the specific movements in the coverage area where edge servers are planted. But in DTN nodes which are designed for high latency application, the mobility factor of the node is unpredicted and hence meeting probability of node with edge server is not perfectly determine. Hence, selection of best edge server on optimal trajectory towards destination is important task. This is the major issue which remains un-addressable in homogeneous and heterogeneous environment.

Residual energy is important factor in DTN node where ultimate source of power is battery. This need to be considered while addressing routing issues in both homogeneous and heterogeneous network. Also prioritizing the message in the intermediate server such as Edge server is also important since the delivery latency is more in DTN as compared to other network. The factors such as resource constraints, bandwidth utilization, congestion, mobility etc. need to be addressed in DTN for both homogeneous and heterogeneous network. Also, the deficiency to route data to destination using optimal trajectory towards destination before establishing contact with Edge server i.e identification of best node on optimal path towards destination need to be addressed in such diverse architectural network.

Related Work

There has been lot of work reported by various researcher in routing strategies of delay tolerant network. There are several routing algorithm available for routing in homogeneous delay tolerant network environment. Most of them are basically classified into Single copy (Forwarding) based and Multi copy (Flooding) based. Some commonly used routing protocols suggested by the researchers and their techniques are described below.

Agila Bitsch Link et al. proposes geographic location based routing (known as Geo-routing) which relies on geographic positional information which provides a promising approach for enhancing the routing efficiency in DTN. Based on geographic locality of destination, source directly sends message instead of establishing route using traditional

MaNet approach. Each node keeps track on own location as well as carefully makes itself aware about the locality of other neighbor nodes. With this piece of information, a message can be routed to the destination without having knowledge of the network topology or a prior route discovery phase.

Gil Eduardo de Andrade et. al. studies geographical routing in the context of velocity and direction and position of nodes. Authors also simulates and effectively addresses the performance by varying these mentioned factors and proves the effectiveness of geo-routing in DTN.

Michael Demmer et al. introduce a new session layer in Delay Tolerant Networks. Authors classifies the routing protocols according to the methodology used to find destinations and whether replicas of messages are transmitted or not into Replica based (flooding) protocols, Knowledge based (store and forward) protocols and coding based protocols. This requires some form of heuristic information of network before data transfer. As a result, data transfer occurs only on the best path instead of replicating it unknowingly. Authors also compare developed protocols in terms of different metrics like hop count, delivery ratio and number of copies to gain an insight into their overall effectiveness with respect to the network environment situations. They design a routing protocol that has efficient resource consumption, low latency and high delivery ratio and that runs a smart algorithm that can select best path and manage buffer in an effective way which help for minimizing energy consumption.

Iqbal, S.M.A. et al. proposes agent aided routing scheme to increase the message delivery ratio by minimizing delays and spreading the messages faster and efficiently . The source of a message is emitted to create unlimited copies of a message and distributes these copies to agents. Agents acting as the representative of the source receive messages from the source based on their node visited list and delivery predictability. Agents forward the copies to the other intermediate nodes or deliver the message directly to the destination. Thus, the agents help in faster and efficient dissemination of a message in the network which helps in reducing message delivery delay. They compare newly designed scheme with other schemes through simulations. Simulation results show that agent based routing scheme performs better in various network scenarios.

Sudip Misra et al. formulates bandwidth redistribution as a utility maximization problem, and provide solution by using a modified descending bid auction. Proposed scheme, named as AQUM, each gateway aggregates the demands of all the connecting mobile nodes and makes a bid for the required amount of bandwidth. Authors investigates the existence of Nash equilibrium (NE) in the proposed solution. Theoretically, they deduce the maximum and minimum selling prices of bandwidth, and proves the convergence of AQUM.

Yuanyuan Zeng et al. proposes an Energy efficient Routing and Rate Allocation (ERRA)scheme based on Q-learning that can optimize the energy efficiency with the constraints of congestion, buffer and delay. ERRA solves the routing and rate allocation together with reinforce learning, and then make decisions on relay selection and rate schedule. ERRA explores the possible strategies, and then exploits the knowledge obtained to adapt its relay and schedule strategies. ERRA achieves the desired overall objective by considering the stochastic non-cooperative game under on-line multi-commodity routing situations. The simulation results show that ERRA achieves good energy efficiency and delivery ratio within the delay bound.

Barun Kumar Saha et al. studies an aspect of human-network interplay in MOONs, by considering the effects of human emotions on the network operations. Authors also uses Meftah's computational model of the emotions, which itself is based on Plutchik's model. Authors models the current emotion of any user as a function of the individual's previous state of emotion and the content of the messages received by her. Authors also models the scenarios in which the users switch off their device's radios, and their effects on the message delivery ratio. Results shows that the delivery ratio of the messages tends to zero due to increased traffic load and inactivity of the device's radios.

Yong Li et al. accurately evaluates the performance of DTN routing protocols to show their advantages and inferiority. At the same time, the energy is very limited in a large number of DTNs, such as sparse mobile sensor networks and emergency ad hoc networks, and it impacts the routing performance significantly. They investigates the performance of routing schemes for the energy-constrained DTNs. First, they models the two-hop relaying, epidemic routing and K-hop forwarding with energy constraint based on a continuous time Markov chain. Then, they evaluate the system performance by measuring message delivery delay and delivery cost using explicit expressions. By both simulation and numerical results, they demonstrate the accuracy of their proposed model and reveals that the energy constraint can actually avoid the message storms.

Yue Cao et al. focus on geographic routing DTN, by considering sparse network density. Authors explore the Delegation Forwarding (DF) approach to overcome the limitation of the geometric metric which requires mobile node moving toward destination, with the Delegation Geographic Routing (DGR) proposed. Besides, they handle the local maximum problem of DGR, by considering nodal mobility and message life time. Analysis and evaluation results show that DGR overcomes the limitation of the algorithm based on the given geometric metric. By overcoming the limited routing decision and handling the local maximum problem, DGR is reliable for delivering messages before expiration lifetime. Meanwhile, the efficiency of DGR regarding low overhead ratio is contributed by utilizing DF.

Sudip Misra et al. addresses the problem of sporadic internodes wireless connectivity by taking an evolutionary theory based approach. Authors proposes a Distributed Information Based Cooperation Ushering Scheme (DISCUSS) to promote cooperation in message forwarding between nodes where the nodes maintains and exchanges information with one another during contacts about the messages created and delivered in the network. These nodes evaluate their own performance and compare that with the approximated network performance to adapt the most successful forwarding strategy. Results claim that the message delivery ratio in the network improves up to 31 \%, when the nodes dynamically switch their strategies, as compared to the case when they do not.

T. I. Ghosh et al. proposes an algorithm for computing bandwidth guaranteed paths for traffic engineering in WiMAX IEEE 802.16 standard based networks using the mesh topology. Authors claims that the underlying algorithm fulfills routing requests \lq on the fly \rq without a priori knowledge of future requests. The problem was motivated by the need for efficient handling of traffic and network resource utilization. Proposed algorithm performs well in comparison with the selected benchmarks on metrics such as the number of rejected requests and the active links present in the network.

P. V. Krishna et al. proposes a routing protocol capable of detecting intrusions, while undertaking the tasks of routing in Wireless Mobile Networks. Authors base the routing tasks in the existing protocol on the existing optimized link-state routing protocol. The protocol uses the sampling mechanism for the detection of malicious information. Concepts of learning automata have been introduced. Authors claim that the performance of proposed protocol achieved an intrusion detection rate of 94%.

During study of related research work given in above, more emphasis is given on finding location of destination and not on optimizing the delivery path to the destination. The path optimization by considering geographical location and area of destination will prominently saves a lot of energy in transmitting data. Definitely, routing overhead will be reduced to significant extent. Proposed protocol in this chapter tries to take an advantage of known location of destination and associated angle to optimize path to the destination by effectively utilizing its resources.

Proposed Routing Protocol

Before proposing a new routing protocol, study of the layer and node communication is an important issue. The DTN bundle protocol assumes that contact between nodes is not always available. The communication is only possible at certain times, which is different from the assumption done in the Internet. Hence, traditional TCP/IP protocol suit needs to be modified to configure a node in DTN. During the contact, certain number of bits can be transmitted, which defines the contact's volume. If the data to be transferred exceeds the volume of any known contact, the bundle has to be fragmented. Two types of fragmentation are explored: proactive and reactive. In proactive fragmentation, a DTN node may divide a block of application data into multiple smaller blocks and transmit each block as an independent bundle. This approach is called proactive fragmentation because it is used primarily when contact volumes are known (or predicted) in advance. In the reactive fragmentation, the DTN nodes in communication may fragment a bundle cooperatively when a bundle is only partially transferred. The receiving bundle layer modifies the incoming bundle to indicate the data received is a fragment and forwards it normally; the sender node truncates the successfully transferred part from the original bundle and keeps the undelivered fragment for a subsequent contact. This approach is called reactive fragmentation because the fragmentation process occurs after an attempted transmission has taken place.

LAYER AND NODE COMMUNICATION IN DTN

DTN uses store and forward message switching technique by introducing a new protocol layer called the "Bundle Layer" on top of the transport layer. The bundle layer is a major responsible for storing and forwarding entire bundles called messages hereafter, between a source node and destination node. Nodes wish to send data must have the bundle layer support on top of the TCP layer. Transport layer protocol either TCP or UDP below the bundle layer is selected based upon the reachability and accessibility properties estimated about each region in and around the intended node in the network. The Figure 1

Figure 1. Layer communication in delay tolerant network

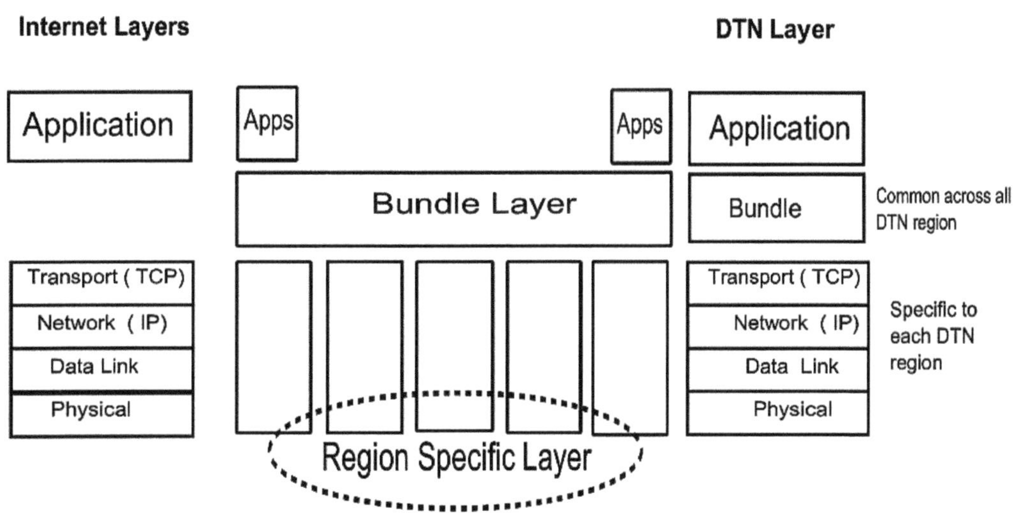

shows the bundle layer and its position in the existing TCP/IP Internet protocol suit. The TCP protocol is responsible for end-to-end communication. DTN layer at the source node communicates with other node called destination using the communication unit as message called bundle. Therefore, DTN communication is also called as "Bundle-to-Bundle communication".

Figure 2 suggests the layer-to-layer communication in DTN where bundles of messages

or simply bundle is sent from source node to the destination node. Acknowledging depends on the selected transport layer protocol, either UDP or TCP and therefore it is optional. Lower layers are protocol dependent which depends on the network layer protocol. So, it is called protocol dependent transfer with optional acknowledgment.

ROUTING PROSPECTIVE IN DELAY TOLERANT NETWORK

Routing is the heart of communication. Taking into account mobility of node in delay tolerant network, DTN register short contact duration and experience rapid changes in topology. The node mobility pattern directly influences transmission range. Physical obstacles and interferences contribute to intermittent connectivity and high error commonly observed in these network. The main objectives of this work are to provide limited broadcasting and minimize the energy consumption in delay tolerant network. The major research on DTN protocols can be roughly Categorized in three categories:

- To derive scaling laws for the delay-capacity tradeoff due to mobility;
- To design and evaluate new DTN protocols for routing and later on solving associated problems such as congestion control, limiting broadcasting areas data dissemination etc;
- Formally analyzing and evaluating the expected performance of DTN protocols under various environments according to parameters considered.

Figure 2. Node communication in delay tolerant network

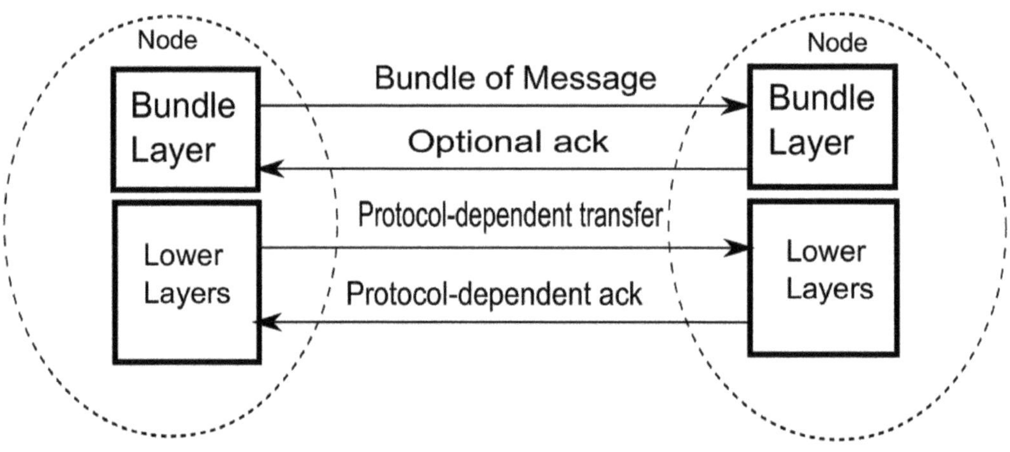

Routing Protocol as SEDAR

Direction Aware Routing Protocol for Delay Tolerant Network is proposed which is based on category two. Proposed protocol is designed and investigated in homogeneous and heterogeneous network. The main goal is focusing on increasing the security level of message scheduled for transmission consuming less amount of energy in the DTN. SEDAR belongs to *Single Copy Replication Routing* category where the original copy of the received message is used. SEDAR uses a unicast approach for relaying messages.

SEDAR Algorithm

SEDAR algorithm is given here. This algorithm used direction based concept to find the best suitable node on the optimal path towards the destination.

Let *G(V,E)* represent a graph where *'V'* represents set of vertices' called nodes and *'E'* represents set of edges called connecting paths. Let *S(Xs, Ys)* & *D(Xd, Yd)* be the source and destination respectively. Let *N1 (Xn1,Yn1)· · · Nn (Xnn,Xnn)* be the immediate neighbors on of node *S*.

Step 1: Compute distance 'DS' between S(Xs, Ys)& D(Xd, Yd) using equation 1.
Step 2: Compute Direction angle, θ using equation 2.
Step 3: Identify Sector using Traditional 360° coordinate system.
Step 4: Find cooperative neighbors within identified sector.
Step 5: Compute angle $\theta N1 ... \theta Nn$ w.r.t. source, S.
Step 6: Compute relative angle difference $\alpha N1 ... \alpha Nn$ w.r.t source, S.
Step 7: Compute minimum relative angle difference, α_{min} using equation 3.
Step 8: Identify best neighbor node Nn and deliver data.
Step 9: Shift origin and repeat step 1 to 7 till all data packets reach to final destination, D.

Design of Secured Energy Efficient Direction Aware Routing Protocol for Delay Tolerant Network (SEDAR)

The working of SEDAR is classified into *FIVE* major phases as follows:

1. Gathering Distance information from GPS;
2. Direction Angle Computation;
3. Node determination on Optimal Trajectory;
4. Relaying messages;
5. Shifting of Origin (Optional)/

Gathering Distance Information from GPS

The mobility nature of nodes in DTN often alters location. Major assumption in SEDAR is that all nodes aware about the location of all other nodes i.e. network topology. The lower MAC layer is assumed to be responsible for collection of GPS coordinate and some useful information from other neighbor nodes. Consider location coordinates of the source as *(Xs, Ys)* and destination as *(Xd, Yd)* respectively. Several methods can be used to estimate the position of a node. SEDAR uses the known locations of two or more reference nodes. Reference nodes are those nodes whose GPS is enabled. The distance between source and destination is estimated by using traditional distance calculation formula given in equation 1 as

$$DS = \sqrt{(Xd - Xs)^2 - (Yd - Ys)^2} \tag{1}$$

All nodes stores this estimated distance about others in their routing table called Direction Routing Table (DRT) whose structure is shown in Table 3. DRT is shown with one row (entry) about neighboring node with node id and its corresponding coordinate: *3(100,100)*. Node with id *3* assume to meet with source node, SN at time $t_{en} = 12.0$, $NEL = 80\%$. Node's *X*-Coordinate and *Y*-Coordinate can be obtained using their Global Positioning System(GPS) Coordinate. Message format used in transmission in SEDAR is shown in Figure 1.

Figure 3, gives SEDAR datagram which contains two major parts: SEDAR Header and SEDAR Payload. SEDAR Header is slightly modified by introducing extra fields. The rest of the header is unmodified. The Brief meaning of SEDAR datagram is given in Table 4.

Table 3. Structure of Direction Routing Table (DRT)

Sr. No.	Node ID	X_i	Y_i	Θ	t_{en}	NEL
1	3	100	100	45°	12.0	80%
...	...	...	...	...	...	...

Figure 3. SEDAR datagram format

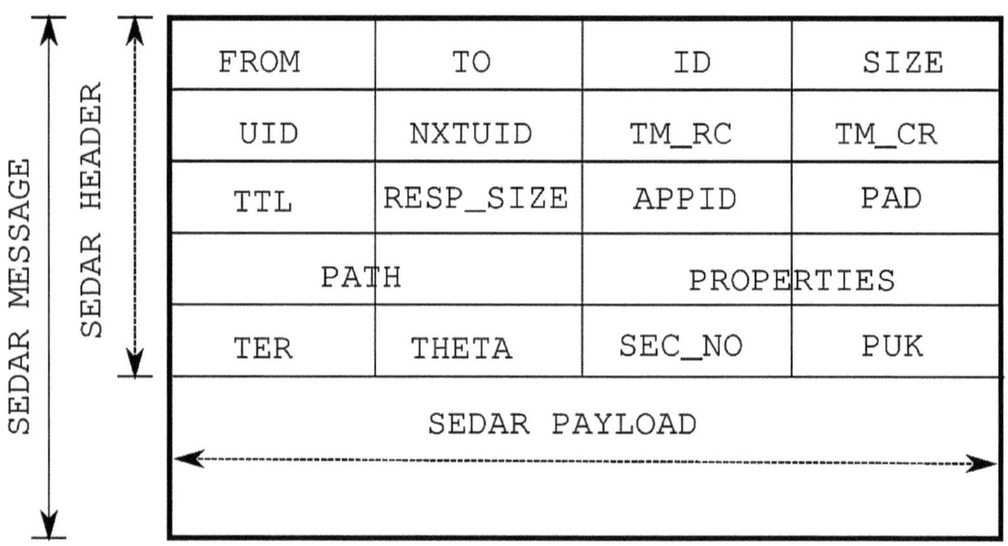

Table 4. Brief description of SEDAR fields

Field	Description
From	Source Node,SN
To	Destination Node,DN
ID	Message ID
SIZE	Message Size
UID#	Message Unique ID
NXTUID#	Message Next Unique ID
TM RC	Time at Message Received
TM CR	Time at Message Created
TTL	Time to Live
RESP SIZE	Response Size
APPID#	Application Identifier
PAD#	Pad Data Bits
PATH	PATH, Message Traversed from SN
PROPERTIES#	Message Properties
TER*	* Total Energy Required
THETA*	Estimated angle of Destination
SEC NO*	Sector Number, Destination Lies
PUK*	Public Key of Receiver

Fields relevant for Simulator
* Fields explicitly introduced

SEDAR Extra Header

The basic protocol format of sending data is insufficient for proposed SEDAR. Hence SEDAR explicitly introduce fields marked as * given in Table 4. Modified fields are listed as follows:

- TEL(βTEL)
- THETA(Θ)
- SEC_NO
- PUK
- *TEL* (βTEL *)*

Node requires energy for sending, receiving and processing messages. The total energy required for sending all messages from source to destination denoted by βTEL is given as

$$\beta TEL = \sum_{i=1}^{N} \left(\beta trans^{i} + \beta process^{i} + \beta recv^{i} \right)$$

where $\beta trans^i$, $\beta process^i$, and $\beta recv^i$ sent energy required for sending, receiving and processing messages respectively in all.

THETA (Θ) is the angle between the source and destination.

SEC_NO gives corresponding sector in which destination lies.

PUK is the public key of the destination node. SEDAR assumes all nodes should maintain pair of keys, PUK and PRV called public key and private key. Public key is used for encryption and private key is used for description using traditional public key cryptography. Therefore, if a message is represented by SEDAR DATAGRAM as $SEDAR_{DATAGRAM}$ is to be encrypted then using traditional public key encryption, encrypted $SEDAR_{DATAGRAM}$ is given as

$$ENCRYPTEDSEDAR_{DATAGRAM} = EA_{PUK}\,[SEDAR_{DATAGRAM}]$$

and

$$T_{thresh} = DA_{PUK}\,[ENCRYPTEDSEDAR_{DATAGRAM}]$$

where EA_{PUK} represents any suitable algorithm for Public Key Encryption and DA_{PUK} represent corresponding decryption algorithm. *PUK* and *PRV* are public and private keys used for encryption and decryption. SEDAR works in Network layer and therefore a particular application layer imposing SEDAR assume to provide suitable public key encryption model for securing the message. Global clock is assumed with every node, synchronized at the start. All nodes will update DRT after every interval of T_{THRESH} time. In case if new node encountered into the existing network, corresponding neighboring nodes update their DRT based on information gathered during the encounter. Lower MAC Layer is assumed to take care about encountering and relevant processing of new node. This layer also takes care of mobility nature of nodes and their subsequent consequences arise. As soon as a new entry is made in the DRT, time of entry says ten is inserted in the DRT Table. MAC layer also exchanges information about the Node's Residual Energy Level (NEL) during MAC Layer transmission.

Direction Angle Computation

The geographic area is partitioned into various sectors. Sectors are created as per traditional 360° quadrant system as shown in Figure 4. The transmission range of each node is divided into Four sectors as per traditional quadrant system. With the help of GPS coordinates of the destination node, an angle Θ is estimated using the formula given in equation 2 as

The geographic area is partitioned into various sectors. Sectors are created as per traditional 360° coordinate system as shown in Figure 4. The transmission range of each node is divided into *4* sectors as per traditional quadrant system. With the help of GPS coordinates of the destination node, an angle Θ is estimated using formula given in equation 2

$$\theta = \tan^{-1}\frac{Ysd}{Xsd} \quad (2)$$

Figure 4. Traditional 360° coordinate system

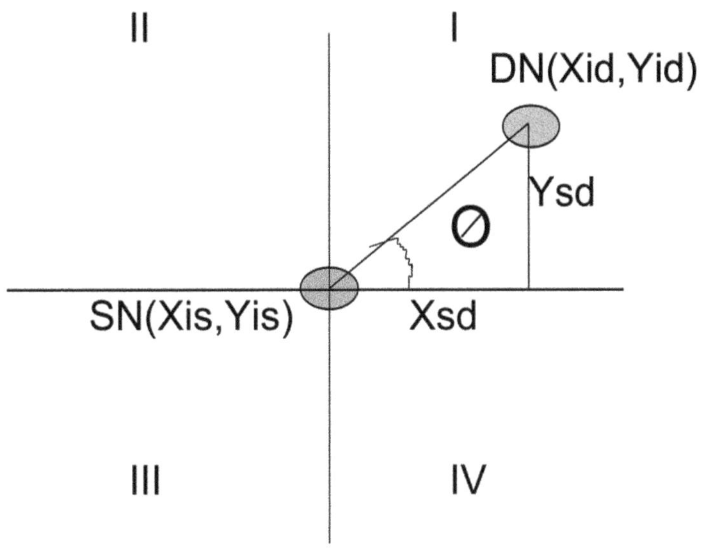

Figure 5. Network topology nodes along with their coordinates

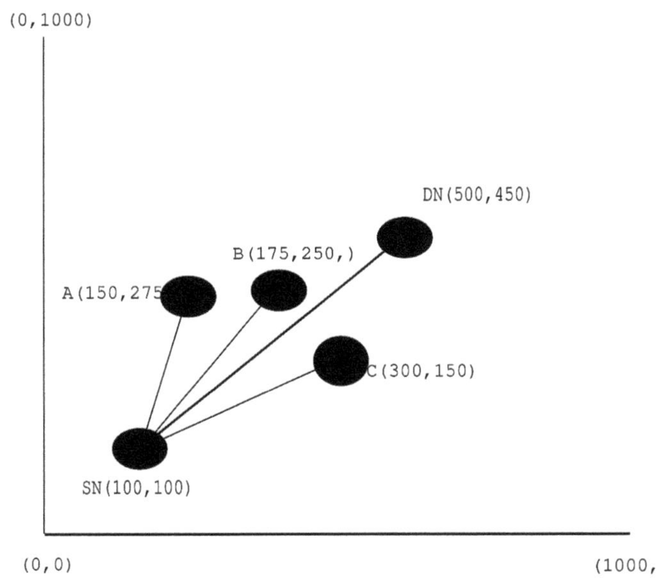

where $Xsd = Xs - Xd$ and $Ysd = Ys - Yd$.

Node Determination on Optimal Trajectory

After computing, θ, the decision of finding an optimal trajectory the destination node is to be taken. The best intermediate node on optimal trajectory is to be identified. To explore this consider Figure 3. Simple Network topology of Five nodes is considered for evaluation. The source node SN, willing to send messages to destination node DN. Nodes A, B and C is immediate neighbors of SN. Immediate neighbors indicate neighbors which lie in the coverage area of SN. SN uses its own DRT and computes an angle Θ using equation 1 and 2. During DRT lookup, SN finds immediate neighbors as node A,B and C. SN computes an angle Θ for each neighbor say θa, θb and θc. Nodes A,B and C are assumed to have sufficient residual energy level(NEL) required to cooperate. Also, it observes the TEL field of incoming message which indicate total amount of energy required by source to send all messages. During DRT lookup, if SN finds a specific neighbor whose NEL is high compared with other nodes, then that neighbor node is selected as an intermediate cooperative node on the trajectory toward DN. If all neighbors have enough energy level and ready to cooperate to SN, localization of a particular neighbor toward a destination is a big question.

As stated earlier, let's consider *A(Xa, Ya)*, *B(Xb, Yb)*, and *C(Xc, Yc)* are the neighboring nodes. Angle Θ for each of those neighbor nodes is estimated using equation 2 as

$$\theta a = \tan^{-1}\frac{Ya}{Xa}, \theta b = \tan^{-1}\frac{Yb}{Xb} \text{ and } \theta c = \tan^{-1}\frac{Yc}{Xc}$$

These sets of estimates give us the exact angle from source belief toward the destination. But this information fails to localize neighbor node in an optimal trajectory toward destination node. For this another set of estimate is required in terms of relative angle estimates. Minimum Relative Angle, $\alpha_{\min}$ between source to destination is estimated using equation 3.

$$\alpha_{\min} = Min(\alpha a, \alpha b, \alpha c)\ldots \tag{3}$$

where αi is the relative angle between SN and neighbor *node i*. This relative angle αi is computed as

$$\alpha_i = \left(\theta - \theta i\right)$$

where Θ is the angle between SN and DN and Θi is the angle between SN and neighbor *node i* willing to cooperate. The example given in Figure 4 has *n=5* nodes. Neighbors *A,B* and *C* are ready to extend theirs cooperation to source node SN. The equation 3 is used to compute α for each of those neighboring nodes *A, B* and *C* as follows

$$tx = \left|Xs - Xi\right|, \ \alpha b = \left|\theta - \theta b\right| \text{and } \alpha c = \left|\theta - \theta c\right|.$$

Relaying Messages

Minimum relative angle among these three neighbor nodes is computed using equation 3 as

$$\alpha_{\min} = Min\left(\alpha a, \alpha b, \alpha c\right)$$

After computation of Minimum Relative Angle $\alpha_{\min}$, neighbor, i.e. *node i* whose relative angle α is equivalent to $\alpha_{\min}$ is considered as the Best neighbor node on optimal trajectory toward the destination. The source starts relaying messages to the Best neighbor node selected so far.

Shifting of Origin (Optional)

This phase is used if Best neighbor selected in previous phase does not find an entry of destination in DRT. If destination node, DN is unreachable, the original coordinate system is shifted from source node, SN to the selected neighboring node, *node i*. The transmission factor *tx* and *ty* are computed as

$$tx = \left|Xs - Xi\right| \text{and } ty = \left|Ys - Yi\right|$$

After shifting origin all above *FIVE* phases are repeated till completion of message transmission. As a result, destination node *DN* will receive all the messages originated at source node, SN relayed by intermediate cooperative neighbor nodes on the optimal trajectory.

Exploring SEDAR Using Example

This section analyzes SEDAR using the example given in Figure 4. Consider the initial content of Direction Routing Table of source node, *SN (100,100)* as given in Table 4. This content of DRT is gathered by the MAC layer at the source during neighbor discovery phase. We assume topography as *1000 by 1000*, where bottom-left corner of the topography at *(0,0)* i.e. the origin and top-right corner at *(1000,1000)*. Source node *SN*, wish to send messages to Destination node, *DN*. *SN* look up its DRT and finds entry in DRT table that satisfied the criteria for route freshness using route entry time during neighbor discovery phase. DRT's entry is updated i.e. T_{EN}. Prior to communication, it is assumed that the neighbor nodes viz. A, B and C have conveyed their full interest message transmission. This is possible only if the destination node is reachable from each of them. On the other hand route to the destination is available at the DRT. For simplicity, the content and details of route entry are not shown in the DRT.

Next, the area under consideration is divided into *4* zones called sectors as per the traditional coordinate system. The sector classification is done at newly shifted origin. These are *Sector I*, *Sector II*,

Table 5. DRT at source node, SN

Sr. No.	Node ID	X_i	Y_i	θ	T_{EN}	NEL
1	SN	100	100	45°	12.0	85%
2	A	150	275	61.38°	12.0	83%
3	B	175	250	55.00 °	12.0	84%
4	C	300	150	26.56 °	12.0	85%
5	DN	500	450	41.98 °	12.0	85%

Sector III and *Sector IV*. *Origin* is shifted from *(0,0)* to *(100,100)* and the content of the DRT is modified accordingly as given in Table 5. By observing angle θ of DRT, *SN* will find that the destination lies in the *Sector I*. So it only focuses *Sector I*. Later some heuristic and predictive information is used to select best neighbor node on the optimal trajectory toward destination. Figure 6 gives the network topology at the initial stage.

Origin is shifted in *45°*, the angle θ with respect to newly shifted origin remains unaltered. Only coordinates of the neighboring nodes will get changed by the transmission factor tx and ty related to earlier origin. Updated entries are given in Figure 7.

In stage II, equation 3 is used to estimate values of α which gives $\alpha a = 19.40$, $\alpha b = 13.02$, $\alpha c = 15.42$. Then, Minimum relative angle among neighbor nodes *A, B* and *C* is

Figure 6. Network topology at initial stage

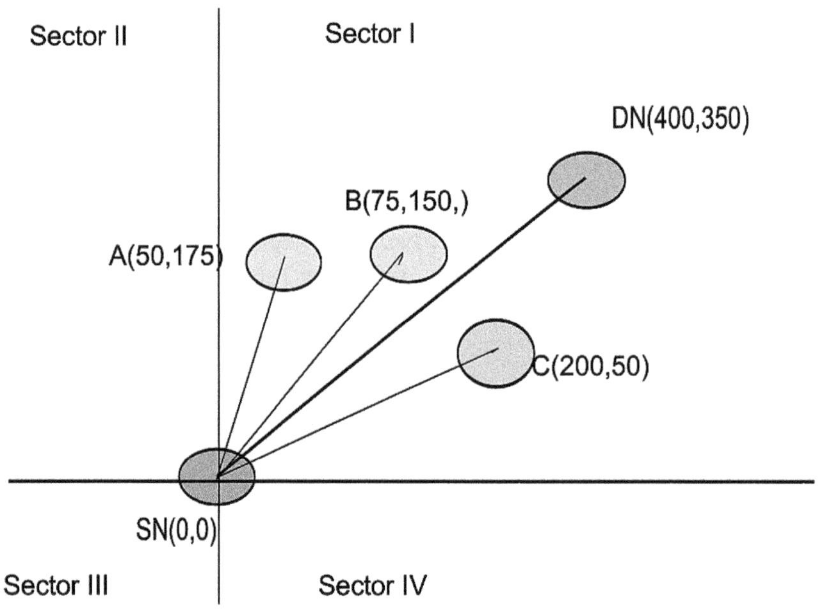

Table 6. Modified DRT at source node, SN

Sr. No.	Node ID	X_I	Y_I	θ	T_{EN}	NEL
1	SN	0	0	0°	14.0	85%
2	A	50	175	61.38°	14.0	83%
3	B	75	150	55.00 °	14.0	84%
4	C	200	50	26.56 °	14.0	85%
5	DN	400	350	41.98 °	12.0	85%

Figure 7. Network topology updated after shifting origin

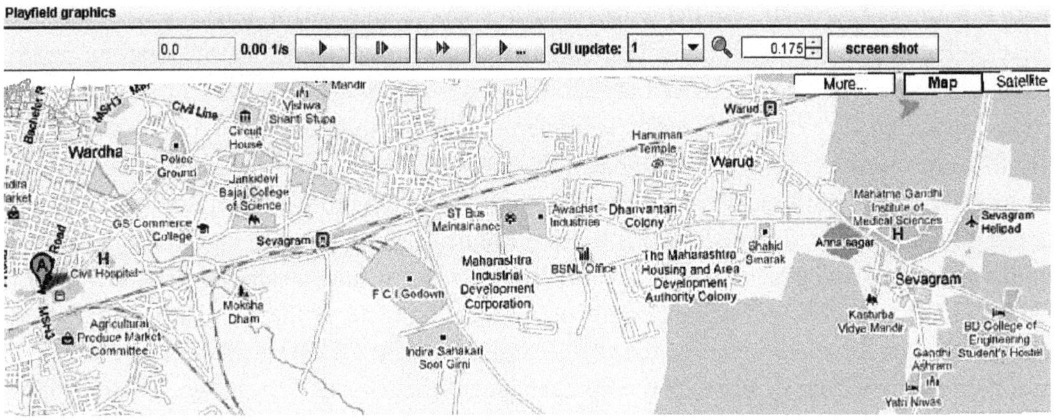

$$\alpha_{\min} = Min(19.40, 13.02, 15.42)$$

This clearly provides the *Best Neighbor node* on the optimal trajectory from *Source node, SN* to *Destination node, DN*. In the next iteration, origin is shifted from *Source node, SN* to *Best Neighbor Node* i.e. Node B in this example and same procedure is repeated till all messages get routed to final destination.

EXPERIMENTAL EVALUATION IN HOMOGENEOUS ENVIRONMENT

To provide a homogeneous environment in which the proposed SEDAR and the conventional routing protocols can be evaluated for their performance, an Opportunistic Networking Environment (ONE) is configured to use in homogeneous environment. This environment consists of a memory of unlimited size and a set of data structures defining the processing to be performed by the simulated system.

The ONE Simulator Framework

The ONE simulator is a Java based simulator offering a broad set of DTN protocols simulation capabilities in a single framework, which has been specifically designed for evaluating DTN routing and application protocols. This simulator helps to create scenarios based upon different synthetic movement models and real-world traces and offers a framework for implementing routing and application protocols. Interactive visualization and post-processing tools support evaluating experiments and emulation mode allows the ONE simulator to become part of a real-world DTN test bed. It has an extensive set of ready-to-use modules which include six synthetic mobility models that can be parameterized and combined to approximate real-world mobility scenarios, six configurable, well known DTN routing schemes, a set of base primitives to design application protocols, a basic battery and energy consumption model, several input/output filters for interacting with other simulators, and a mechanism for the integration with real-world test beds.

At each simulation step the engine updates a number of modules implementing simulation functions. The main functions of the ONE simulator are modeling of node movement; inter node contacts, routing and message handling. The result collection and analysis are done through visualization, reports and post-processing tools. The connectivity between the nodes is based on their location, communication range and the bit rate. The routing function is implemented by routing modules that decide which messages to forward over existing contacts. Finally, the messages themselves are generated through event generators. The simulation results are collected primarily through reports generated by report modules during the simulation run. Report modules receive events (e.g. message or connectivity events) from the simulation engine and generate results based on them. The results generated may be logs of events that are then further processed by the external post-processing tools, or they may be aggregate statistics calculated in the simulator. Secondarily, the graphical user interface (GUI) displays a visualization of the simulation state showing the locations, active contacts and messages carried by the nodes. Node movement capabilities are implemented through mobility models. The mobility models define the algorithms and rules that generate the node movement paths. It also includes a framework for defining the algorithms and rules used in routing and come with ready implementations of well known DTN routing protocols.

OpenJUMP Tool

To create a map based movement model, OpenJUMP tool is used. The OpenJUMP is a Geographical Information System (GIS) open source tool. It is a Java based unified mapping platform, hence the suffix JUMP. The OpenJUMP window contains a list of layers and a graphic display in which the layers are displayed. These layers are used to display geographic objects (features). A Layer usually contains objects of specific geographic theme (e.g. water bodies, forest, buildings, soils). The layers are used to display vector data. To create a simulation model to simulate proposed, the various layers are created in accordance with the various groups. A group layer indicates a particular entity moving on the map. After configuration of layer model, a vector data file is generated. This vector data file is passed as input to ONE Simulator for simulating given map based movement model.

For simulating SEDAR, ONE simulator described in subsection is used. The route map between two places viz. Wardha (District in Maharashtra State, India) & Sevagram (Village in Wardha District, Maharashtra State, India) is considered by configuring ONE simulator as shown in Figure 8.

Simulation Results

SEDAR is simulated in the homogeneous mode for simulation time of 12 hours. After simulation, the results of simulation are collected and analyzed and message statistics are collected. Result analysis is formulated in terms of graphical representation given in this subsection. Simulation parameters are given in Table 7, which are kept constant during comparison of existing routing protocols. Figures 8 shows the simulation of SEDAR routing in the simulator environment described in this section. Each run specifies the different instant of simulation. These instances are obtained during working of the SEDAR routing protocol in the described scenario considering homogeneous nodes.

Proposed SEDAR is simulated in the homogeneous mode for simulation time of 12 hours. After simulation, the results of simulation are collected and analyzed and message statistics are collected. Result analysis is formulated in terms of graphical representation.

Figure 8. Route map between Wardha and Sevagram

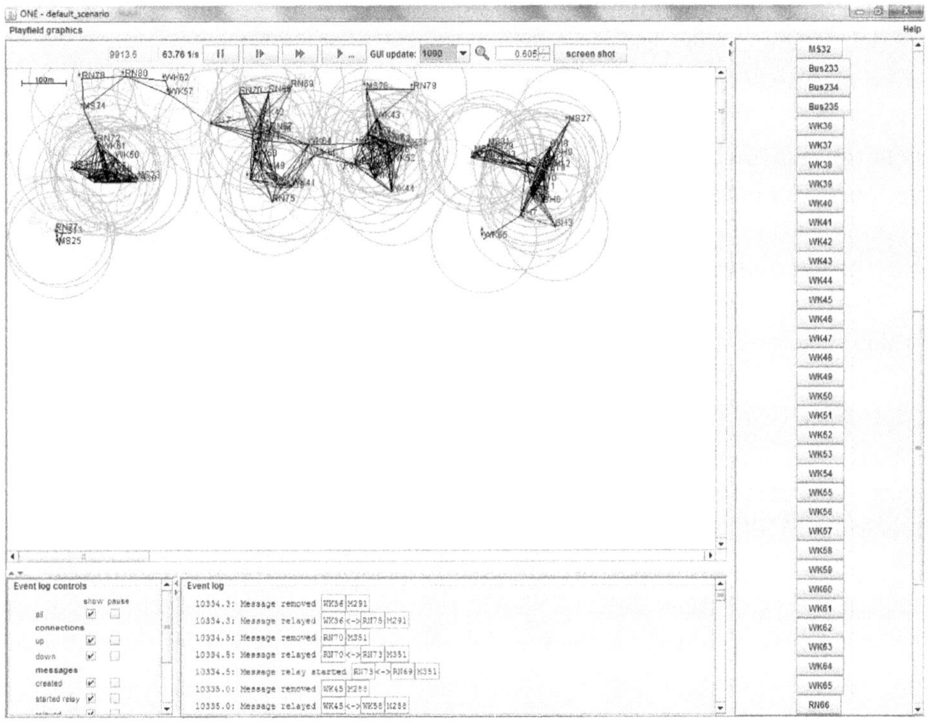

Table 7. Simulation parameters

Sr. No.	Parameter	Value
1	Simulation Time	43200s [12 Hours]
2	Number of host groups	8
3	Number of Nodes	81
4	Office Pause Time	0.5s
5	Working Time	28000 seconds
6	Schedule difference	20s
7	Transmit Speed	250 kbps
8	Transmit Range	100 M
9	Buffer Size	50 Messages
10	Message TTL	300 Minutes[05 Hours]

Packet Delivery Ratio in SEDAR

Packet Delivery Ratio (PDR) is defined as the ratio of the number of messages delivered to the number of messages created. In SEDAR the PDR is 0.8354. PDR is less during early stage of simulation and reaches to almost highest value in the middle and then nearly becomes constant to its highest value. This is because in the early stage, the packets are created and delivered and message buffer is found to be empty. But the SN does not meet with a DN which delays the delivery. Later stage, enough relay nodes cooperates SN which results into high PDR. Analysis of the results obtained proves that the SEDAR gains PDR of 0.834. This can be elaborated from the graph given in Figure 9.

By observation, Number of Packets created is found to be almost equal at every instance of pause time, whereas Number of Packets delivered is found more in the initial case, decreases in middle hours and increase at the end of simulation. The simulation is particularly considered using Working Day Model where more rush in traffic is found at the middle part of pause time. As a effect of this heavy traffic, network congestion occurs which results in decreasing PDR. On the other hand, the early morning and evening period (rather than the office hours), the network is mostly free from congestion results in increasing PDR. The control of traffic in this Map Based Movement Model is done using various probability values

Energy Consumption in SEDAR

To reduce energy consumption, a model which utilizes less energy during transmission and reception of the message is formulated. Some of the parameters used to simulate energy efficient SEDAR as given in Table 8.

For computing energy consumption in SEDAR, Java code is designed which initialize residual energy at start to every node considered in the scenario. Node requires energy for each process mention as Sr.No. 2 to 4 in Table 8. Results obtained after simulation of SEDAR is collected and analyzed. From the result, Average Residual Energy is computed for all participating nodes. The graph of Average Residual Energy available at particular instance is plotted against Pausetime by considering granularity of 1 Hour. From the results, it can be foreseen that 15% energy is consumed during 12 Hours of simulation. This is clearly understood from the graph given in Figure 11.

Figure 9. Message transmission in SEDAR

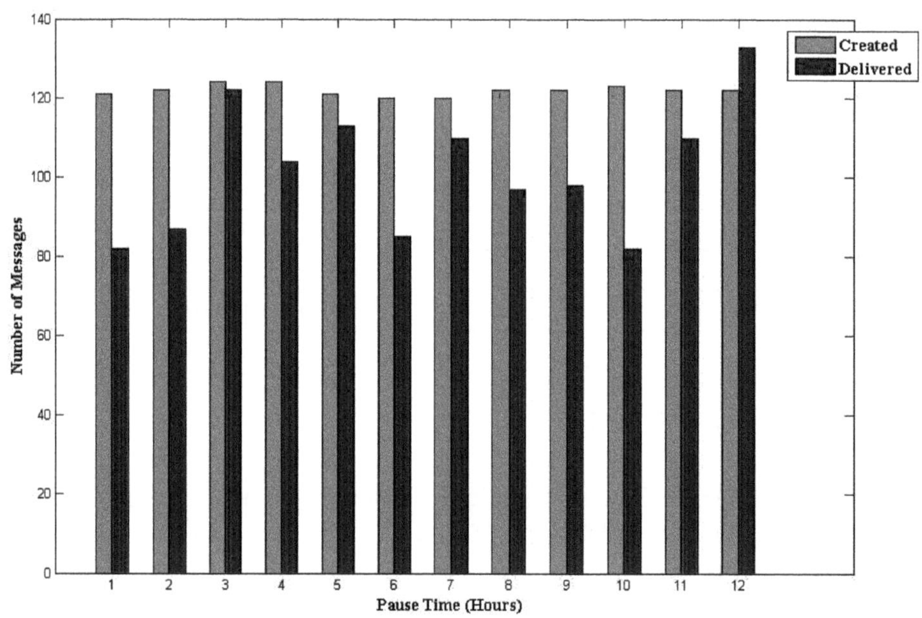

Figure 10. Plot showing number of messages (packets) created and delivered in SEDAR

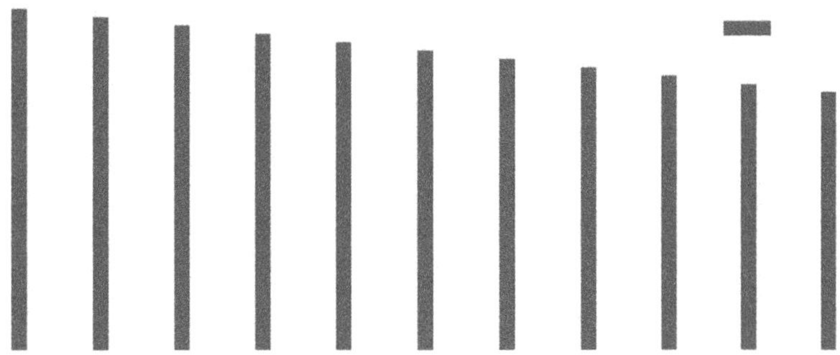

Table 8. Parameters used to simulate energy efficient SEDAR

Sr. No.	Parameter	Value
1	Initial Residual Energy	100000 unit
2	Scan Energy	0.61 unit
3	Transmit Energy	0.80 unit
4	Scan Response Energy	0.80 unit

Figure 11. Plot showing average residual energy available

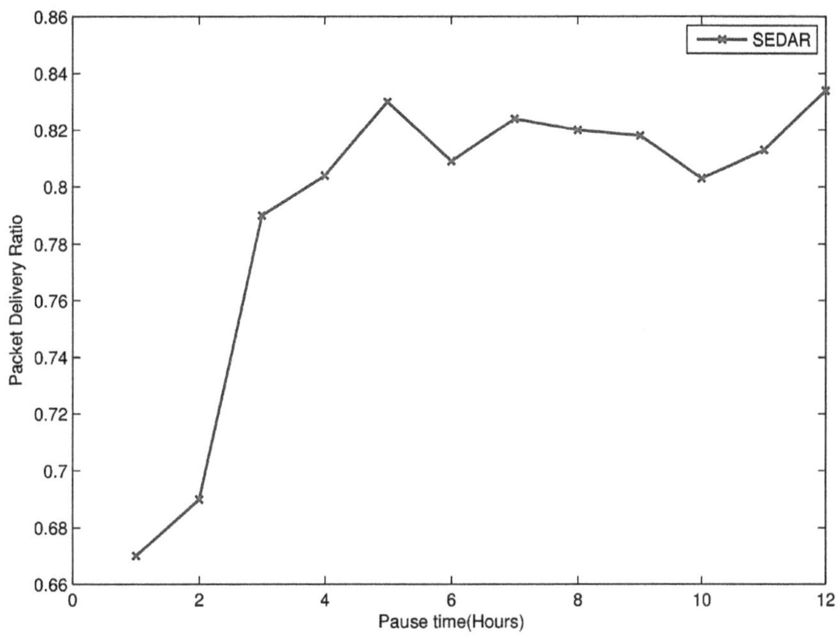

Comparative Analysis of SEDAR With Existing Protocols

In this section, the proposed SEDAR is compared with other existing routing protocols in DTN viz. First Contact Routing (FCR), Direct Delivery Routing (DDR), Prophet Routing (PROPHET) and Spray and Wait Routing (SPW).

Comparative Analysis Using Packet Delivery Ratio

For this comparative analysis, the parameters listed in Table 8 are kept constant with the same model given in Figure 7. The graph of PDR verses Pausetime(Hours) is plotted as given in Figure 10 for various scenarios.

From the graph, it can be observed that the PDR in proposed SEDAR is found to be more than other existing routing protocols. The reason is that in SEDAR, the transmission area is restricted to focus on the particular sector in which destination lies. As a result, limited broadcasting is achieved, which significantly reduces Routing Overhead and increases the PDR. This also helps to avoid congestion to moderate level. Thus, relay nodes get a better chance to deliver the messages within a limited period of time.

Comparative Analysis Using Average Residual Energy

The comparative analysis of average residual energy is carried out. The graph of Average Residual Energy within the nodes verses Pausetime(Hours) is plotted as given in Figure 13.

From the graph, it can be observed that th proposed SEDAR consumes less energy as compared with other DTN routing protocols. Because of the restriction on flooding and sectorization, the delivery of

Figure 12. Plot showing PDR in various scenario

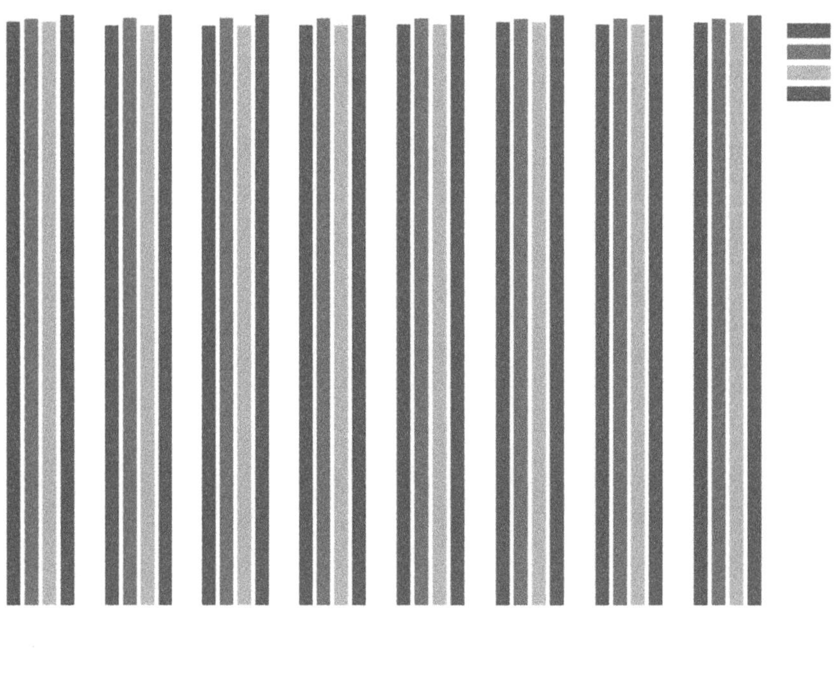

Figure 13. Plot showing energy consumption in various DTN routing protocols

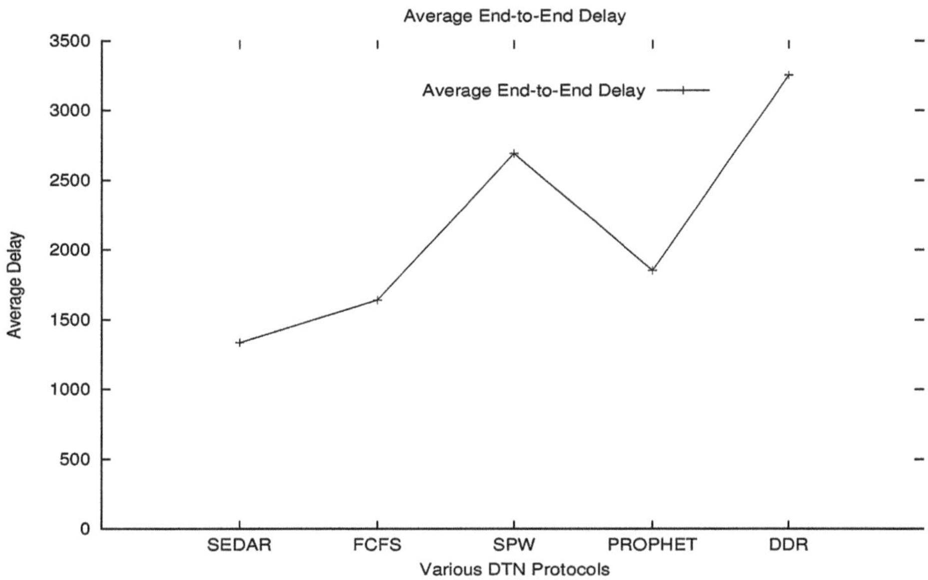

message is almost unicast. Hence it reduces energy consumption. It is inferred that 3% to 5% energy is saved in SEDAR comparative to other routing protocols.

Comparative Analysis Using End-to-End Delay

The End-to-End delay is nothing but the time required from transmission of message from source to reception of same message at the destination. This analysis is performed by considering sample messages for proposed and existing DTN routing protocols. After simulation, report of message delivery and its details are generated. This report is used to compute average delay value for all the protocol under consideration. This computed average end-to-end delay is plotted against various routing protocols as given in Figure 14. Thus, SEDAR is found to be optimum in terms of end-to-end delay compared with other DTN routing protocols. In this comparison, it is observed that the Direct Delivery Routing protocol found to be worst as it delivers messages only when source contact with destination. Rest of the protocols actively contributes to lower down the end-to-end delay.

Comparative Analysis using Variation in Buffer Size

Every node needs storage space called message buffer to store incoming messages till they becomes ready for further transmission. Hence, it is also required to observe the performance of proposed system by varying message storing capacity. This section compares proposed SEDAR with various DTN routing protocols by observing the effect of variation in message buffer as stated earlier. Here, message buffer size is varied from 50 messages to 200 messages with step of 50. It is concluded that the performance of First contact routing and proposed SEDAR is almost similar and both outperforms others existing protocols under consideration.

Figure 14. Plot showing end to end delay in various DTN routing protocol

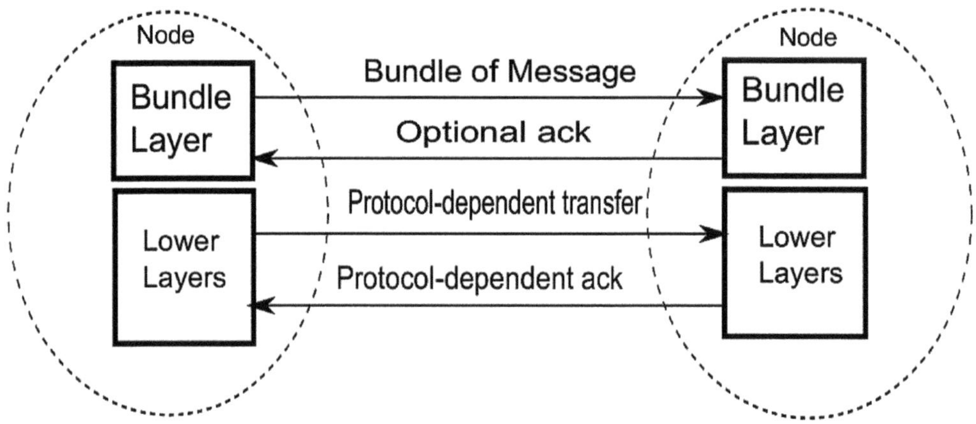

EXPERIMENTAL EVALUATION IN HETEROGENEOUS ENVIRONMENT

Heterogeneous system differs considerably from the homogeneous system in several ways; more importantly in that the performance metric is not the congestion, but rather than the delivery rate. Heterogeneity arises in a many scenarios in delay tolerant networks and is one of the key issues that control the performance of network while evaluating packet forwarding capabilities of the routing protocol. There has been incredible amount of research conducted in effective routing protocol implementation in heterogeneous system. This section is concerned with the issues conceived due to heterogeneous nature of network. Author's principal objective is to develop a novel secured and energy efficient routing protocol for heterogeneous environment. The issues with which they deal range from the decision of selecting optimize trajectory for custodian acknowledgment transmission to the complexity arises due to network congestion in heterogeneous environment. In this section, the motivation for the proposed work is presented, discussing heterogeneous network and associated issues.

There are several analytical studies on the overall performance of a few routing protocols based on both flooding and forwarding approaches have been presented in the literature. However, these works are mainly based upon a homogeneous model in which any mobile node is making contacts with others according to a Poisson process. The homogeneous networking model fails to potentially exploit the capability of delay tolerant network. Heterogeneity is achieved when the communication is done in wired and wireless network. This leads to give birth to special issues about network parameters. Obviously, such heterogeneous network behaves totally different from homogeneous network. Hence, it is necessary to evaluate the performance of proposed SEDAR protocol in heterogeneous network environment.

Heterogeneous Network

Heterogeneous network used in wireless networks indicates the usage of multiple types of access technologies for a wireless system, and these radio access networks RAN can form a motley of macrocells, picocells, and femtocells in order to offer wireless service in an environment with a wide variety of wireless coverage zones, ranging from an open outdoor environment to office buildings, homes, and underground spaces. Therefore, it has a hierarchical structure with overlapped and overlaid wireless service areas and complex interoperation between these macrocells and smaller cells sometimes in the form of WLAN working together to provide a mosaic of coverage.

Research in heterogeneous network can be basically categorized into three groups. The first group focuses on the new architecture and implementation of HetNet system. The second group is trying to find the good and optimized RAN selection algorithms. The last group works on the system performance analysis of heterogeneous network. This section proposes the work belonging to second category where a new routing protocol is proposed which outperforms in highly dense mobile nodes environment.

Related Work

Different instances of network elements that export the same functional capability are, in general, vulnerable to different security attacks. Hence, a network architecture that supports a collection of heterogeneous network elements for the same functional capability offers a greater possibility of surviving security attacks as compared to homogeneous networks.

H. Li et al. propose a distributed scheme for calculating resources available in node vicinity, as a tool to support meaningful routing decisions. A generic model is developed first, and is then applied to individual network assets. The model is based on a sparse network, where resources are potentially not uniformly distributed. It uses recent encounters to estimate resource availability in node vicinity. It has been shown that a store-carry-forward scheme may benefit from accessing vicinity resource estimates. This knowledge allows nodes to implement meaningful custodian election and queue management strategies, approached here from a holistic perspective. It also demonstrates that routing protocols not only use fewer resources overall, but also consume resources preferentially from nodes with higher resource levels, sparing nodes with limited supplies. As a result, disparities in available resources across the node population are significantly reduced, and nodes are less likely to leave the network as a consequence of resource depletion.

Yang Yu et al. proposes a DTN architecture implementing custody transfer (CT) mechanism to improve the reliability of DTN. Authors also proposed a mathematical model of the DTN data link. Author developed developed a NS2 based DTN simulator and conducted three simulations. The first two simulations are conducted on links whose hops are from 2 to 10, where all relay nodes accept or deny CT. Authors claims that the simulation results shows the destination node receives far more data when CT is available; furthermore, the storage consumption of the source node decreases. Also authors proves that the value of retransmission timer that is greater than the maximum round-trip-time between adjacent nodes and less than the minimum disconnection interval is a good candidate for DTN data transfer.

T. Venu Madhav et al. proposes Energy Efficient Cluster Routing Protocol for Heterogeneous Wireless Sensor Networks. Authors suggested an improved clustering strategy. Authors also propose Energy Efficient Stable Election Protocol (EESEP) is proposed. The protocol found to be efficient in increasing network lifetime. Simulation results prove that the proposed protocol outperforms when compared to LEACH. They also show the impact of heterogeneity in terms of node energy.

Yong Li et al. design an optimal relaying scheme for DTNs, which takes into account node's heterogeneous contact rates and delivery costs when selecting relays to minimize the delivery cost while satisfying the required message delivery probability. Here, Extensive results based on real traces demonstrate that the proposed relaying scheme requires the least delivery cost and achieves the largest maximum delivery probability, compared with the schemes that neglect nodes' heterogeneity.

V. K. Chaithanya Manam et al. proposes a message driven based algorithm add on to the existing routing protocols that enables intelligent forwards based on individual message lifetime and delivery requirements, thereby improving energy efficiency. Further, authors analytically model the performance of those protocols in a realistic DTN setting that comprises of nodes with different transmission radii. The results show that the routing protocols with the proposed energy efficient add-on improve the energy efficiency up to 60% and 75%. Their results are validated by extensive simulations.

Milena Randkovic et al. proposes CafRep that uses a combined local social, buffer and delay metrics for congestion aware message forwarding and replication. Authors also claims that the proposed technique maximizes message delivery ratio and availability of nodes while minimizing latency and packet loss rates as congestion levels get increases. Proposed model is a combined relative utility driven heuristic that allows highly adaptive forwarding and replication policies by managing to detect and offload congested parts of the network and adapting the sending/forwarding rates based on resource and contact predictions. Authors also empirically investigated a number of weighting models for analyzing the impact of different utilities on the CafRep performance. They prove that CafRep outperforms five other state of the art DTN adaptive and non-adaptive routing protocols across the majority of metrics across all the traces. Authors also shows that CafRep maintains performance levels in two application scenarios: publish subscribe constant bit rate podcasting and facebook traffic application.

Design of SEDAR-H

This section briefly describe the design space for SEDAR and shows different dimensions of possible approaches that can be useful for message dissemination in heterogeneous DTN environment. Earlier proposed SEDAR found to work efficiently to increase the performance in DTN. The evaluation of proposed SEDAR is done using ONE Simulator and it is also compared with few well known protocols under study. ONE simulator fails to provide support for simulating SEDAR in heterogeneous environment which is a hybrid network of both wired and wireless nodes to evaluate its performance. For simulation of SEDAR in heterogeneous environment, authors need to modify some of the phases proposed in working of SEDAR. For relevant modification in SEDAR, authors need to provide custodian acknowledgment to every intermediate node encountered on optimal trajectory to increase security level. For this, we need to modify the proposed SEDAR message format.

Recalling the working phases of SEDAR listed in chapter 3 as

- Gathering distance information,
- Direction angle computation,
- Node determination on optimal trajectory,
- Relaying messages,
- Shifting of origin (optional).

The design approach in SEDAR-H modifies Relaying messages phase of above list of phases so as to send custodian acknowledgment to immediate back relay node either sender or intermediate node on upstream trajectory. This custodian acknowledgment is used to increase the level of confidence between participating nodes, increasing security

Figure 15 shows the communication model considered for design, implementation & simulation of SEDAR-H in heterogeneous environment. Wired node W is connected with Mobile node M by using intermediate Base station node BS as gateway. The traffic towards M is routed through BS. As discussed earlier, Relaying messages phase is modified to find Best Neighbor Node on optimal trajectory toward DN and DN is expected to send custodian message as acknowledgment to immediate previous node.

This custodian acknowledgment carries information to instruct neighbor about status of received message. Here, custodian report is sent by every intermediate node to the immediate previous node about status of received message called "bundle". Finally, destination node sends acknowledgment of bundles received to the last but one node as custodian acknowledgment.

Simulation of SEDAR-H

To provide an environment in which proposed SEDAR-H can be evaluated for its performance, an open source, NS2 Simulator has been modified as a complete hybrid environment. For this, Packet Format and Routing Table class has been newly designed in relevance with working of proposed SEDAR-H. A groups of sending and receiving functions have been coded and implemented. The destination aware computation methodology for finding optimal trajectory has been used in TCL script and corresponding variable binding has been provided to link TCL and C++.

Figure 15. Node configuration while communication in DTN

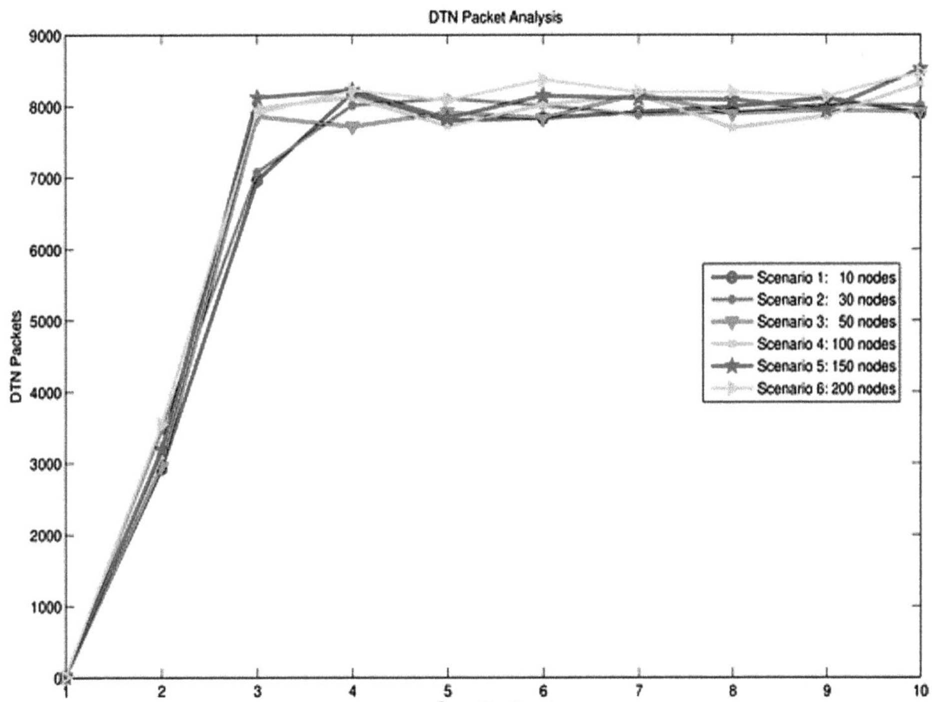

Simulation Framework for SEDAR-H

Network Simulator (NS) is a discrete event simulator targeted at networking research. NS2 provides substantial support for simulation of TCP, routing, and multicast protocols over wired and wireless (local and satellite) networks. It is an object oriented simulator, written in C++, with an OTCL interpreter as a frontend. NS is implemented using two languages viz. C++ and TCL. This is because simulator has two different kinds of things it needs to do. On one hand, detailed simulations of protocols require a systems programming language which can efficiently manipulate bytes, packet headers, and implement algorithms that run over large data sets. For these tasks run-time speed is important and turn-around time (run simulation, find bug, fix bug, recompile, re-run) is less important. On the other hand, a large part of network research involves slightly varying parameters or configurations, or quickly exploring a number of scenarios. C++ is fast to run but slower to change, making it suitable for detailed protocol implementation. OTCL runs much slower but can be changed very quickly (and interactively), making it ideal for simulation configuration. In addition to these, a scenario file defining the destination of mobile nodes along with their speeds and a connection pattern file (CBR file) defining the connection pattern, topology and packet type are also used to create the trace files and nam trace files. Also the network parameters can be explicitly mentioned during the creation of the scenario and connection-pattern files using the library functions of the simulator.

Simulation Setup

Simulation parameters are precisely written in following tables. These parameters are divided into two parts. Table 9, gives constant parameters for entire scenario considered. Table 10 gives variable parameters which changes according to the scenario under consideration.

Performance Evaluation

For evaluating performance, entire simulation with various scenarios is considered. The results are collected in tracefile using Event trace generator tool in NS2 Simulator. The useful information is extracted using GAWK script. First, the number of data packets (DTN Packets) sent in the network for all scenarios are extracted as given in Table 11.

Overall result obtained for all scenarios under consideration is plotted against Pausetime(s) verses number of DTN packets in all scenarios together given in Figure 16.

Traffic Analysis Using Δ and Ø

Poisson distribution is widely used by many researchers for modeling data traffic. But, it is shown by many authors that Poisson distribution is not adequate for modeling local-area and wide area network traffic. Present day traffic in DTN, Internet and in LANs is mostly bursty in nature. Therefore, in the simulation we consider random Map based traffic distribution. To analyze the effect of traffic, two parameters viz. Δ and $\Lambda 1$ are used which are computed using current knowledge of network. The parameter, $\Lambda 1$ is computed using Equation 4.

Table 9. Constant simulation parameters

Sr. No.	Parameter	Value
1	Channel	Wireless Channel
2	Propagation	Tworayground
3	Physical Interface	Wirelessphy
4	Interface Queue	Queue/Droptail/Priqueue
5	Interface Queue Length	100 Packets
6	Adhoc Routing Protocol	Sedar
7	Antena	Omniantenna
8	Topography	800 Meter By 800 Meter
9	Total Simulation Time	100 Seconds
10	Wired Nodes	2
11	Base Station Nodes	2
12	Ieee Model	802.11
13	Data Rate	11 Mega Bytes (Mb)
14	Basic Rate	1mb
15	Tcp Window Size	100
16	Size Of Data To Be Transmitted	10 Mega Bytes (Mb)
17	Size Of Tcp Segment	1460 Bytes

Table 10. Variable simulation parameters

Sr. No.	Parameters
1	Number of Mobile Nodes
2	Number of File Transfer Protocol Connection
3	Source and Destination pair
4	Start time and End time of FTP

$$\Lambda 1 = \frac{T\Delta}{1000} \tag{4}$$

To compute Δ, Equation 5 is used. These values are given in Table 12.

$$\Delta = \frac{Fc}{N} \tag{5}$$

The graph given in Figure 17 is plotted between Δ values and various Number of Nodes considered for simulation. From the graph, it has been observed that as the number of nodes are increases, Δ values decreases.

Table 11. Data packets for all scenarios

Sr. No.	Pause Time(s)	Scenario					
		1	2	3	4	5	6
1	10	0	0	0	0	0	0
2	20	2894	3415	2912	3050	3114	3484
3	30	6943	7051	7839	7951	8094	7884
4	40	8176	8004	7711	8119	8210	8214
5	50	7792	8087	7899	7703	7804	8063
6	60	7818	7992	7820	8001	8127	8358
7	70	7914	7875	8169	8157	8108	8199
8	80	7991	7914	7892	7690	8093	8201
9	90	8116	7991	7942	7860	7950	8139
10	100	7901	8013	7920	8312	8513	8477

Figure 16. Plot showing DTN packets in all scenarios

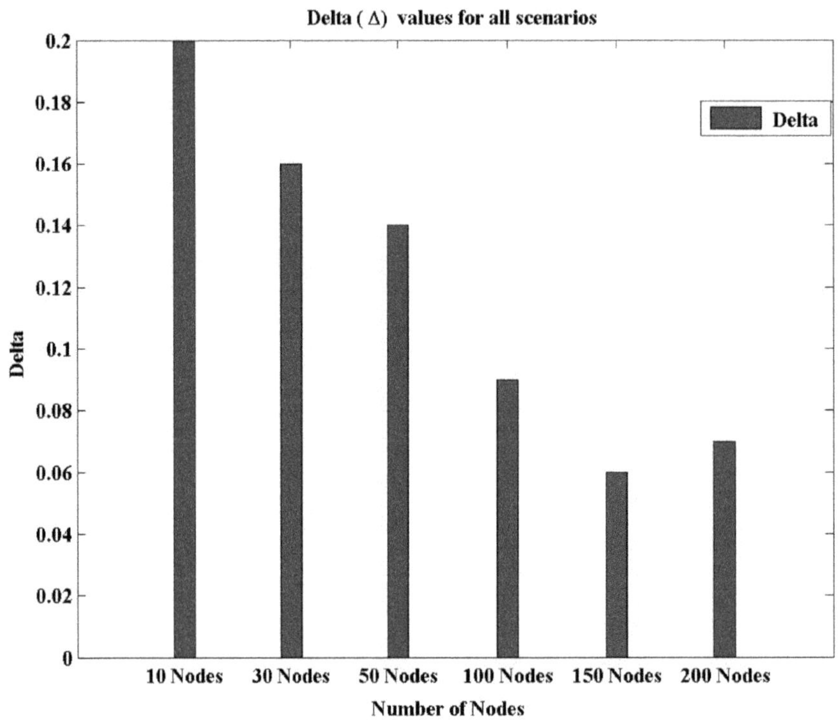

This Λ1 is calculated for all DTN packets sent in the network for given FTP connection as per scenario considered. Factor T is computed using Equation 6.

$$T = \sum_{i=1}^{10} X(i, j) \tag{6}$$

Table 12. Δ and Λ1 values for DTN packets

Sr. No.	Number of Nodes	Number of FTP Connection Fc	Total DTN Packets	Δ	Λ1
1	10	02	65545	0.20	13.109
2	30	05	66366	0.16	10.618
3	50	07	66104	0.14	9.254
4	100	09	66843	0.09	6.015
5	150	10	68013	0.06	4.080
6	200	11	69019	0.07	4.831

Figure 17. Plot showing Δ values for all scenarios

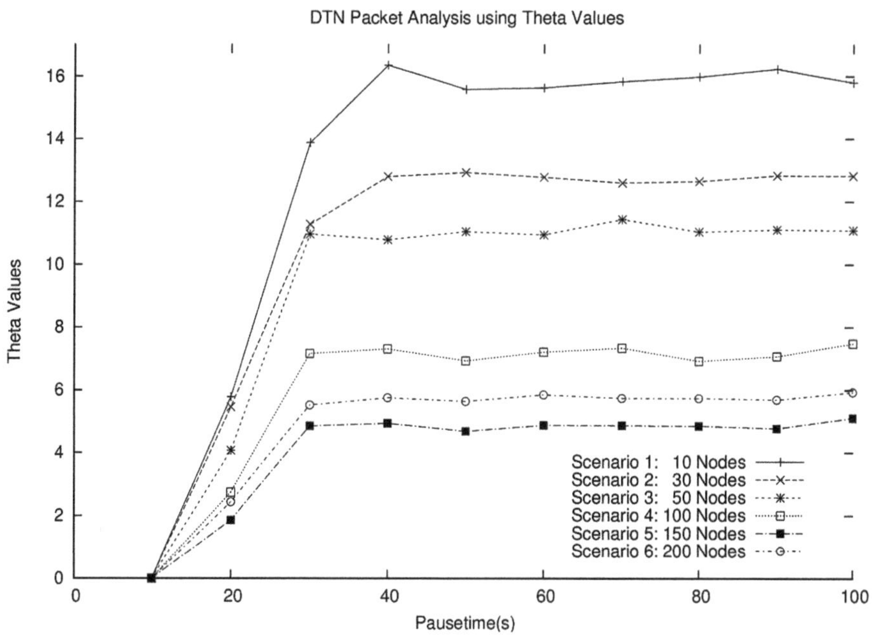

where X(i, j) represent row i column j intersection value where i = 1, 2, · · ·, 10 and j = 2

For each of those results. The Λ1 and Δ values are observed to be decreased when number of FTP connection i.e. network traffic and number of mobile nodes gets increased. Another observation is that the number of DTN messages sent is directly proportional to the number of FTP connections setup in the network. More the number of FTP connections, more will be the traffic and therefore more DTN packets will be sent. Another parameter, θ is computed using Equation 7

$$\theta i = \frac{PDTNi\Delta j}{100} \tag{7}$$

where PDTNi represent the values of total DTN packets sent at Pausetime k for i = 1, 2, · · ·, 10, j = 1, 2, · · ·, 6 & k = 10, 20, · · ·, 100. The graph given in Figure 18 is plotted against various θ values verses Pausetime(s) for all the scenarios.

From the graph, it is inferred that as the number of mobile nodes get increases, θ values get decreases. Further, θ becomes almost constant to average θ value for all scenario. As per the graph given in Figure 18, $\Lambda 1$ values are found to be more when number of mobile nodes are less. As soon as, number of mobile nodes are increased in subsequent scenarios, $\Lambda 1$ values decreases. This is because of the network congestion.

Dropped Packet Analysis Using $\Lambda 2$

Another observation can be formulated by analyzing the parameter $\Lambda 2$ as given in Equation 8.

$$\Lambda 2 = \frac{T\Delta}{10} \quad (8)$$

where Δ is same as computed in Equation 5. This parameter $\Lambda 2$ is calculated for the total number of DTN packets which are dropped in the network. As DTN packets are sent over reliable data stream protocol called TCP, DTN packets are rarely dropped in the network and even if, it will be retransmitted due to properties of TCP Protocol. Therefore, instead of analyzing parameter $\Lambda 2$ for Transport layer, Network layer and Data link layer are considered. The Network layer and Data link layer packets are labeled as RTR and MAC.

Figure 18. Plot showing Θ values for all DTN packets

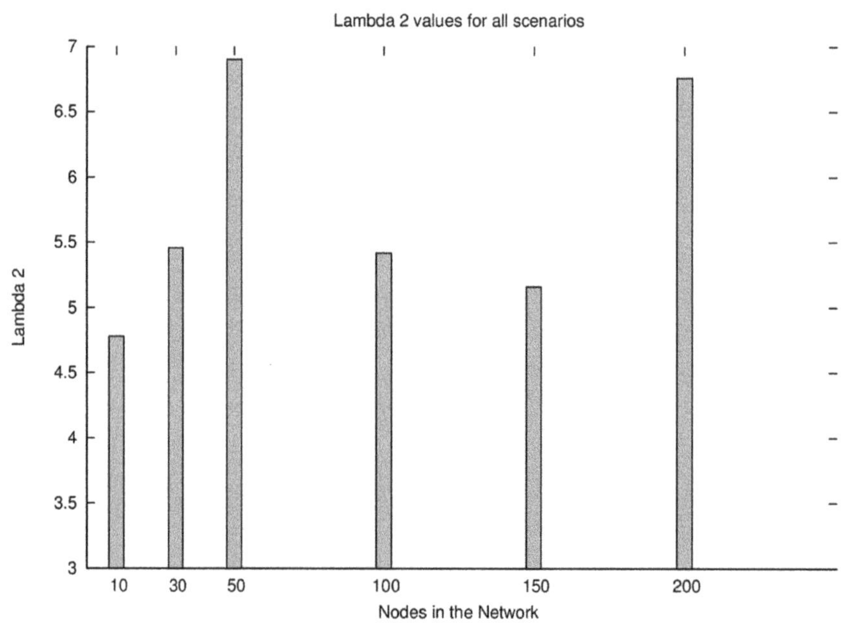

From the Table 13, it has been observed that the Λ2 values increases when the number of mobile nodes i.e mobile node density and number of FTP connection increases. This becomes clear if we observe the Figure 19.

However, Λ2 ≤ 7 and varies around 6 for all scenario under consideration. This also foreseen from the graph in Figure 20.

From the graph, frequent packets drop occurs in the early stage of simulation whereas it becomes almost constant during later stage. The comparison of Λ1 and Λ2 is given in Figure 21. The Λ2 is inversely proportional to Λ2 i.e. Λ1 decreases when Λ2 increases. Crossover point is observed when number of nodes becomes 110 approximately. After this crossover point, the corresponding parameters gives constant results.

Table 13. Δ and Λ2 values for RTR and MAC packets dropped

Sr. No.	Number of Nodes	Number of FTP Connection Fc	RTR and MAC Packets Dropped	Δ	Λ2
1	10	02	239	0.20	13.109
2	30	05	341	0.16	10.618
3	50	07	493	0.14	9.254
4	100	09	602	0.09	6.015
5	150	10	860	0.06	4.080
6	200	11	966	0.07	4.831

Figure 19. Plot showing Λ2 values for all scenario

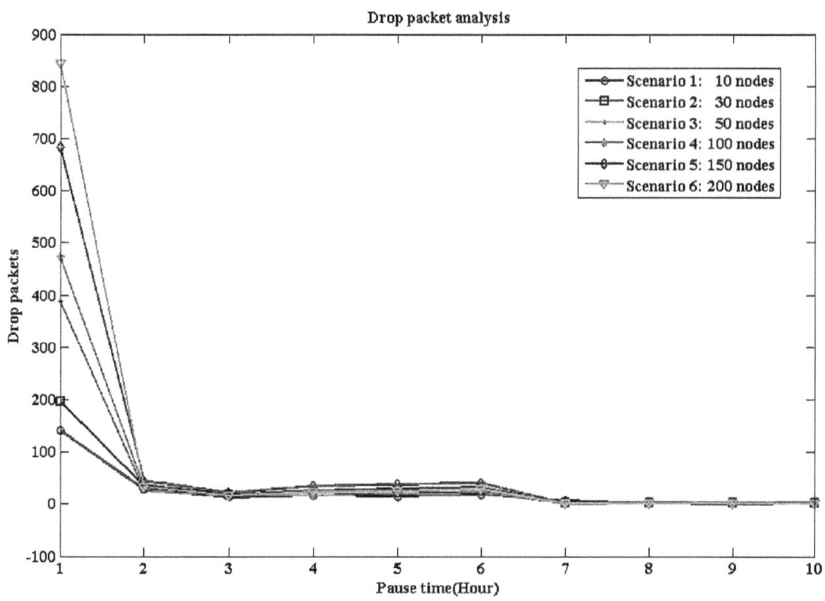

Figure 20. Plot showing packets dropped in all scenario

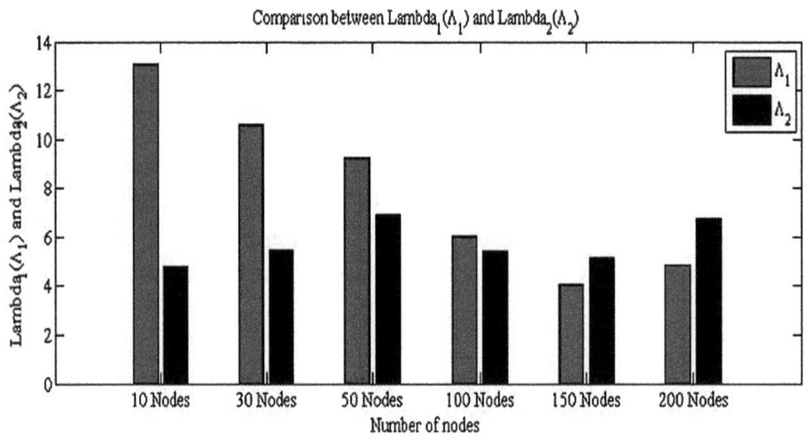

Throughput Analysis

Throughput is defined as the total number of bits sent per unit time. To perform throughput analysis, a graph of Pausetime(s) verses Number of DTN packets received by the destination within the interval of 0.5s is plotted in Figure 22.

For this, the packet size is considered as 1540 bytes as the length of protocol data unit which includes 1460 bytes as data size received from higher layer and 80 bytes as header size added by corresponding layer. It is observed that the throughput values are high for scenario with mobile nodes 50 and 100. Whereas, it reaches to lower values for rest of the scenario with mobile nodes 10,30 and also for the

Figure 21. Plot showing comparison of $\Lambda 1$ and $\Lambda 2$

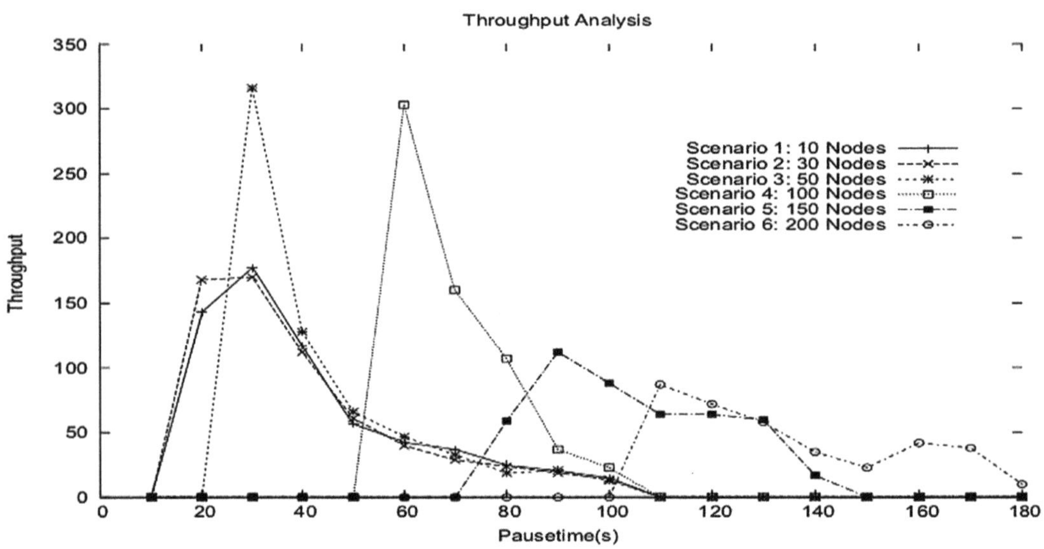

Figure 22. Plot showing throughput values of all DTN packets

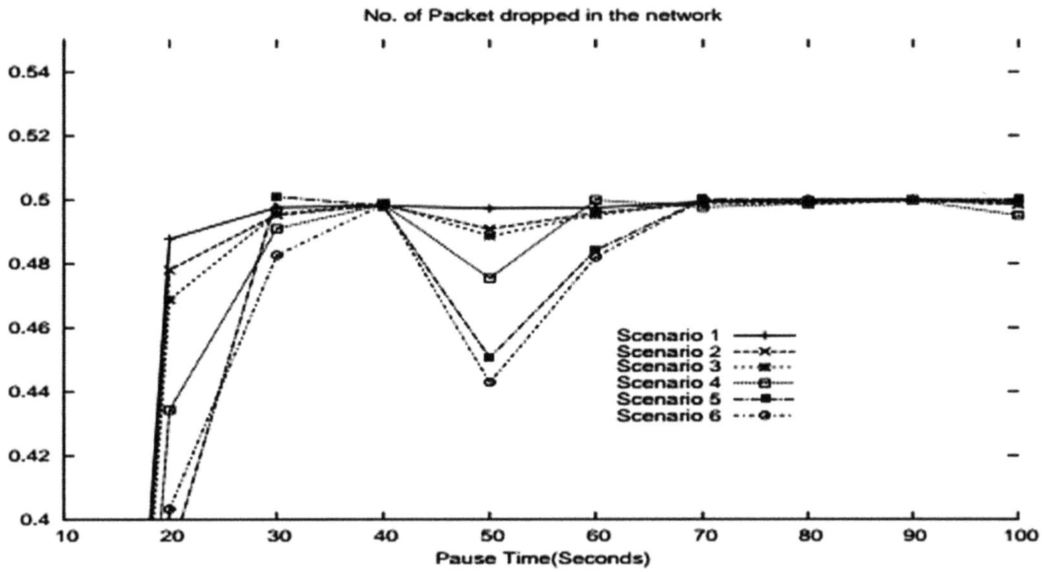

scenario with mobile nodes 150,200. Thus, it is inferred that the increasing number of mobile nodes and FTP connection definitely affect throughput beyond threshold values. However, proper selection of number of mobile nodes and number of FTP connection in proposed SEDAR routing protocol results to increase overall throughput.

Routing Overhead

Routing of control packets during route setup phase consumes network resources such as network bandwidth. Hence, computation of routing overhead becomes important factor. Routing overhead is defined as the ratio of total data packets sent to total control packets sent in the network.

CONCLUSION

This chapter gives the introduction of delay tolerant network. Initially, layer architecture and node communication is given. State of the art routing protocols are categorized in to flooding and forwarding strategies. The comparisons of these two categories of protocols are given. Detailed design of direction routing protocol is given. The issue of relaying data packets towards destination is solved. Extra header is provided to in Direction Aware Routing protocol. The methodology of proposed SEDAR has been described. Basic theme is to restrict the message broadcasting by reducing transmission area. This is achieved using the direction angle θ of the destination node. This also reduces network congestion which results into reducing routing overhead. Proposed SEDAR uses public key cryptography approach to secure message transmission. Proposed SEDAR also implements energy aware model to consume less energy during message transmission and reception. For relaying message, proposed SEDAR select best relaying node on an optimal trajectory toward the destination. This is achieved by introducing origin shifting and dividing given geographical area into various sectors. Performance of proposed SEDAR is evaluated using a map based model with 81 mobile nodes grouped into various 8 groups. The proposed SEDAR succeed to achieve the PDR of 83.4%. This PDR in proposed SEDAR is also compared with various protocols under study. The proposed SEDAR also found to save 3 to 5% energy. The performance of proposed SEDAR-H has been evaluated in heterogeneous highly dense and dynamic mobile node environment. Heterogeneous environment has been created using with the help of stationary Wired nodes, Base station nodes and Mobile nodes. Various scenarios have been created for this evaluation by varying the number of mobile nodes and the traffic routed between Wired nodes and Mobile nodes thorough Base station nodes. Results have been analyzed by introducing four factors Δ , θ and › 1 and › 2 .From the results, it has been inferred that increase in the number of mobile nodes affect overall throughput of the network. Routing overhead has constant effect on the delay tolerant network traffic and almost 50% routing overhead is observed to be created for data traffic. Increase in the number of nodes and number of FTP connection rarely affect the overall performance of the network. Here, the proposed protocol found to be effective while evaluating its performance in heterogeneous highly dense mobile node environment. Thus, it has been concluded that in both homogeneous and heterogeneous environment destination aware routing protocol (SEDAR) increase the packet delivery ratio, decrease the overall energy consumption and routing overhead.

FUTURE RESEARCH DIRECTIONS

This research work has attempted to explore all the major components of a delay tolerant network to achieve the stated goals. However, much remains to be done. During the current research work, various avenues for future research related to this dissertation can be envisaged. The major area of immediate research is the investigation of new routing protocol which can handle multiple attacks, intelligent DTN routing protocol which is capable of taking decision on varying environments. Other possible areas are discussed and briefly outlined below.

1. In case of packet format of proposed protocol, we have not prioritize the message. However, the proposed work can be extended to include the multiple level of priority. On the basis of time-criticality and loss sensitivity, traffic may be divided into following prioritized class: time-critical and loss-sensitive traffic as priority 1; time-critical and loss-insensitive traffic e.g., video and voice as priority 2; non time-critical and loss-insensitive traffic e.g., down-play multimedia and business critical traffic as priority 3; and non time-critical and loss-sensitive traffic e.g., file-transfer and emails as priority 4.
2. In case of routing, the sectorization method had been used to restrict confined area and then unicast message delivery one by one in destination's direction. However, this work can be extended to apply sectorization technique in multiple direction to further minimize delay on account of increasing bandwidth consumption.

3. In case of routing, majority of the protocols in DTN are categorized as either flooding or forwarding have their merits and de-merits. However both This work attempted on flooding category. However, extension of this research work can be done in evaluating the performance of proposed protocol in forwarding category and making firm decision about best category for implementing proposed protocol.
4. In case of simulation, the proposed research work focused on using TCP and FTP protocols. The evaluation is suitable for homogeneous and heterogeneous environment. However this work can be extended to test and evaluate the effect of other similar layer protocols on proposed protocol to strengthen the claims.

ACKNOWLEDGMENT

We highly acknowledge the reviewing work done by the editor and reviewer of the IGI global for their enormous support. Thank you.

REFERENCES

Agila Bitsch Link, J. I., Schmitz, D., & Wehrle, K. (2011). GeoDTN: Geographic Routing in Disruption Tolerant Networks. *Proceedings of IEEE, Global Telecommunications Conference (GLOBECOM-2011)*, 1 - 5.

Cao, Y., Sun, Z., Wang, N., & Cruickshank, H. (2013). Converge-and-diverge: A geographic routing for delay/disruption-tolerant networks using a delegation replication approach. *IEEE Transactions on Vehicular Technology*, *62*(5), 2339–2343. doi:10.1109/TVT.2013.2238958

Chaithanya Manam, V. K., Mahendran, V., & Siva Ram Murthy, C. (2012). Message-driven based energy efficient routing in heterogeneous delay tolerant networks. *Proceeding of HP-MOSys'12, 1st ACM workshop on High performance mobile opportunistic systems*, 39 - 46. doi:10.1145/2386980.2386989

D'souza, J., & Jose, J. (2010). Routing approaches in Delay Tolerant Networks: A Survey. Proceeding of International Journal of Computer Applications, 17, 9-15.

Fall, K. (2003). A Delay –Tolerant Network Architecture for challenged internets. *Proceeding of annual conference of the special Internet Group on Data Communication (ACM Siggomm'03)*, 27-34.

Frenkiel, R. H., Badrinath, B. R., Bores, J., & Yates, R. D. (2000). The infestations challenge: Balancing cost and ubiquity in delivering wireless data. *Proceeding of IEEE Personal Communications*, *7*(2), 66–71.

Grossglauser, M., & Tse, D. N. C. (2002). Mobility increases the capacity of adhoc wireless networks. *Proceeding of IEEE/ACM Transactions on Networking*, 10, 477-486. doi:10.1109/TNET.2002.801403

Jain, S., Fall, K., & Patra, R. (2004). Routing in a delay tolerant network. *Proceeding of ACM SIGCOMM*.

Jian, S., Sangman, M., & Chung, L. (2008). Routing Protocols in Delay Tolerant Networks: A Comparative Survey. *Proceeding of 23rd International Technical Conference on Circuits /Systems, Computer and Communications (ITC-CSCC 2008)*, 1577-1580.

Jones, E., & Paul, A. S. W. (2008). Routing Strategies for Delay – Tolerant Networks. *Journal of Computer Communication.*

Li, Y., & Wang, Z. (2011). Optimal relaying in heterogeneous delay tolerant networks. *IEEE International Conference on Communications ICC 2011*, 1 - 5. doi:10.1109/icc.2011.5963142

Mangrulkar & Atique. (2012). Design and development of delay tolerant network: A case study. *Proceedings of IEEE International conference Nuicone 2012*, 1 - 5.

Mangrulkar, R. S., & Mohammad, A. (2010). Routing Protocol for Delay Tolerant Network: A Survey and Comparison. *Proceeding of IEEE International Conference on Communication, Computing and Control* during, 210-215.

Mapp, G., Shaikh, F., Aiash, M., Vanni, R. P., Augusto, M., & Moreira, E. (2009). Exploring efficient imperative hand over mechanisms for heterogeneous wireless networks. *Proceedings of the 12th International Conference on Network-Based Information Systems NBiS09*, 286 - 291.

Mathurapoj, A., & Pornavalai, C. (2009). Fuzz-Spray: Efficient Routing in Delay Tolerant ad-hoc network based on Fuzzy Mechanism. *Proceeding of FUZZ-IEEE 2009*, 104-109.

Mishra, S., Krishna, P. V., & Abraham, K. I. (2010). Adaptive Link State Routing and Intrusion Detection in Wireless Mesh Networks. Information Security, 4(4), 374-389.

Misra, Das, Khatua, & Obaidat. (2014). Qos-guaranteed bandwidth shifting and redistribution in mobile cloud environment. *IEEE Transaction on Cloud Computing, 11*(99), 1 - 10.

Radenkovic & Grundy. (2012). Efficient and adaptive congestion control for heterogeneous delay-tolerant networks. *Journal of Adhoc Networks, 10*(7), 1322 - 1345.

Sandulescu & Nadjm-Tehrani. (2008). Opportunistic DTN Routing with Window-aware Adaptive Replication. *Proceeding of AINTEC'* 08'.

Shah, R. C., Roy, S., & Brunette, W. (2003). Data Mules: Modeling three tier architecture for sparse sensor networks. Proceeding of Sensor Network Protocols and Applications, 30-41.

Spyropoulos, T., Psounis, K., & Raghvendra, C. S. (2003). Spray and Wait Efficient routing in intermittently connected Networks. Proceeding of Mobile Computer and Communication Review, 7(3).

Spyropoulos, T., Psounis, K., & Raghvendra, C. S. (2007) Spray and Focus Efficient Mobility Assisted Tolerant Networks. *Proceeding of workshop on PerCom apos*, 79-85.

Vahdat, A., & Becker, D. (2000). Epidemic Routing for partially connected adhoc networks. Tech Report CS- 200006, Duke University, April 2000.

Venu Madhav, & Sarma. (2012). Energy efficient cluster routing protocol for heterogeneous wireless sensor networks. *LNICST*, (108), 452 - 455.

Chapter 7
A Review of Trust Management for Mobile Ad Hoc Networks

Sarbjeet Singh
Panjab University, India

ABSTRACT

Establishing trust in Mobile Ad hoc Networks (MANETs) is a challenging task. Trust management consists of various activities like prediction of trust, computation of trust, propagation of trust, aggregation of trust etc. Trust prediction can be done based on past behavior of node, internal parameters of the node, or it can be induction-based or neural-network based. Trust computation can be centralized or distributed. Trust propagation can be social networks based, small world phenomenon based, web of trust based, distributed hash table based. Trust aggregation can be weighted average based, probability based, sequential, conditional-sequential, parallel or subjective logic based. This work presents the review of various activities pertaining to trust management to enable researchers, academicians and practitioners to identify and address trust related issues in a better way.

INTRODUCTION

Trust is a fascinating subject and has been studied by different researchers in various fields like philosophy, psychology, sociology, economics, law, political science, computer science etc. Trust mean different things to different people and vary from person to person and system to system. MANET is a temporary network of wireless mobile nodes without centralized infrastructure. Trust determination of nodes becomes more important as a node can enter or leave the system any time. Moreover the network is dynamic, short-term and short-lived so determination of trust of nodes is also a challenging task. This chapter presents the notion of trust, its characteristics and various activities pertaining to trust management. Chapter also highlights the unique characteristics of MANET and the need and importance of addressing trust in MANET.

DOI: 10.4018/978-1-5225-2342-0.ch007

The work is organized as follows: Section 2 explains the concept of trust in various disciplines and its characteristics. Section 3 presents MANET and its characteristics. Section 4 discusses trust management activities. Section 5 highlights important work in the area of trust management in MANET. Section 6 summarizes the chapter and concludes with future scope.

CONCEPT OF TRUST AND ITS CHARACTERISTICS

Each discipline has defined and considered trust from a different perspective. For example in psychology, trust is described by state of mind of an individual. It has three perspectives viz. cognitive, emotive and behavioral. (Deutsch, 1962) defined trust as "confidence that an individual will find, what is desired from another rather than what is feared". If an individual finds an entity's behavior good and according to expectations under different situations, then individual starts building psychological trust in his mind about that entity and disseminates this trust among others when it interacts with other individuals. Trust between a child and mother, husband and wife, brother and sister, cousins and friends takes different form and depends upon the psychological state of an individual. In sociology, trust can be defined and considered from two viewpoints: individual and societal. Trust from individual viewpoint is close to psychological trust. Societal trust is described by the societal setup and collective societal state of mind. Societal trust is important to understand the building and establishment of trust in social networks. Psychological and societal trusts if lost are difficult to regain. In philosophy, trust has been considered important but dangerous (Cho, Swami, & Chen, 2011). Trust is the foundation of birth, survival and end of relationships. In economics, trust is generally considered from utility point of view. From utility point of view, an entity may put more trust on another entity even in the presence of risk and uncertainty if the action maximizes utility. For example while giving credit card numbers for online purchase of items an individual puts trust on various stakeholders, which is close to economic trust. In economics, prisoner's dilemma games are used to model trust and cooperation in systems (Sherchan, Nepal, & Paris, 2013). In law, trust is taken as a relationship between two parties in which one, the trustee, has the power to manage property and the other, the beneficiary, has the privilege of receiving benefits from that property. In the area of computer science also, researchers have made attempts to define trust in various disciplines like computer networks, internet applications, computing systems, security systems, operating systems etc. In social networks, trustworthiness of an individual is derived from its virtual presence. It depends upon several parameters like honesty, attitude, interests, behaviour, beliefs, reliability, courage, patience, compatibility, adaptability, responsibility, expectations, dependency etc. One of the popular definitions of trust which has been adopted by many computer scientists is the one given by Diego Gambetta as "a particular level of the subjective probability with which an agent assesses that another agent or group of agents will perform a particular action, both before he can monitor such action (or independently of his capacity ever to be able to monitor it) and in a context in which it affects his own action" (Chakrabarty, 2007). (Grandison & Sloman, 2000) also provides a working definition of trust for Internet applications as "the firm belief in the competence of an entity to act dependably, securely, and reliably within a specified context". From the concept of trust in various disciplines, following characteristics of the trust are derived:

- It is context specific.
- It may or may not be transitive.

- It is multi-dimensional.
- It changes with time.
- It is event sensitive.
- It may or may not be symmetric.
- It is subjective.
- It is dynamic.
- It changes with experience.
- It is hard to regain if lost.

Context specific means that trust depends on the context in which it is being investigated. For example, a car mechanic can be trusted to repair a car but not to repair a computer. Symmetric characteristic means that if X trusts Y then Y trusts X also but this may or may not be the case. Similarly if X trusts Y and Y trusts Z then X may or may not trust Z. Subjective property means that it varies from individual to individual and system to system. Multi-dimensional property means that it depends on several factors and it is a multi-faceted concept. Dynamic property means that it is not a fixed value and changes with time and experience. Event sensitive property means that certain events can raise it significantly while certain other events can lower it significantly. Also, trust is hard to regain if lost.

Other concepts related to trust are trustworthiness, reputation, feedback and opinion (Chakrabarty, 2007). Many a times the terms trust and trustworthiness are used interchangeably but these are different. If an entity is really trustworthy then it will behave as expected with high probability. Trust computed by an entity must be close to actual trustworthiness of the entity in the environment otherwise it indicates some problem with the accuracy of trust computation method. Reputation is an indication of general perception about the trustworthiness of an entity as seen by others as a whole in the environment. Feedback is a piece of information given by an entity about the trustworthiness of other entity and an opinion is general impression of an entity regarding the trustworthiness of other entity. Trust computation involves the use of trust functions which calculate the trust of an entity. (Zhang, Yu, & Irwin, 2004) proposed following classification scheme for trust functions:

- *Subjective or Objective.*
- *Complete information based or localized information based.*
- *Transaction-based or Opinion-based.*
- *Rank-based or threshold based.*

A trust computation method is said to be objective if the parameters affecting trust are quantitatively measurable. It is subjective if trust is calculated from a specific entity's point of view. Similarly, if the trust computation method is making full use of all the global information available in the environment to compute trust then it is complete information based otherwise localized information-based if it making use of only local information available with the node to determine trust. Trust computation method is said to be transaction based if trust is computed after going through a number of transactions with the entity on which trust is computed. It is opinion based if it is computed by requesting opinion from other entities. Rank based trust functions present entities trustworthiness in ranks by comparing them with each other. Threshold based trust functions declare an entity trustworthy if its trustworthiness is greater than a predefined threshold. The choice of trust function depends on the information available and the application or problem in hand. (Ruohomaa & Kutvonen, 2005; Ruohomaa, Kutvonen, & Koutrouli,

2007) presented the survey of trust management and reputation management. Trust management in cloud computing systems has been discussed by (Firdhous, Ghazali, & Hassan, 2011; Noor et al., 2013). (Sherchan, Nepal, & Paris, 2013) presented the survey of trust in social networks.

CHARACTERISTICS OF MANET

MANET is a self configuring network without centralized infrastructure consisting of autonomous mobile nodes connected temporarily through wireless channel. MANET is a kind of wireless ad hoc network. The unique characteristics of MANET make trust management a difficult task. Following is a list of important characteristics of MANET:

- It is self configuring network.
- It is without centralized infrastructure.
- It consists of mobile devices.
- Mobile devices are free to move in any direction.
- Each device is supposed to be a router.
- It is temporary network.
- Mobile devices change their links with other devices frequently.
- Network is dynamic having autonomous topology.
- It has resource constraints.
- It has limited physical security.
- Communication channel is unreliable.

In order to maintain a secure environment, following security concerns need to be addressed:

- Confidentiality,
- Integrity,
- Non-repudiation,
- Authentication,
- Authorization,
- Access Control,
- Availability,
- Privacy,
- Trust.

Confidentiality requirement means that nodes should not be able to see or interpret each other's sensitive data or information. It is met with encryption. Integrity requirement means that unauthorized changes to messages are detectable. It is met with hashes. Non-repudiation means that entity cannot deny performed a particular action if it has actually performed. Authentication requirement deals with verifying the identity of the entity and authorization concerns what an entity is authorized to do in the system. Access control deals with policies that allow or restrict access to resources. Availability means services should be available whenever required. Privacy deals with protection of private and sensitive information. Trust requires its presence among various nodes for confident and faithful interaction.

MANETs are easily susceptible to attacks. (Wu et al., 2007; Jawandhiya et al., 2010) discussed the survey of attacks in mobile ad hoc networks. One way of classifying attacks is on the basis of whether they disrupt normal functionality of MANET i.e. active or passive attacks. An active attack disrupts normal functionality of MANET by interrupting, fabricating or modifying information whereas passive attacks do not disturb the normal communication operations (Wu et al., 2007). Passive attacks include traffic analysis, eavesdropping and monitoring of information over communication channel. Active attacks include modification, repaying, spoofing, jamming etc (Wu et al., 2007). Another classification is whether attacks are coming from internal nodes (internal attacks) or from outside the network (external attacks). Attacks have been classified on the basis of layers also, for example, (Wu et al., 2007) categorizes attacks on the basis of layers as follows:

- ***Physical Layer:*** *Jamming, interceptions, eavesdropping.*
- ***Data Link Layer:*** *Monitoring, traffic analysis, WEP weakness etc.*
- ***Network Layer:*** *Wormhole, Blackhole, flooding, resource consumption etc.*
- ***Transport Layer:*** *Session hijacking, SYN flooding.*
- ***Application Layer:*** *Repudiation, data corruption.*

In jamming, radio signals get jammed and message gets corrupted or lost. In interception and eavesdropping, attacker gets access to shared communication medium and can play with information or data being transmitted over communication channel. Monitoring and traffic analysis allow attacker to observe and analyze data for the presence or absence of certain events or actions. Attackers can also exploit the weaknesses of wired equivalent privacy (WEP). In Wormhole attack, malicious entities pretend to connect two points in the network (Cho, Swami, & Chen, 2011) and routing gets disrupted when routing control messages are tunneled (Wu et al., 2007). The tunnel between two colluding attackers is referred as a wormhole (Wu et al., 2007). In Blackhole attack, a node pretends having a valid route, intercept messages and then drops them. Grayhole attacks are also there in which a malicious entity selectively drops packets (Cho, Swami, & Chen, 2011). Resource consumption attack is known as sleep deprivation attack where a malicious entity consumes battery life or other resources of the victim by requesting excessive route discovery or by forwarding unnecessary packets (Wu et al., 2007). SYN flooding attack is a denial of service attack where a malicious node creates many half-opened TCP connections but never completes the handshake process (Wu et al., 2007). In session hijacking attack, a malicious entity impersonates the victim entity and continues the session with the spoofed IP address and sequence number. Repudiation refers to denial of participation in an act. Besides launching an attack from a particular layer, attacks can be launched from multiple layers also, for example, denial of service (DoS) attack, man-in-the-middle attack etc. The concept of trust can play major role in designing solutions to counter attacks. If trustworthiness information of each node is available then this information can play vital role in identifying malicious users and intruders.

TRUST MANAGEMENT

Trust Management Systems are responsible for managing trust in the environment. Thrust management systems can be classified as

- Credentials and Policy based Trust Management Systems.
- Reputation based Trust Management Systems.
- Social-Network based Trust Management Systems.

In credentials and policy based trust management systems, trust is established on the basis of predefined policies and by exchange of credentials among participants of the system. If credentials are valid and policies are satisfied, entity is assumed to be trustworthy and access is granted. In reputation-based trust management systems, trust is established by evaluating the reputation of the system in the environment. Social-network based systems evaluate trust by exploiting social relationships among participants.

The different phases and tasks of trust management systems can be categorized as:

- Trust Information Collection and Initialization.
- Trust Negotiation.
- Trust Determination.
- Trust Prediction.
- Trust Evaluation.
- Trust Aggregation.
- Trust Evolution.
- Trust Updation.
- Trust Dissemination.
- Trust Revocation.
- Trust Storage.
- Trust Visualization.

Following is a brief description of each of these tasks:

Trust Information Collection and Initialization

The first step in the usage of any trust management system is the selection of type of trust management system and the sort of information on the basis of which trust will be determined. The way information will be collected is decided. The logic behind initialization of information is defined. We are into this phase when nothing has happened in the environment and transactions are yet to begin. If trust management system to be used is credentials and policy based then credentials and policies are defined and fixed. If it is reputation based then reputation information is collected or the mechanisms to collect initial reputation information are identified and used. The types of trust functions to be used are also decided. In social networks based trust management systems, social relationships are identified and information pertaining to these relationships is collected and initialized.

Trust Negotiation

This phase comes into picture when a new entity joins the system. In this phase credentials exchange and policy exchange operations take place. Reputation information is also exchanged and shared. New entrant's trust metrics are initialized and the initial trust of other participants with the new entrant is also computed. Information regarding new entrant is also disseminated among all the participants in the environment.

Trust Determination

This is the core phase of trust management system where trust is actually determined. Trust determination can be centralized or decentralized. In centralized trust management systems, a central agent helps nodes in determining trust on other nodes. In decentralized systems, each node determines trust on other nodes at its own without the help of centralized agent (Govindan & Mohapatra, 2012). Trust can be determined either by prediction or by computation. Following is a brief description of each of these.

Trust Prediction

Trust prediction can be done based on past behavior of node, internal parameters of the node, or it can be induction-based or neural-network based. Trust prediction depends on the type of trust management system being used and the information available with the node. (Jonker & Treur, 1999) propose a trust prediction scheme based on mathematical inductions. A reputation prediction scheme in MANET using Radial Basis Function-Neural Network has been proposed by (Ham et al., 2009).

Trust Evaluation

The way trust is evaluated depends on the type of trust management system. In credentials and policy based system trust is assumed if the entity possesses valid credentials and defined policies are satisfied. In reputation based systems, feedback about reputation is asked from participants (or centralized agent) and then received information is aggregated to compute reputation trust of other entity. Opinions from specific entities can also be called and trust can also be evaluated based on recommendations. In social network based systems, existing relationships among entities are examined using various mechanisms and trust is evaluated. Various trust evaluation schemes exist depending on the type of trust functions (Subjective vs. Objective, Complete information based vs. Localized information based, Transaction-based vs. Opinion-based, Rank-based vs. Threshold based) being used by trust management system.

Trust Aggregation

In trust aggregation phase, trust received from different sources is aggregated. It can be weighted average based, probability based, sequential, conditional-sequential, parallel or subjective logic based. As trust value is received from various sources, one of the sources may be malicious also. Trust aggregator must be able to dissolve the effect of false or wrong information received. Trust aggregator must also not be computationally very complex. (Damiani et al., 2006) proposed a Weighted Ordered Weighted Averaging (WOWA) based aggregation technique. (Pardo, 2009) has presented aggregation of trust in probabilistic logic. (Huang & Nicol, 2009) have presented calculus of trust, describing sequential and parallel aggregation operators (Govindan & Mohapatra, 2012). Gossip-based aggregations of trust and computations have been described by (Bachrach & Parnes, 2009) and (Kempe, Dobra, & Gehrke, 2003). (Wang & Singh, 2006) presented trust aggregation using subjective logic (Govindan & Mohapatra, 2012). Reputation aggregation using fuzzy logic is described by (Song et al., 2005). Aggregation techniques can also be utilized to reduce resource consumption in obtaining information from distant nodes (Cho, Swami, & Chen, 2011).

Trust Evolution

Trust evolution deals with building of trust in the environment. Successful transactions with an entity increase the trust and reputation of the entity in the environment whereas unsuccessful or malicious transactions decrease the trust and reputation of the entity. Trust evolution also depends on the type of trust management system being used. In social network based trust management system, trust is evolved with the formation and fall of relationships in the environment. (Wang et al., 2007) have presented trust evolution model for P2P systems.

Trust Updation

Trust updation deals with update of trust metrics. Trust needs to be updated periodically or with the occurrence of certain events. Event based updates may take place at the end of transactions (Chakrabarty, 2007). Various parameters get changed after the execution of transactions depending upon the experience resulted of transactions. Successful and unsuccessful experiences update trust metrics accordingly. Updates must be propagated in the environment (or to the concerned entities) in a timely manner in order to correctly determine the trustworthiness of an entity.

Trust Dissemination

Trust dissemination deals with propagation or distribution of trust information in the environment. Trust need to be propagated among nodes which are not in direct contact with each other. Trust need to be disseminated to all or concerned entities. Trust propagation can be web of trust based, small world phenomenon based, social net based or distributed hash table based (Govindan & Mohapatra, 2012).

Trust Revocation

Trust revocation deals with reversal of opinion formed by an entity about the trustworthiness of other entity. Trust revocation is particularly important in MANET because of their inherent properties of self organization (Baras & Jiang, 2005).

Trust Storage

It deals with the storage aspects of trust information and parameters. Storage of trust can be centralized or decentralized depending upon whether the trust management system is centralized or decentralized.

Trust Visualization

Trust visualization deals with visualization of trust in the environment from various perspectives. As trust is a subjective notion, an entity can have different trustworthiness figures with different entities with respect to different contexts. Trust visualization deals with analyzing the spread of trust in the environment from various viewpoints.

Trust management systems are also expected to possess certain characteristics, termed as desirable characteristics, as follows:

- Scalability:
  - Message overhead,
  - Storage overhead,
  - Computation overhead.
- Reliability.
- Security.

Scalability deals with the ability of the system to scale with increase in number of nodes in the environment. Message exchanges, storage requirements and computation complexity must not increase too much with the increase in number of nodes in the environment. Similarly reliability and security features are also expected from trust management systems which ensure that trust value obtained is reliable enough to interact with the node and system is not vulnerable to threats.

TRUST MANAGEMENT IN MANET

This section highlights the work done in the area of trust management in MANET. Survey of trust management for mobile ad hoc networks has been discussed by (Cho, Swami, & Chen, 2011). (Suryanarayana & Taylor, 2004) presented survey of trust management and resource discovery in peer-to-peer applications. (Agarwal & Motwani, 2009) presented survey of clustering algorithms for MANET. Survey of trust computations and trust dynamics in MANETs has been discussed by (Govindan & Mohapatra, 2012).

(Eschenauer, Gligor, & Baras, 2004) discussed a framework for establishing trust in MANET. Authors presented three components of their scheme: generation, distribution and evaluation of trust evidence. Trust evidence can be generated by any node about any other node and evidence can be an identity, a public key, or any information required by the policy and evaluation metric. Evidence can be obtained on-line or off-line. For distribution, a principal can distribute trust evidence in the network and can disconnect itself afterwards. Evidence can be replicated to ensure availability. Authors have also proposed a swarm based evidence distribution scheme. Simulation results of freenet in MANET to distribute trust evidence using NS-2 have also been presented by authors. Their network has 50 nodes, having wireless range of 250 meters, randomly distributed over a 1km x 1km zone. The network is highly connected and average number of hops between two nodes is two. Experiments have been conducted to understand the effect of mobility and routing on the performance of freenet and to measure the impact of request routing on the diversity of evidence stored in the network. (Liu & Issamy, 2004) presented an enhanced reputation model for MANETs. Model incorporates reputation formation, evolution and propagation along with two important dimensions: time and context. The proposed reputation mechanism ensures reliability against attacks by obtaining true reputation of an agent by effectively distinguishing trust-telling and lying agents. Reputation formation has been implemented through experience manager, recommendation manager and reputation manager components. Experience manager records previous interactions. Reputation manager store recommendations, exchange reputation information and manage table of recommender reputations of recommenders. Reputation manager administers and calculates the service reputation of a peer by taking into consideration the inputs from experience manager and recommendation manager. Different weights are assigned to experiences and recommendations. Reputation evolution involves evaluation of subjective degree of satisfaction. Authors have also presented the mechanism to obtain the degree of satisfaction. In order to evaluate the effectiveness of their approach authors carried out three sets of simulation tests. (Pirzada & McDonald, 2004) presented a trust-based communication model in ad-hoc networks. Under this model, a trust agent performs three functions viz. trust derivation, quantification and computation. In trust computation, weights (range 0 to 1) are assigned to events that were monitored and quantified. Type of application demanding trust determines the value of weights. Trust is established without the existence of a central trust authority. The model provides a dynamic measure of reliability and trustworthiness in an ad-hoc network.

In the work presented by (Baras & Jiang, 2005), trustworthiness of agents has been decided based on neighbors' opinion. Voting approach has been used in which value of each vote represents the opinion of the voter. In this approach state information is not required from all the agents but rather from neighbors.

(Velloso et al., 2010) presented trust management in MANETs using a scalable maturity-based model. Trust is based on previous interactions and recommendations. A Recommendation Exchange Protocol (REP) has also been presented which enables nodes to exchange recommendation information. The key concept introduced is relationship maturity which helps improving efficiency. Trust information is not disseminated over the entire network instead trust information is exchanged among nodes within radio range. This characteristic makes model suitable for large networks. Simulations have been performed to evaluate the performance of proposed REP protocol. Results show that implementation of REP significantly reduces number of messages. Other results indicate that model detects behavior changes of nodes and is robust to slander and colluding attacks. The proposed model has ability to tolerate up to 35% of liars. The effectiveness of relationship maturity parameter is also shown which in certain scenarios reduces trust error level by 50%.

(Subramanian & Ramachandran, 2012) provide trust based scheme for QoS assurance in MANETs. Authors focused on a trust management framework that is intended to cope with the misbehavior problem of node and increases the performance of MANETs. Trust value for each node is implemented by Trust Based Reliable AODV (TBRAODV) protocol. The reliability gets enhanced as nodes are allowed to participate in routing based on trust value else node is identified to become a misbehaving node. NS-2 simulator has been used to analyze the performance of proposed TBRAODV protocol. Trust based routing increases the performance and improves QoS parameters.

(Bar, Mandal, & Singh, 2013) presented trust based AODV routing protocol and its utility in MANET. Depending on the packet forwarding ability of the node, a trust value is obtained to generate the rank of the node. During route discovery, a path involving more trusted nodes is chosen. Un-trusted nodes can be excluded from the route. The packet transmission takes place through more trusted path rather than shortest path. Simulation of the approach has been carried through the use of OMNeT++ software and the results show that higher threshold values give less packet drops and results in more reliable communication. In this approach trust value is calculated every 0.07 second and updated. The approach helps identifying black hole node and is excluded from the route.

CONCLUSION

This chapter presents a survey of trust in Mobile Ad Hoc Networks highlighting the notion of trust in various disciplines and its characteristics. Chapter also lists unique characteristics of MANETs which make trust management difficult. Different phases and tasks of trust management have also been presented and discussed highlighting trust information collection and initialization, trust negotiation, trust determination, trust prediction, trust evaluation, trust aggregation, trust evolution, trust updation, trust dissemination, trust revocation, trust storage and trust visualization. Major work done by different researchers in the area of trust management in MANET has also been presented.

REFERENCES

Agarwal, R., & Motwani, M. (2009). Survey of clustering algorithms for MANET. *International Journal on Computer Science and Engineering*, *1*(2), 98–104.

Bachrach, Y., Parnes, A., Procaccia, A. D., & Rosenschein, J. S. (2009). Gossip-based aggregation of trust in decentralized reputation systems. *Autonomous Agents and Multi-Agent Systems*, *19*(2), 153–172. doi:10.1007/s10458-008-9073-6

Bar, R. K., Mandal, J. K., & Singh, M. M. (2013). QoS of MANet Through Trust based AODV Routing Protocol by Exclusion of Black Hole Attack. *Procedia Technology*, *10*, 530–537. doi:10.1016/j.protcy.2013.12.392

Baras, J. S., & Jiang, T. (2005, February). Managing trust in self-organized mobile ad hoc networks. In N. D. S. S. Workshop (Ed.), Extended abstract. Academic Press.

Chakrabarti, A. (2007). *Grid Computing Security*. Springer Berlin Heidelberg.

Cho, J. H., Swami, A., & Chen, R. (2011). A survey on trust management for mobile ad hoc networks. *IEEE Communications Surveys and Tutorials*, *13*(4), 562–583. doi:10.1109/SURV.2011.092110.00088

Damiani, E., De Capitani di Vimercati, S., Samarati, P., & Viviani, M. (2006). A WOWA-based aggregation technique on trust values connected to metadata. *Electronic Notes in Theoretical Computer Science*, *157*(3), 131–142. doi:10.1016/j.entcs.2005.09.036

Deutsch, M. (1962). Cooperation and trust: Some theoretical notes. *Nebraska Symposium on Motivation*. Nebraska University Press.

Eschenauer, L., Gligor, V. D., & Baras, J. (2004, January). On trust establishment in mobile ad-hoc networks. In *Security Protocols* (pp. 47–66). Springer Berlin Heidelberg. doi:10.1007/978-3-540-39871-4_6

Firdhous, M., Ghazali, O., & Hassan, S. (2011). *Trust and Trust Management in Cloud Computing–A Survey*. Internetworks Research Group, Universiti Utara Malaysia, Technical Report.

Govindan, K., & Mohapatra, P. (2012). Trust computations and trust dynamics in mobile adhoc networks: A survey. *IEEE Communications Surveys and Tutorials*, *14*(2), 279–298. doi:10.1109/SURV.2011.042711.00083

Grandison, T., & Sloman, M. (2000). A survey of trust in internet applications. *IEEE Communications Surveys and Tutorials*, *3*(4), 2–16. doi:10.1109/COMST.2000.5340804

Ham, F. M., Imana, E. Y., Ondi, A., Ford, R., Allen, W., & Reedy, M. (2009). Reputation Prediction in Mobile Ad Hoc Networks Using RBF Neural Networks. In Engineering Applications of Neural Networks (pp. 485-494). Springer Berlin Heidelberg. doi:10.1007/978-3-642-03969-0_46

Huang, J., & Nicol, D. (2009, April). A calculus of trust and its application to PKI and identity management. In *Proceedings of the 8th Symposium on Identity and Trust on the Internet* (pp. 23-37). ACM. doi:10.1145/1527017.1527021

Jawandhiya, P. M., Ghonge, M. M., Ali, M. S., & Deshpande, J. S. (2010). A survey of mobile ad hoc network attacks. *International Journal of Engineering Science and Technology*, *2*(9), 4063–4071.

Jonker, C. M., & Treur, J. (1999). Formal analysis of models for the dynamics of trust based on experiences. In *Multi-Agent System Engineering* (pp. 221–231). Springer Berlin Heidelberg. doi:10.1007/3-540-48437-X_18

Kempe, D., Dobra, A., & Gehrke, J. (2003, October). Gossip-based computation of aggregate information. In *Foundations of Computer Science, 2003. Proceedings. 44th Annual IEEE Symposium on* (pp. 482-491). IEEE. doi:10.1109/SFCS.2003.1238221

Liu, J., & Issarny, V. (2004). Enhanced reputation mechanism for mobile ad hoc networks. In *Trust management* (pp. 48–62). Springer Berlin Heidelberg. doi:10.1007/978-3-540-24747-0_5

Noor, T. H., Sheng, Q. Z., Zeadally, S., & Yu, J. (2013). Trust management of services in cloud environments: Obstacles and solutions. *ACM Computing Surveys*, *46*(1), 12. doi:10.1145/2522968.2522980

Pardo, P. (2009). Aggregation of trust for iterated belief revision in probabilistic logics. In *Scalable Uncertainty Management* (pp. 165–179). Springer Berlin Heidelberg. doi:10.1007/978-3-642-04388-8_14

Pirzada, A. A., & McDonald, C. (2004, January). Establishing trust in pure ad-hoc networks. In *Proceedings of the 27th Australasian conference on Computer science-Volume 26* (pp. 47-54). Australian Computer Society, Inc.

Ruohomaa, S., & Kutvonen, L. (2005). Trust management survey. In *Trust Management* (pp. 77–92). Springer Berlin Heidelberg. doi:10.1007/11429760_6

Ruohomaa, S., Kutvonen, L., & Koutrouli, E. (2007, April). Reputation management survey. In *Availability, Reliability and Security, 2007. ARES 2007. The Second International Conference on* (pp. 103-111). IEEE. doi:10.1109/ARES.2007.123

Sherchan, W., Nepal, S., & Paris, C. (2013). A survey of trust in social networks. *ACM Computing Surveys*, *45*(4), 47. doi:10.1145/2501654.2501661

Song, S., Hwang, K., Zhou, R., & Kwok, Y. K. (2005). Trusted P2P transactions with fuzzy reputation aggregation. *IEEE Internet Computing*, *9*(6), 24–34. doi:10.1109/MIC.2005.136

Subramanian, S., & Ramachandran, B. (2012). *Trust Based Scheme for QoS Assurance in Mobile Ad-Hoc Networks*. arXiv preprint arXiv:1202.1664

Suryanarayana, G., & Taylor, R. N. (2004). *A survey of trust management and resource discovery technologies in peer-to-peer applications*. Academic Press.

Taneja, S., & Kush, A. (2010). A Survey of routing protocols in mobile ad hoc networks. International Journal of Innovation. *Management and Technology*, *1*(3), 2010–0248.

Velloso, P. B., Laufer, R. P., de O Cunha, D., Duarte, O. C. M. B., & Pujolle, G. (2010). Trust management in mobile ad hoc networks using a scalable maturity-based model. *Network and Service Management, IEEE Transactions on, 7*(3), 172-185.

Wang, Y., & Singh, M. P. (2006, July). Trust representation and aggregation in a distributed agent system. In AAAI (Vol. 6, pp. 1425-1430). AAAI.

Wang, Y., Tao, Y., Yu, P., Xu, F., & Lü, J. (2007). A trust evolution model for P2P networks. In Autonomic and Trusted Computing (pp. 216-225). Springer Berlin Heidelberg.

Wu, B., Chen, J., Wu, J., & Cardei, M. (2007). A survey of attacks and countermeasures in mobile ad hoc networks. In Wireless Network Security (pp. 103-135). Springer US. doi:10.1007/978-0-387-33112-6_5

Zhang, Q., Yu, T., & Irwin, K. (2004, November). A Classification Scheme for Trust Functions in Reputation-Based Trust Management. In *ISWC Workshop on Trust, Security, and Reputation on the Semantic Web*.

Chapter 8
A logistic-Map-Based PN Sequence for Stochastic Wireless Channels

Deb Sunder Swami
Gauhati University, India

Kandarpa Kumar Sarma
Gauhati University, India

ABSTRACT

The work in this chapter mainly focusses on a pseudo-noise (PN) sequence generator which is constructed exploiting the features of one-dimensional chaotic systems such as a logistic map. The use of logistic map to generate strong cryptographic sequences is novel in approach in terms of its use with a range of transmission techniques for wireless communication because it is easy to conceive and requires simple devices to generate the sequence. The generated sequence is used as a spreading sequence in a DS-SS modulation multipath environment with AWGN and Rayleigh fading channel, under various channel fading conditions. The BER curves and channel impulse responses are obtained and compared with the existing linear feedback shift register (LFSR) based PN sequence and Kasami sequence. Results from this comparative analysis indicate that the proposed method generally yields a greater number of reliable, unpredictable and random bits than existing techniques under the same conditions and can be practically implemented for DS-SS scheme as a spreading sequence.

INTRODUCTION

In the era of modern telecommunication technologies, people exchange information with each other via wired or wireless networks. To avoid unauthorized access and illegal usage of information being transferred over the insecure communication channel, securities system are needed to be build and deployed. Most of the security systems rely on the use of cryptographic techniques to provide necessary security to the user's sensitive data against illegal usage. The challenging task in the design of cryptographic techniques is to generate sequences with high randomness and statistical properties. The quality of the

DOI: 10.4018/978-1-5225-2342-0.ch008

pseudo-random sequence generated by encryption technique determines its strength from a cryptographic viewpoint. Apart from cryptography, PN sequences are equally applicable and significant in the areas of DS-SS modulation, statistical sampling, computer simulation, (Cruselles, Soriano, and

Melus, 1995) etc. The PN sequence generators may be realized in hardware and as well as in software. LFSR based PN sequence generators are well suited for hardware realizations. LFSR based generators are simpler in implementations, have high speed performance and good statistical properties. However, the disadvantage with LFSR based design is that the feedback tapings can be determined under Berleykamp Massey attack (Ahmad and Farooq, 2011). The importance of a careful design of cryptographic PN sequence generators cannot be underestimated as these generators are becoming particularly useful to ensure secure data transmission over an insecure communication channels. Generating high-quality randomness is a vital part of many cryptographic operations.

In the past decade, the behavior of chaotic systems has been much studied and analyzed. According to chaos theory, chaotic systems are nonlinear dynamical systems whose state evolves with time. The future dynamics of these systems are fully defined by their initial conditions. As a result, the behavior of these systems appears random. It has been determined that the chaotic systems have some interesting inherent characteristics such as high dependency on its initial conditions, unstable periodic orbits with long period, ergodicity etc. Due to these characteristics, the chaotic systems are adopted as promising candidates while designing the security systems for text, image, voice, video, etc. (Ahmad and Farooq, 2011). Chaos based encryption methods provides cryptographically better protection than the conventional cryptographic techniques. Keeping these points under consideration, the characteristics of the chaotic systems are exploited to build chaos based PN sequence generator, which can ascertain the excellent statistical and randomness performance. The simple one-dimensional chaotic 'logistic map' is integrated to generate the real-valued chaotic sequence, which on pre-processing and quantization gives a PN sequence that has noise-like characteristics. To justify the claim, the statistical, randomness and encryption performance of the proposed chaos-based PN sequence generator is compared with existing LFSR based PN sequence and Kasami sequence.

BACKGROUND

A PN sequence can be analyzed for quantitative performance measurements for wireless communication applications. The analysis of a PN sequence includes generation of the sequence, testing its inherent properties and verifying the sequence under some existing criteria. A lot of work has been done on chaos based systems and sequences.

In (Ahmad and Farooq, 2011), the features of one-dimensional chaotic systems are exploited to construct a PN sequence generator by the integration of chaotic logistic maps and cubic maps to generate a strong cryptographic sequence. The proposed system includes the preprocessing and quantization of the real-valued chaotic sequences. The statistical performance of the proposed generator is performed which ascertain that the generated PN sequence has noise-like characteristics. In addition, it is also examined that the proposed chaos-based generator has better statistical and encryption performance than the existing LFSR based generator. In (Cruselles, Soriano, and Melus, 1995) a nonlinear PN generator is proposed that can produce a great number of uncorrelated random sequences with good properties of auto and cross-correlation. Each of these shows good behavior to what is expected for this type of sequence, that is, high unpredictability and good statistical behavior. Moreover, it shows itself to be advantageous for high speed generation of PN codes, with low cost, low power consumption, small volume and simplicity of implementation, and exhibits a structural parallelism useful for VLSI implementations. All these features make it potentially suitable for the generation of independent spreading pseudo random codes in code division multiple access (CDMA) systems.

The use of chaotic sequences as spectral spreading sequences in DS-SS communication systems is proposed in (Heidari-Bateni, and McGillem, 1994). The error probabilities of such systems are investigated and shown to be, for practical purposes, identical to the conventional DS-SS systems which use binary signature sequences. Among the advantages of the use of chaotic sequences in DS-SS is the availability of a great number of them, the ease of their generation, as well as their inherent improvement in the security of transmission. In (tang and tang, 2005), the performance of synchronous DS-CDMA systems over multipath fading channel and AWGN Channel is done. The synchronous DS-CDMA system is well known for eliminating the effects of multiple access interference (MAI) which limits the capacity and degrades the BER performance of the system. This paper investigates the BER performance of a synchronous DS-CDMA system over AWGN and Rayleigh channel, which is affected by the different number of users, as well as different types spreading codes. In [R7], the BER performance under Rayleigh fading channel conditions of MC-CDMA (Multicarrier Code Division Multiple Access) in presence of AWGN for different number of subcarrier, different number of users, and different path gains system analysis is performed by simulating the MC-CDMA using MATLAB program, and a comparison between simulated results is provided. In [R11], a quadrature modulation based complex digital processing (CDP) DS-SS system which exploits the 2-D complex chaotic Ikeda map (IM) as the spreading sequence is proposed. With the proposed system, the mitigation of multiple access interference (MAI) by using complex-valued processing method for the spreading and despreading operations. The performance of the proposed CDP system is evaluated in multiuser environment numerically and analytically under AWGN and Rayleigh flat fading channel conditions. Simulation results reveal that the proposed CDP DS-SS system with the IM spreading sequence significantly outperforms the conventional Gold code based inphase-quadrature (I-Q) DS-SS system. In [R13], a new optimal design method of chaotic spreading sequences with arbitrary period is shown, which optimizes the binary-quantization thresholds with different fractal parameters and initial values. This article simulates and analyzes the design results. The results show that the method can generate a good amount of high-performance chaotic spreading sequences with arbitrary period.

PN SEQUENCE GENERATORS

In cryptography, pseudo random noise is a signal similar to noise which satisfies one or more of the standard tests for statistical randomness. Although it seems to lack any definite pattern, pseudo random noise consists of a deterministic sequence of pulses that will repeat itself after its period. In cryptographic devices, the pseudo random noise pattern is determined by a key and the repetition period can be very long, even millions of digits.

A pseudorandom number generator (PRNG), also known as a deterministic random bit generator (DRBG), is an algorithm for generating a sequence of numbers that approximates the properties of random numbers. The sequence is not truly random in that it is completely determined by a relatively small set of initial values, called the PRNG's state, which includes a truly random seed. Although sequences that are closer to truly random can be generated using hardware random number generators, pseudorandom numbers are important in practice for their speed in number generation and their reproducibility. PRNGs are central in applications such as simulations (e.g. of physical systems via the Monte Carlo method), in procedural generation, and in cryptography. Cryptographic applications require the output to be unpredictable, and more elaborate algorithms, which do not inherit the linearity of simpler solutions, are needed. Common classes of PRNG algorithms include linear congruential generators, lagged Fibonacci generators, and linear feedback shift registers. More recent instances of PRNGs with strong randomness guarantees are based on computational hardness assumptions . Proper statistical properties are a central requirement for the output of a PRNG. In general, the mathematical analysis of PRNG must guarantee complete randomness in generating PN sequences (Cruselles, Soriano, Melus, 1995).

The Logistic Map

The logistic map is a polynomial mapping (equivalently, recurrence relation) of degree 2, often cited as an archetypal example of how complex, chaotic behavior can arise from very simple non-linear dynamical equations. The map was popularized in a seminal paper published in 1976 by Robert May, who by profession then was a biologist. It was in part as a discrete-time demographic model analogous to the logistic equation first created by Pierre François Verhulst (Weisstein, 2012). Mathematically, the logistic map is written as

$$x\left(n+1\right) = \lambda x\left(n\right)\left(1-x\left(n\right)\right) \tag{1}$$

where, x(n) is a number between zero and one, and represents the ratio of existing population to the maximum possible population at year n, and hence $x\left(0\right)$ represents the initial ratio of population to maximum population (at year 0).

The term λ is a positive number, and represents a combined rate for reproduction and starvation.

The logistic map is one of the simplest forms of a chaotic process. Basically, this map, like any one-dimensional map, is a rule for getting a number from a number. The parameter λ is fixed, but if one studies the map for different values of λ (up to 4, else the unit interval is no longer invariant) it is found that λ is the catalyst for chaos.

Population Map

Population biologists are interested in the long term behavior of the population of a certain species. Given certain observed or experimentally determined parameters the biologists set up a mathematical model to describe the fluctuations in population. The problem facing the population biologist is a typical dynamical system question: given initial population $x(0)$, can one predict the long term behavior of the population ? The easiest mathematical way to deal with the above problem is as follows-

Let $x(n)$ be the Population after n generations, where n is a natural number.

The simplest growth law one can imagine is that the population in the next generation is directly proportional to that in the present generation i.e.

$$x(n+1) = \lambda\, x(n) \tag{2}$$

where λ is a constant.

It means that we have the ultimate fate of the population given if

$n > 1$, $x(n) \rightarrow \propto$ indicating overcrowding.

and if

$0 < n < 1$, $x(n) \rightarrow 0$ indicating extinction.

Thus there are essentially only two possibilities. First is unchecked growth and the second is extinction. But experiment tells us that if the population goes on increasing, there will be shortage of food. Its result might lead to extinction of the species.

So the biologists try to incorporate an additional constraint or limiting factor hoping for better reflection of reality. A common practice in such a situation is to assume that there is some limiting value L for the population (since practically it cannot be infinite). If $x(t)$ (population at time t say, if considered the continuous case) exceeds L, the population should tend to decrease. On the other hand, if $x(t) < L$, there is room for more of the species, so $x(t)$ should increase.

The simplest model leading to such behavior which tends to show better realistic results is

$$x(n+1) = \lambda x(n)(L-x(n)) \tag{3}$$

Little simplifications are made on the model. We assume that L=1 is the limiting value of the population. In that case $x(n)$ represents the percentage of the limiting population in the generation. The corresponding difference equation then becomes

$$x(n+1) = \lambda x(n)(1-x(n)) \; \lambda > 0, \; x(n) \varepsilon [0,1] \tag{4}$$

This is the logistic map equation in its natural form (Weisstein, 2012).

The relative simplicity of the logistic map makes it an excellent point of entry into a consideration of the concept of chaos. A rough description of chaos is that chaotic systems exhibit a great sensitivity to initial conditions. As a practice related to the logistic map, most values of λ exist between about 3.57 and 4 (as noted above). A common source of such sensitivity to initial conditions is that the map represents a repeated folding and stretching of the space on which it is defined. In the case of the logistic map, the quadratic difference equation (1) describing it may be thought of as a stretching-and-folding operation on the interval (0, 1) (Weisstein, 2012).

This stretching-and-folding does not just produce a gradual divergence of the sequences of iterates, but an exponential divergence, evidenced also by the complexity and unpredictability of the chaotic logistic map. In fact, exponential divergence of sequences of iterates explains the connection between chaos and unpredictability. A small error in the supposed initial state of the system will tend to correspond to a large error later in its evolution. Hence, predictions about future states become progressively (indeed, exponentially) worse when there are even very small errors in our knowledge of the initial state. This quality of unpredictability and apparent randomness led the logistic map equation to be used as a PRNG in early computers.

Bifurcation

The word 'Bifurcation' literally means splitting into two parts. The term is commonly used in the study of non-linear dynamics to describe any sudden change in the behavior of the system as some parameter is varied. Bifurcation then refers to the splitting of the behavior of the system into two regions, one above, the other below the particular value at which the change occurs. The parameter value at which the change occurs is called a bifurcation point (Weisstein, 2012).

The onset of chaos is often studied by bifurcation theory. In certain parameterized families of one dimensional map, chaos occurs by infinitely many period doubling (PD) bifurcations. PD bifurcation, as a universal route to chaos, is one of the most important phenomenon implying a series of bifurcations in any of which a periodic orbit loses its stability and is replaced by a new stable one with the double period. These bifurcations accumulate at certain values where chaotic behavior begins to appear.

The Figure 1 shows a bifurcation diagram of the logistic map obtained by plotting as a function of λ a series of values for x(n) obtained by starting with a random value x(0), iterating many times, and discarding the first points corresponding to values before the iterates converge to the attractor. In other words, the set of fixed points of x(n) corresponding to a given value of λ are plotted for values of λ increasing to the right.

An enlargement of the previous diagram around λ= 3.5 is also illustrated in Figure 1, with value of λ at which a PD cycle first appears indicated by blue lines.

In order to study the fixed points of the logistic map, let an initial point x(0) lie in the interval [0, 1]. Now find appropriate conditions on λ which keep points in the interval. The maximum value x(n+1) can take is found by differentiating eq. (1), which is found to be 0.5. Plugging this in, Max.[x(n+1)]= λ/4. Therefore, to keep the map in the desired region, we must have $\lambda \in [0, 4]$ (Weisstein, 2012).

Figure 1. Bifurcation in logistic map

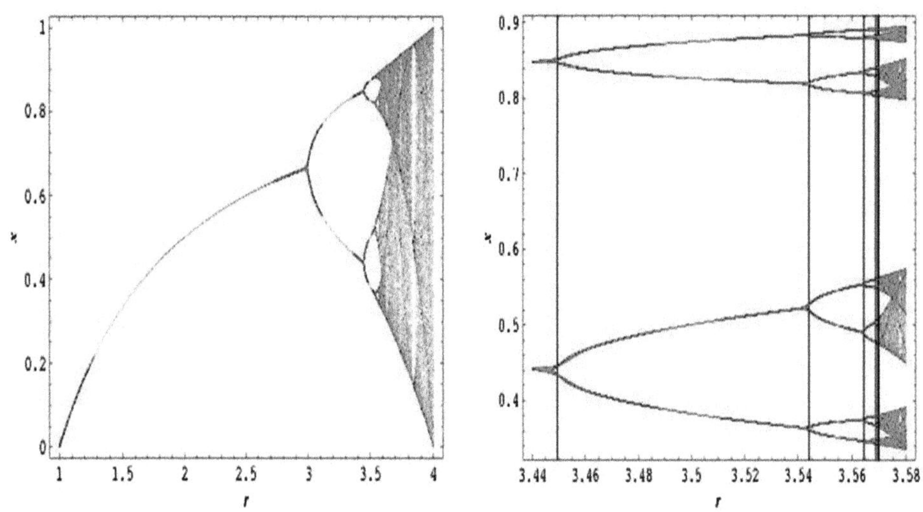

Initial Condition Sensitivity

By varying the parameter λ, the following behavior is observed:

- With λ between 0 and 1, the population will eventually die, independent of the initial population, as shown in Figure 2(a) below.
- With λ between 1 and 2, the population will quickly approach the value (λ-1/λ), independent of the initial population, as shown in Figure 2(b) below.
- With λ between 3 and $1+\sqrt{6}$ (approximately 3.44949), from almost all initial conditions the population will approach permanent oscillations between two values. These two values are dependent on λ, shown in Figure 3(a).
- When $\lambda > 1+\sqrt{6}$ from almost all initial conditions the population will approach oscillations among 8 values, then 16, 32, etc. and it will have infinite oscillation points. Then, it will show chaotic behavior as shown in Figure 3(b).

From almost all initial conditions we can no longer see any oscillations of finite period. Slight variations in the initial population yield dramatically different results over time, a prime characteristic of chaos. Most values beyond 3.56995 exhibit chaotic behavior, but there are still certain isolated ranges of λ that show non-chaotic behavior; these are sometimes called islands of stability. For instance, beginning

Figure 2. Logistic map plots: (a) population extinction (b) population stagnant

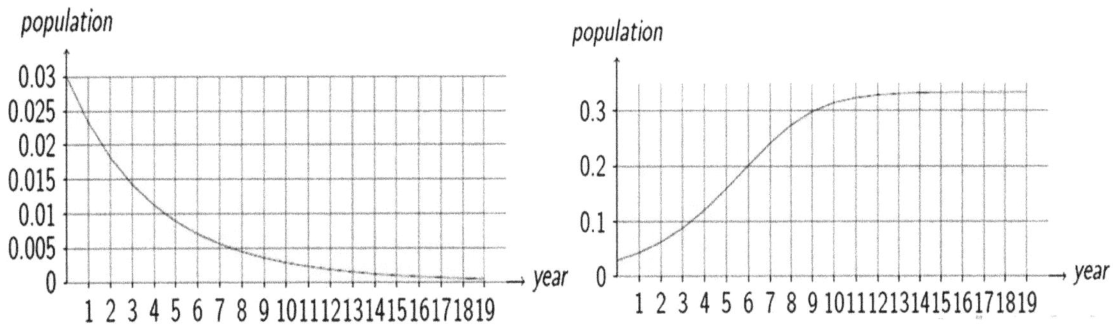

Figure 3. Logistic map plots: (a) population oscillation (b) chaotic oscillation

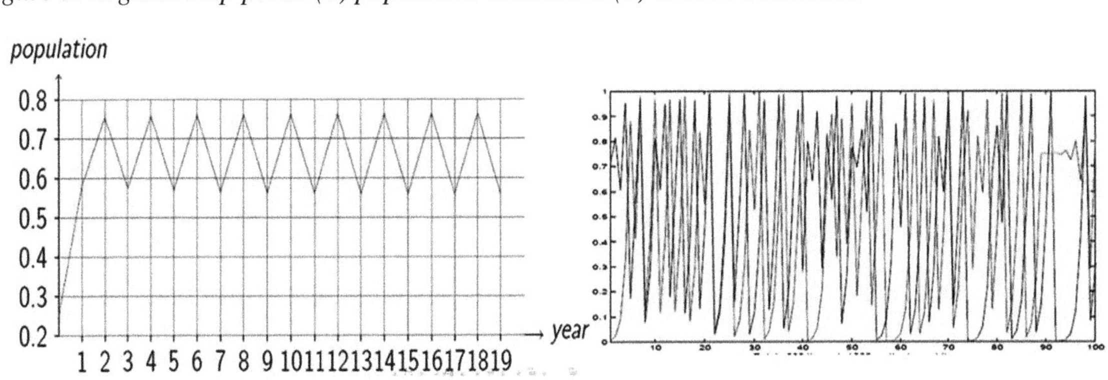

at $1+\sqrt{8}$ (approximately 3.82843) there is a range of parameters λ which show oscillation among three values, and for slightly higher values of λ oscillation among 6 values, then 12 etc.

The development of the chaotic behavior of the logistic sequence as the parameter λ varies from approximately 3.56995 to approximately 3.82843 is sometimes called the Pomeau–Manneville scenario, which is characterized by a periodic (laminar) phase interrupted by bursts of aperiodic behavior. Such a scenario has an application in semiconductor devices (Weisstein, 2012). There are other ranges which yield oscillation among 5 values etc.; all oscillation periods occur for some values of λ. Beyond $\lambda = 4$, the values eventually leave the interval [0, 1] and diverge for almost all initial values.

LOGISTIC MAP BASED PN SEQUENCE GENERATOR

Next, the elaboration on the aspects of logistic map's chaotic behavior and to develop a scheme to create a PN sequence with proper correlation properties and randomness is done. Using MATLAB, a logistic map based PN sequence is created. The characteristics of the sequence so generated are then observed and plotted. The details of the system model for the proposed logistic map PN sequence generator are given.

The system model for the proposed sequence is shown in Figure 4. Here we see that the system is simplistic in nature with respect to wireless applications. As shown in Figure 4, $x(0)$ and λ are feed to the system as initial seeds. On each iteration, the real valued state variable $x_i(n)$ of chaotic map is captured and preprocessed/quantized into a 3-bit value. So, for each real-valued chaotic sequence $x_i(n)$, we get a binary random sequence $Z_i(n)$. After the quantization step the sequence of a predefined length is used for spreading of modulated data. The research shows that the logistic map is chaotic for $3.57 < \lambda < 4$ and $0 < x(n) < 1$ for all $n \geq 0$. As clear from their state equations, the value of state variable obtained on n^{th} iteration will acts as seed for the next $(n+1)^{th}$ iteration. The dynamical orbits of chaotic maps highly deviated from their normal trajectories, which impart randomness to the generated chaotic sequences.

Sequence Generation

Using MATLAB, the logistic map equation is used in a iterative form using a loop to get a sequence of random decimal numbers. The process starts with allocating a value to $x(0)$ as 0.1 and λ as 4. A counter for the loop is taken whose value depends on the length of the sequence required. Inside the loop the

Figure 4. Logistic map generator

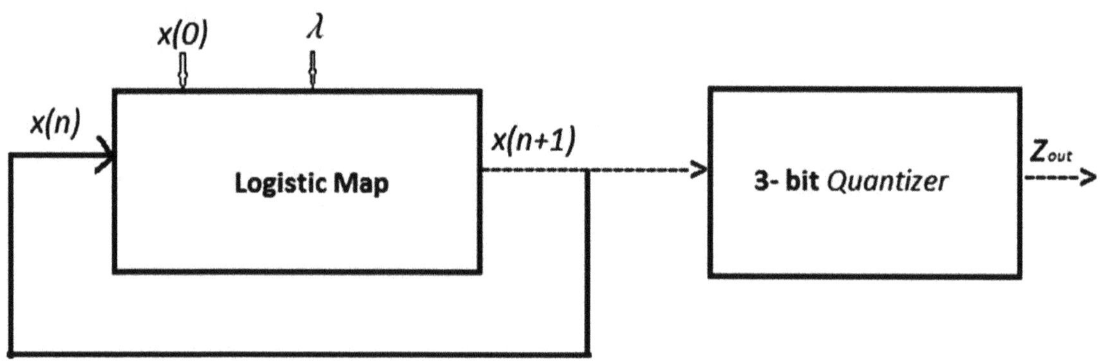

logistic map equation is iterated and a decimal value gets generated. For each decimal value generated, it is quantized into a 3 bit or 4 bit binary value. The process continues until the loop counter exhausts. Then the numbers are concatenated to form a single array. Finally an optional step is carried to convert the sequence into bipolar form. This sequence is then used for further processing and application. The sequence of operation for generation of the sequence is shown in Figure 5.

Initial Condition Sensitivity

The initial conditions of the logistic map determine its properties. Here, we observe the sensitivity of the logistic map to initial conditions. Two sequences are generated with little changes in the initial conditions as is shown in Figure 6. With 0.1 incremental change in $x(n)$ value, the subsequent resulting changes in $x(n+1)$ is obvious from the plots in Figure 6. This property of the sequence is very important because of the randomness it provides to multiple versions of a sequence in a wireless device. With this attribute, while designing the system for multiple users we can change the initial conditions of the sequence to get different sequences being orthogonal to each other.

A shown in the Figure 6 although for the first few values of both the sequences the values are nearly similar values, but in subsequent iterations the values of the number change is diverted to a large extent.

Figure 5. Flow Sequence of generator

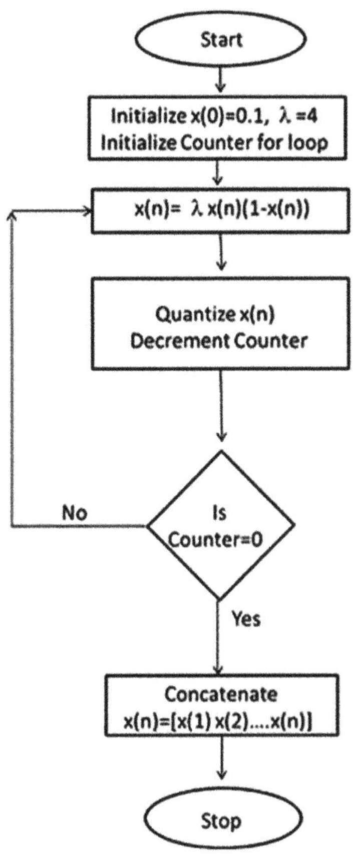

Figure 6. Initial condition sensitivity

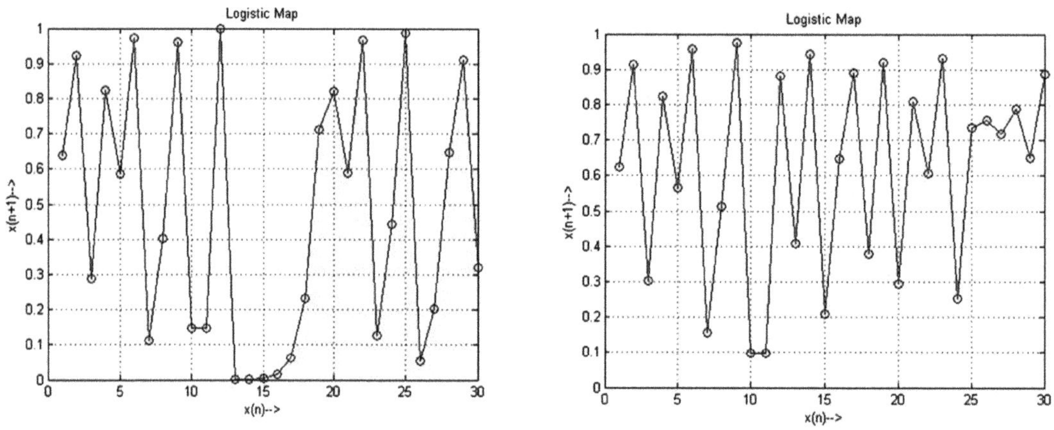

In the long run, the sequences would appear to be orthogonal to each other. So for example, in a multi-user case different sequences with slightly changed initial conditions can produce multiple versions of a sequence which can be used for various applications.

Properties of a PN Sequence

For a sequence to be a pseudo noise or pseudorandom, it should follow the following basic rules.

- **Balance Property:** The relative frequency of 0's and 1's are each 1/2 in the sequence.
- **Run Property:** The run lengths of 0's and 1's are, 1/2 of all run lengths are of length 1, 1/4 are of length 2, 1/8 are of length 3, and so on.
- **Correlation Property:** If a PN sequence is shifted by any non-zero number of elements, the resulting sequence will have an equal number of agreements and disagreements with respect to the original sequence.

Observations

Following the iteration process to produce a sequence of decimal numbers, a quantization process is done to each value. The quantization can be done in 3-bits or 4-bits depending on the length of the sequence required. Also the quantization done was uniform in nature, this ensured that the balance property and the run property of a PN sequence hold.

Figure 7 shows the plot of the quantized version of the sequence both in polar and bipolar format. It was observed to follow the run and balance properties of a PN sequence. Also, the length of the sequence required can be predefined by the number of iterations and the number of bits for used quantization.

Autocorrelation

Correlation is a measure of similarity between two time shifted versions of a sequence. When the two sequences compared are different, it is the 'crosscorrelation' and when they are the same it is the 'auto-

Figure 7. Uniolar and bipolar plot of the generated sequence

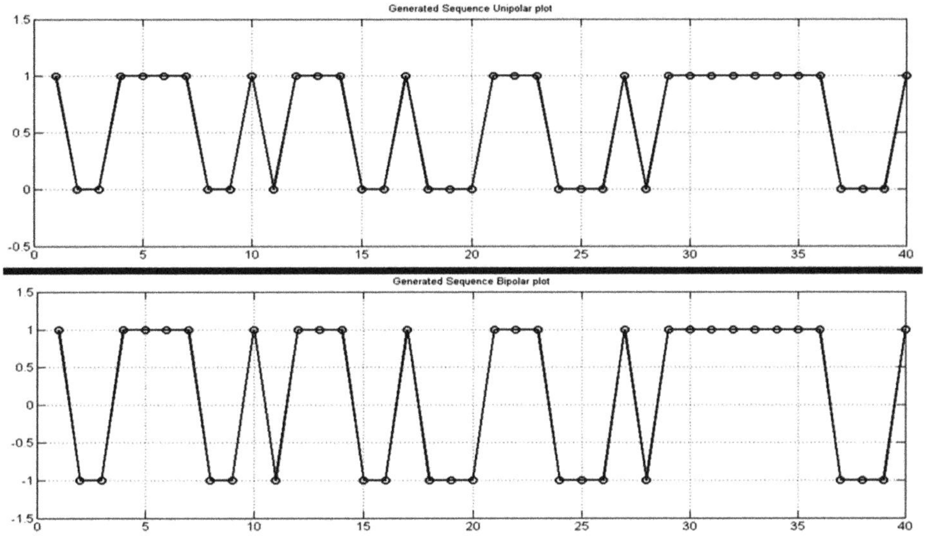

correlation'. The autocorrelation function of the sequence generated is plotted and is shown in Figure 8. For a perfectly random sequence, the autocorrelation function needs to have a delta function like form (Mutagi, 1996), i.e. a PN sequence has an autocorrelation of 1 at zero phase (no time shift), and 0.5 at all other phases. From the autocorrelation function of PN sequence generated, it is clear that it has moderate autocorrelation function form and it satisfies the required condition.

Figure 8. Autocorrelation of generated sequence

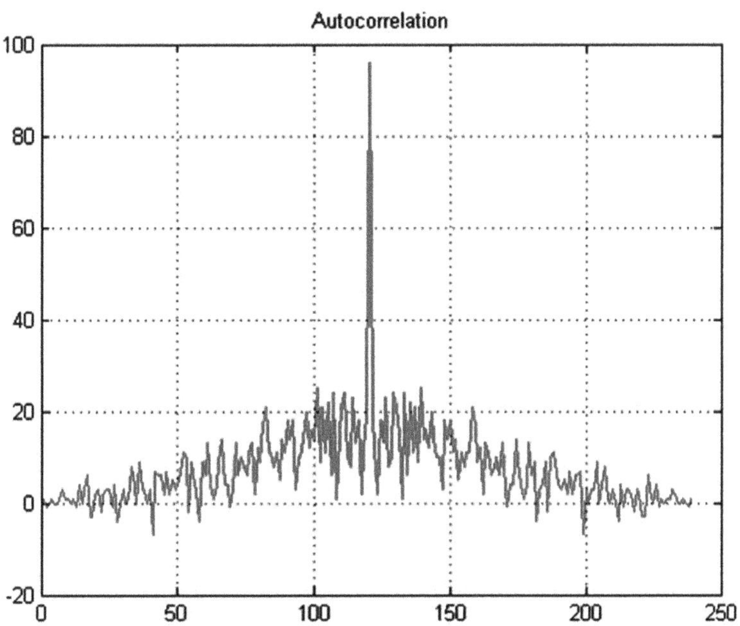

Advantage of Proposed Technique Over Existing Methods

- **Infinite Period:** Traditional LFSR based PN sequence is a semi-random sequence in the sense that it appears random within the sequence length, fulfilling the needs of randomness, but the entire sequence repeats indefinitely. To a casual observer the sequence appears totally random, however to a user who is aware of the way the sequence is generated all its properties are known, (Mutagi, 1996). But for a logistic map PN sequence, the period is infinite. As such the sequence does not repeat within the sequence length.
- **Less Hardware Requirement:** LFSR based PN sequences are constructed using shift registers and modulo-2 adders. An N-stage shift register can generate a maximal length sequence of 2^{N-1} bits, (Mutagi, 1996). So larger the sequence length, more the number of feedback registers required. But in the case of a logistic map based PN sequence, there is minimal hardware requirement. Also the sequence length can be changed in the software by simply changing the number of iterations or the number of bits used for quantization.
- **No Additional Hardware:** LFSR based PN sequence suffer from an issue of 'all-zero state'. This can occur when the initial value given to the shift registers are all zeroes, (Mutagi, 1996). As such, additional circuitry may be needed in them to detect such a state and avoid it. For a logistic map PN sequence this problem does not arise, as the iterated values are always a non-zero.
- **Less Overheads:** LFSR based PN sequence used for spread spectrum application have more overhead. This is because, for an N-stage shift register at the transmitting end, N initial values need to be send at the receiver end for identical sequence generation. But for a logistic map based PN sequence, only one initial value for $x(0)$ needs to be same at both the transmitting and receiving end.

LOGISTIC MAP PN SEQUENCE IN DIRECT SEQUENCE SPREAD SPECTRUM

Direct Sequence Spread Spectrum (DSSS) is a spread spectrum technique whereby the original data signal is multiplied with a pseudo random noise spreading code. This spreading code has a higher chip rate, which results in a wideband time continuous scrambled signal. DSSS significantly improves protection against interfering (or jamming) signals, especially narrowband and makes the signal less noticeable. It also provides security of transmission if the code is not known to the public. The effect is to diffuse the information in a larger bandwidth. Conversely, we can remove the spread-spectrum code (called a despreading operation) at a point in the receive chain before data retrieval. A despreading operation reconstitutes the information into its original bandwidth. Obviously, the same code must be known in advance at both ends of the transmission channel.

The Figure 9 shows a simple set up of a DSSS modulation scheme. Here the logistic map based PN sequence is used for spreading the transmitted data before being sent through a communication link. At the receiver end the same PN sequence is used to de-spread the data before demodulating it.

The operation of the sequence as spreading code in a DSSS modulation scheme has been shown along with BER result plots. Three different modulation schemes have been used to check the performance of the sequence under different conditions. To check the performance of the logistic map PN sequence generated, the spreaded data is tested in a multipath environment with Rayleigh and AWGN channels.

Figure 9. DSSS modulation scheme

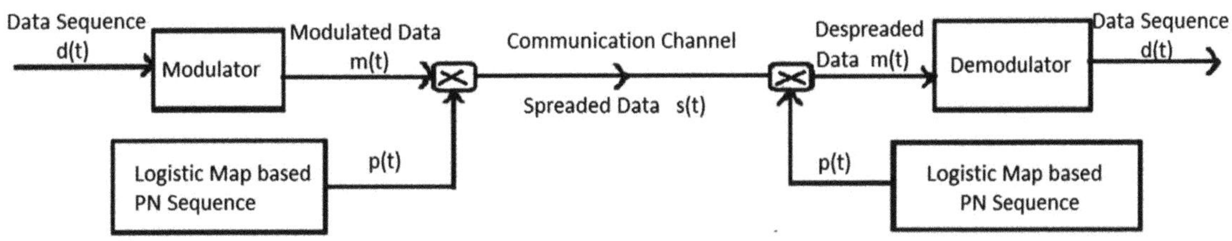

In the next section, the system model for the logistic map PN sequence generator in an AWGN channel and the associated results are shown.

Logistic Map PN Sequence Generator in an AWGN Channel

The block diagram of proposed method for verification of the spreading sequence in a AWGN channel is shown in Figure 10. A comparison is made with Kasami sequence and existing LFSR based PN sequence. In the first simulation, the logistic map based PN sequence is used to spread BPSK modulated data in a multipath AWGN channel. The logistic map based PN sequence has been generated using MATLAB and the simulation has been done on Simulink tool.

BPSK is a method for modulating a binary signal onto a complex waveform by shifting the phase of the complex signal. In digital baseband BPSK, the symbols 0 and 1 are modulated to the complex numbers e(jt) and -e(jt), respectively, where t is a fixed angle or the phase offset. The output from the BPSK modulator is multiplied with the logistic map based PN sequence generated, to spread the data.

Next, the spreaded data is passed through a multipath AWGN channel. The AWGN channel adds white Gaussian noise to a real or complex signal. Then the signal is despreaded and the individual paths are combined in terms of power or gain. The despreaded data is then demodulated using BPSK demodulator

Finally the BER of the signal is calculated. The same procedure is done with the LFSR based PN sequence and Kasami sequence.

The BER curve is plotted for the three sequences in a SNR range of -10 dB to +10 dB. The PN sequence generator generates a sequence of pseudorandom binary numbers using a LFSR. The LFSR is implemented using a simple shift register generator (SSRG, or Fibonacci) configuration. Kasami sequence is also a PN sequence generated by a polynomial method with good correlation properties.

The above mentioned steps are repeated with QPSK and DPSK modulations. In case of QPSK for input m, the output symbol is $e(j\theta + j\pi m/2)$, where θ is the phase offset parameter for input m. The M-

Figure 10. System model 1

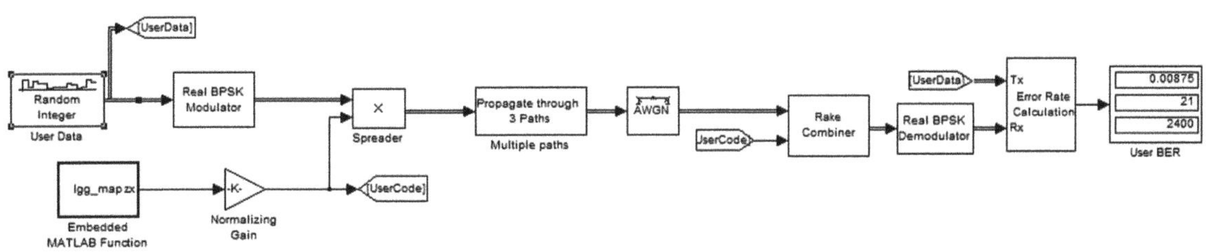

DPSK modulates using the M-ary differential phase shift keying method. The M-ary number parameter, M, is the number of possible output symbols that can immediately follow a given output symbol.

The three spreading sequences are used in both modulation schemes and the results compared. In the following section, the experimental setup is described and the results shown. As will be shown later the logistic map scheme gives better error performance compared to both PN sequence and Kasami sequence.

Results

Using Simulink, the blocks have been arranged and the standard communication criteria has been used and all the three spreading sequences (logistic map PN sequence, LFSR based PN sequence and Kasami sequence) has been tested in same conditions. The BER values for a SNR range of -10dB to +10dB has been obtained. The obtained BER values are then plotted using MATLAB. For calculating BER, 2400 samples have been compared. For each symbol a spreading code of length 64 bits is used to spread the data in the frequency domain. This ensures high chip rate of the data.

The BER curve for BPSK modulation in AWGN channel with comparison between three spreading sequences is provided in Figure 11. It is obvious from the figure that the logistic map PN sequence has better error performance compared to the LFSR PN sequence and Kasami Sequence.

In the next setup the modulation scheme that is used is QPSK. Here also the same conditions have been used for the three spreading sequences. Figure 12 shows the BER curve obtained. Here also it is clear that logistic map PN sequence provides comparatively better error performance and the overall performance remains same. The LFSR PN sequence and the Kasami sequence have similar error performance.

Similarly, in the third setup, DPSK modulation scheme has been used under the same conditions. Figure 13 shows the BER curve obtained from the setup. Again the logistic map PN sequence provides

Figure 11. BER curve for BPSK modulation

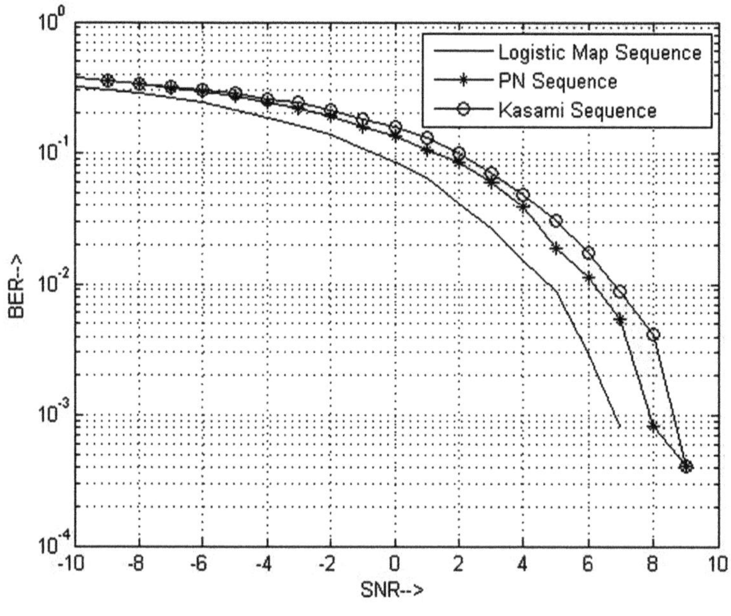

Figure 12. BER curve for QPSK modulation

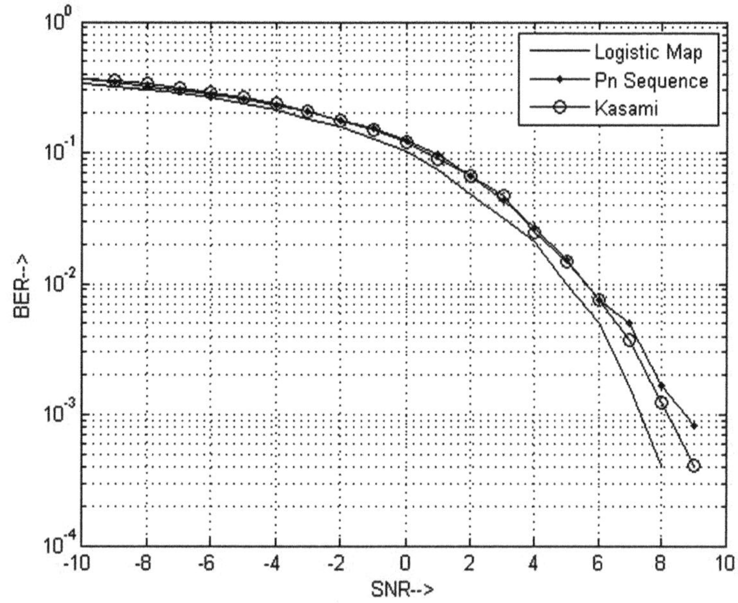

Figure 13. BER curve for DPSK modulation

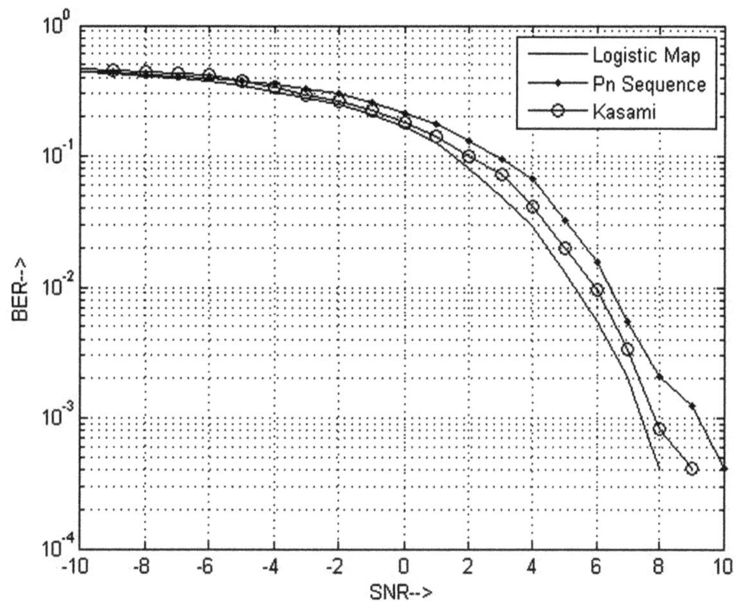

comparatively better error performance then the Kasami and LFSR PN sequence. However, here the overall system performance is slightly degraded due to the differential modulation scheme used.

As is obvious from the curves shown above, the logistic map PN sequence provides better error performance in AWGN multipath environment. Although the error performance is not great, however the difference with respect to existing techniques is significant enough.

Logistic Map PN Sequence Generator in a Rayleigh Fading Channel

In the next setup a Rayleigh fading channel has been added to the model shown in the Figure 14. The Rayleigh channel model in Simulink adds fading noise to the signal, and it accepts only complex inputs. For simulation simplicity the complex phase gains have been removed to maintain uniformity and multipath have been added externally. Rest of the criteria has been kept same as the previous model. The simulations are done for BPSK, QPSK and DPSK modulation schemes. The error performance for logistic map PN Sequence, LFSR PN sequence and Kasami sequence have been compared. Their respective BERs have been plotted. But due to the presence of extra blocks in the communication link, the simulation times were longer.

Results

Figure 15 shows the BER curve for BPSK modulation in Rayleigh channel. It can be seen that the logistic map PN Sequence clearly gives better error performance then the LFSR PN sequence and the Kasami sequence. But due to the presence of Rayleigh channel the fading is flat and the overall system performance degrades to some extent.

In the next simulation, QPSK modulation scheme has been used. Here also the same conditions have been used for the three spreading sequences. Figure 16 shows the BER curve obtained. Here also it is clear that logistic map PN sequence provides comparatively better error performance and the overall performance remains same. The LFSR PN sequence and the Kasami sequence have similar error performance.

In the final simulation DPSK modulation scheme has been used using the same conditions. Figure 17 shows the BER curve obtained from the setup. Again the logistic map PN sequence provides better error performance then the Kasami and LFSR PN sequence.

From the above simulations it is clear that the proposed logistic map based PN sequence gives better error performance then both LFSR based PN sequence and Kasami sequence. But due to the presence of a Rayleigh fading channel, the overall error performance was poor and the curves obtained were flat as in a practical communication link. Another observable thing was that for higher SNR values the logistic

Figure 14. System model 2

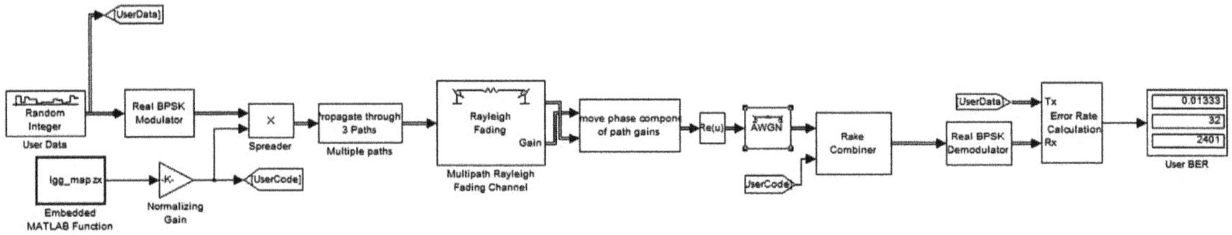

Figure 15. BER curve for BPSK modulation in Rayleigh Channel

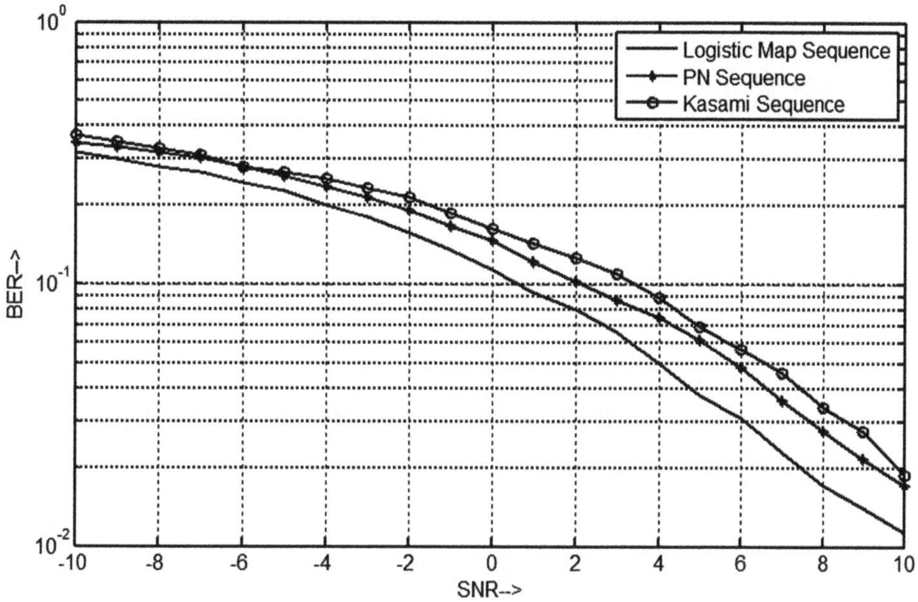

Figure 16. BER curve for QPSK modulation in Rayleigh Channel

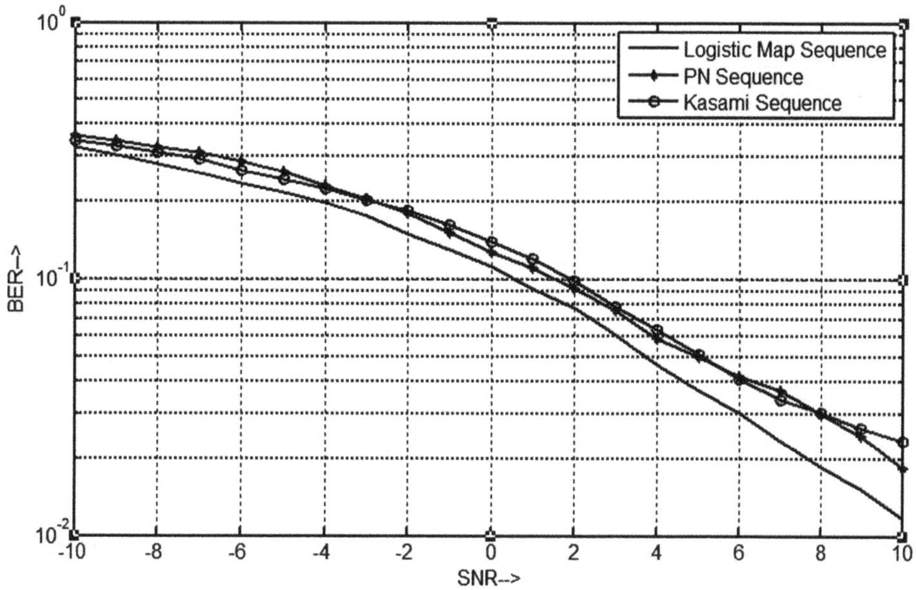

Figure 17. BER curve for DPSK modulation in Rayleigh Channel

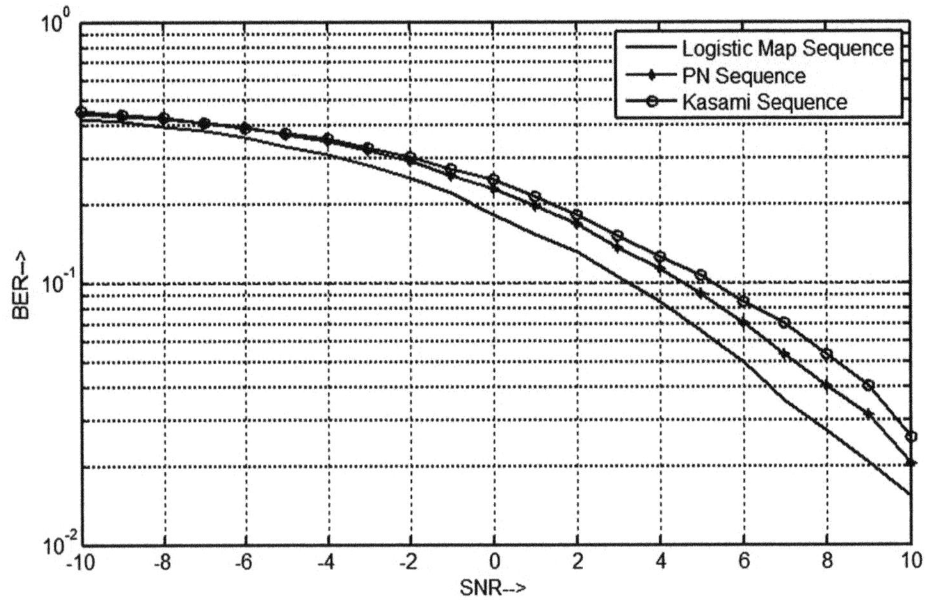

map sequence's performance was significantly better than the other sequences. The performance can be improved by better designing of the Rake combiner where diversity techniques are used for channel estimation.

Orthogonal Code

A comparison is also made with Hadamard code in BPSK modulation scheme and BER performance obtained for the logistic Map PN sequence was found to be much better than the Hadamard code (BER curve shown in Figure 18). A Hadamard code is generated from a Hadamard matrix, whose rows form an orthogonal set of codes. Hadamard matrices are square matrices whose entries are +1 or -1, and whose

Figure 18. BER curve for Hadamard code vs. logistic map PN sequence

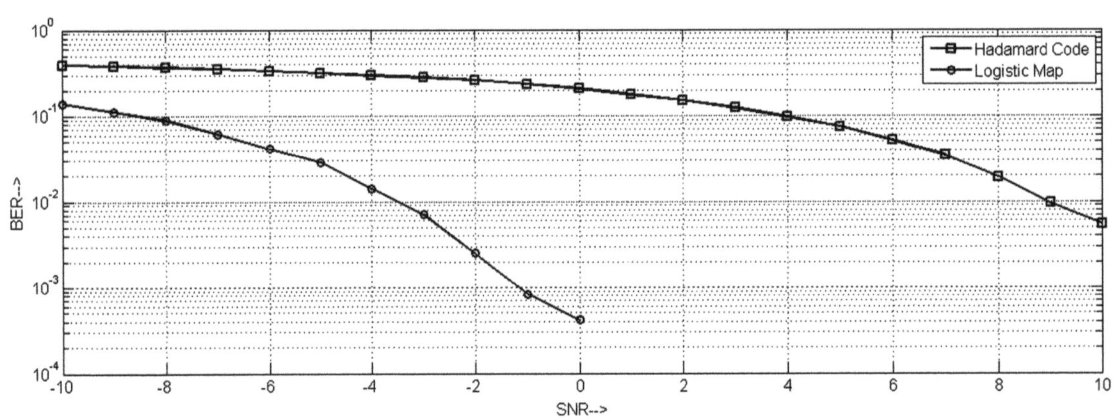

rows and columns are mutually orthogonal. Orthogonal codes can be used for spreading in communication systems in which the receiver is perfectly synchronized with the transmitter.

Fading Channel Model

In the next setup the fading characteristics of a wireless communication channels is verified using the logistic map based spreading sequence. The time dispersion and frequency dispersion mechanisms in a mobile radio channel lead to four possible distinct effects, which are manifested depending on the nature of the transmitted signal, the channel, and the velocity (Rappaport, 2005).

Some fading characteristics of a wireless communication channel are tested using the logistic map based PN sequence in a Rayleigh fading channel with different modulation schemes. The BER curves and the channel impulse responses shown below justify the characteristics for the above mentioned channels.

For modeling slow fading and fast fading channels, Doppler frequencies of 10 Hz and 110 Hz has been used respectively. For modeling a frequency selective fading channel, a delay vector with different delays and path gains has been used in the Rayleigh Fading Channel block.

For BPSK modulation scheme, the BER performance of slow fading channel is marginally better because of the dependence on the lower Doppler frequency value, as shown in Figure 19, whereas for fast fading due to more Doppler Shift, there is marginally more error.

The channel impulse response of both the cases is shown in Figure 20, which showed that in a fast fading channel the channel impulse response changed at a much faster rate compared to a slow fading channel. The impulse responses are generated using MATLAB visualization tool.

The BER curves shown in Figure 21 (left) shows that frequency selective fading is the worst case scenario and the BER curve is very flat, as such an equalization technique has been used to compensate for the poor performance.

For a frequency selective fading case the channel impulse response is shown in Figure 21 (right). The impulse response has a multipath delay spread which is greater than the reciprocal bandwidth of the transmitted message waveform. When it occurs, the received signal includes multiple versions of the transmitted waveform that are attenuated and delayed, and hence the received signal is distorted.

Figure 19. BER curve showing slow fading vs. fast fading for BPSK

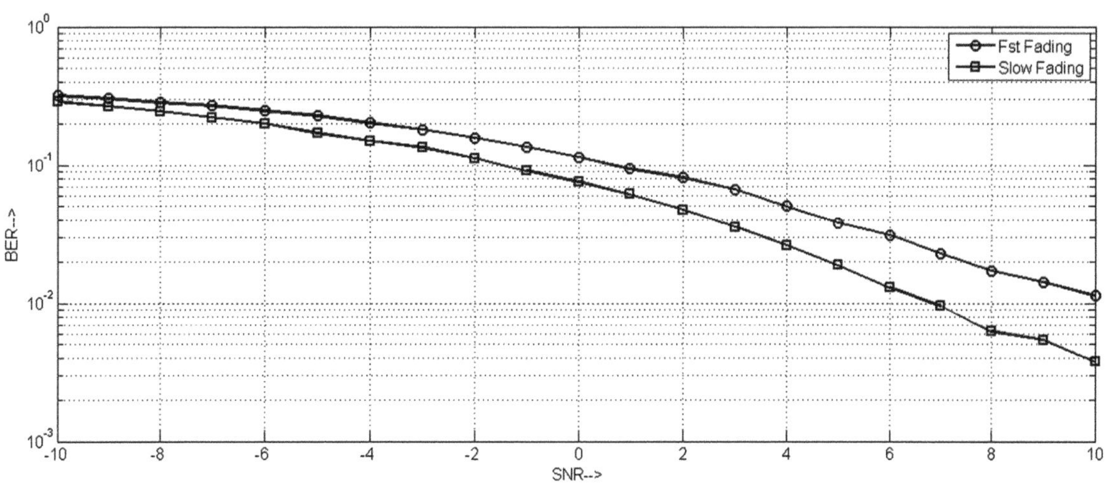

Figure 20. Channel impulse response of slow fading (left) and fast fading channel (right)

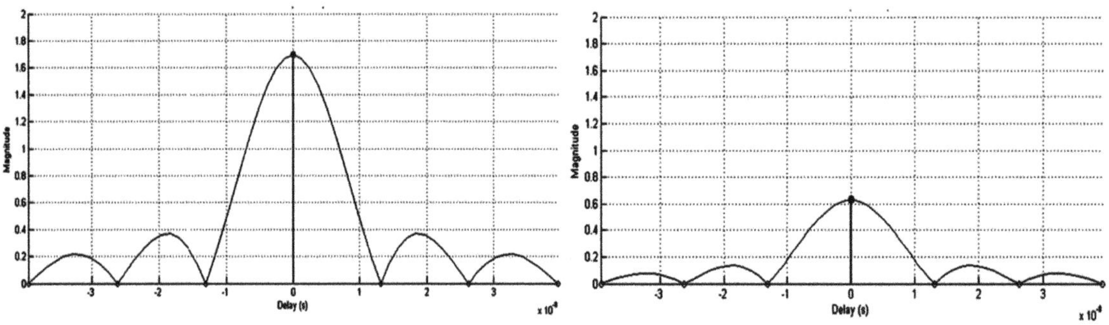

Figure 21. BER curve showing frequency selective fading for BPSK (left) and the channel impulse response (right)

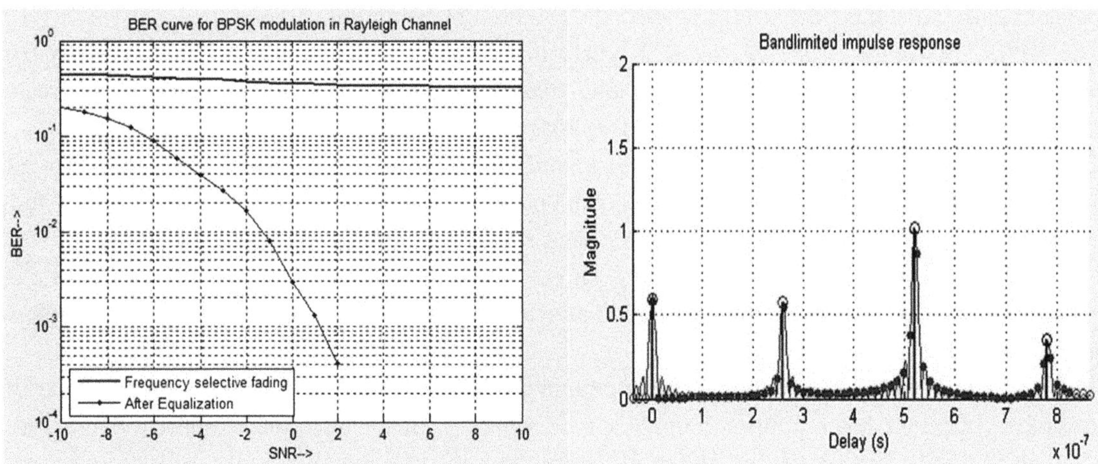

The same procedures are carried for QPSK and DPSK modulation schemes as well. For QPSK modulation, Figure 22 shows slow fading and fast fading channel BER curves. Here as well, the slow fading channel has significantly better error performance compared to a fast fading channel because signal distortion due to fast fading increases with increasing Doppler spread relative to the bandwidth of the transmitted signal. Their respective channel impulse responses are shown in Figure 23 where it was observed that in a fast fading channel the channel impulse response changed at much faster rate compared to slow fading channel.

Similarly for QPSK, the BER curve for frequency selective fading channel is shown in Figure 24 (left) which presents the worst case scenario; as such equalization is used to improve the BER performance. The channel impulse response is shown in Figure 24 (right), under this condition the channel impulse response had multipath delay spread.

The above procedures are carried for DPSK modulation schemes too. For DPSK modulation, Figure 25 shows slow fading and fast fading channel BER curves. Here as well, the slow fading channel has better error performance compared to a fast fading channel because signal distortion due to fast fading increases with increasing Doppler spread relative to the bandwidth of the transmitted signal. Their re-

Figure 22. BER curve showing slow fading vs. fast fading for QPSK

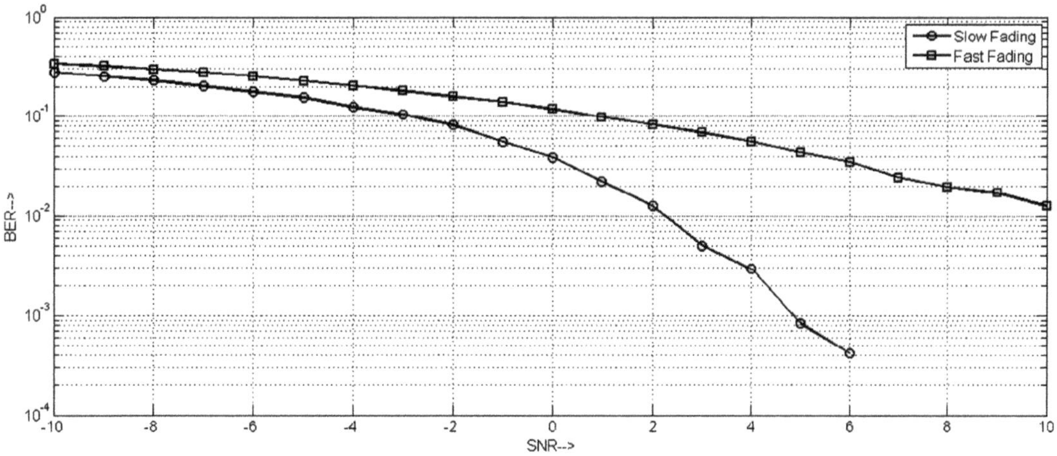

Figure 23. Channel impulse response of slow (left) and fast fading channel (right)

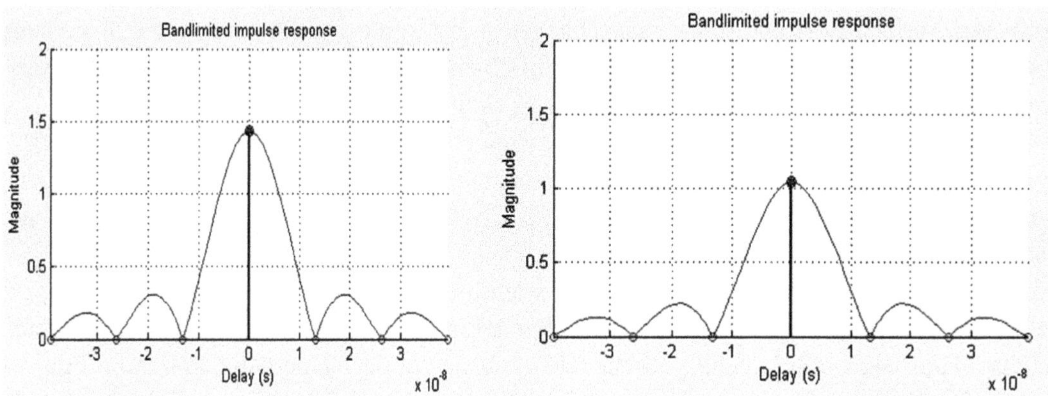

Figure 24. BER curve of frequency selective fading channel for QPSK (left) and the channel impulse response (right)

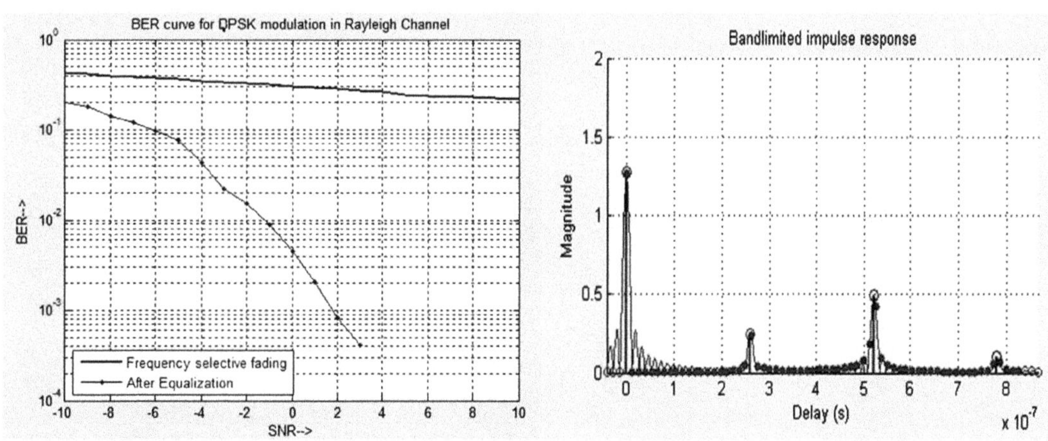

Figure 25. BER curve showing slow fading vs. fast fading for DPSK

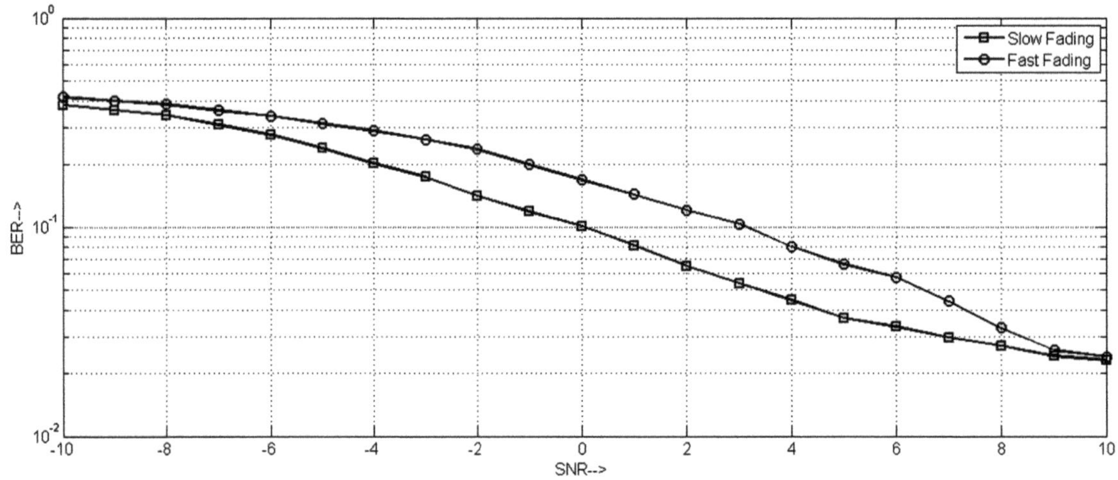

spective channel impulse responses are shown in Figure 26 where here also it was observed that in a fast fading channel the channel impulse response changed at faster rate compared to slow fading channel. It is clear that the velocity of the mobile (or velocity of objects in the channel) and the baseband signaling determines whether a signal undergoes fast fading or slow fading.

Similarly for DPSK, the BER curve for frequency selective fading channel is shown in Figure 27 (left) which presents the worst case scenario; as such equalization is used to improve the BER performance. The channel impulse response is shown in Figure 27 (right), under this condition the channel impulse response had multipath delay spread.

From the above simulations we can say that when a channel is specified as a fast fading or slow fading channel, it does not specify whether the channel is flat fading or frequency selective fading in nature, fast fading channel only deals with the rate of change of the channel due to motion. Hence a flat fading, fast fading channel is a channel in which the channel impulse response is approximated to be a delta function (no time delay).

Figure 26. Channel impulse response of slow and fast fading channels fading channel

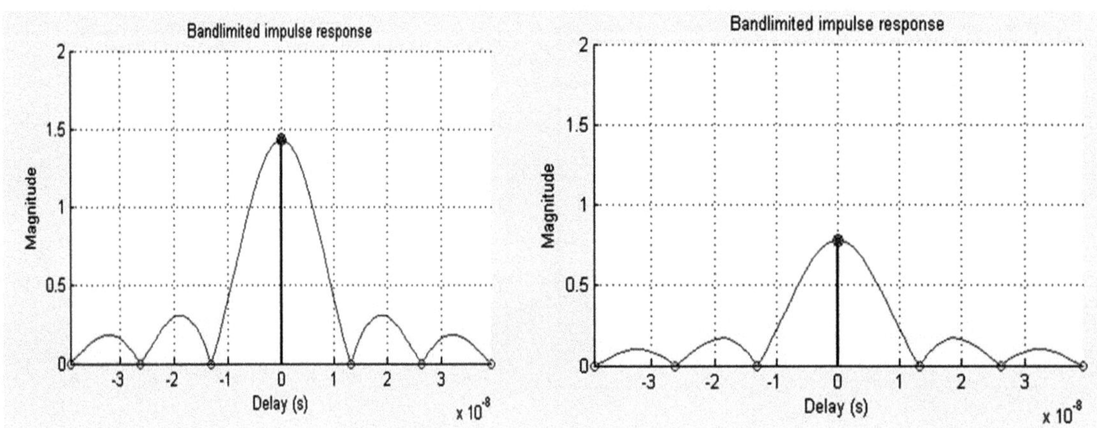

Figure 27. BER curve of frequency selective fading channel for DPSK (left) and the channel impulse response (right)

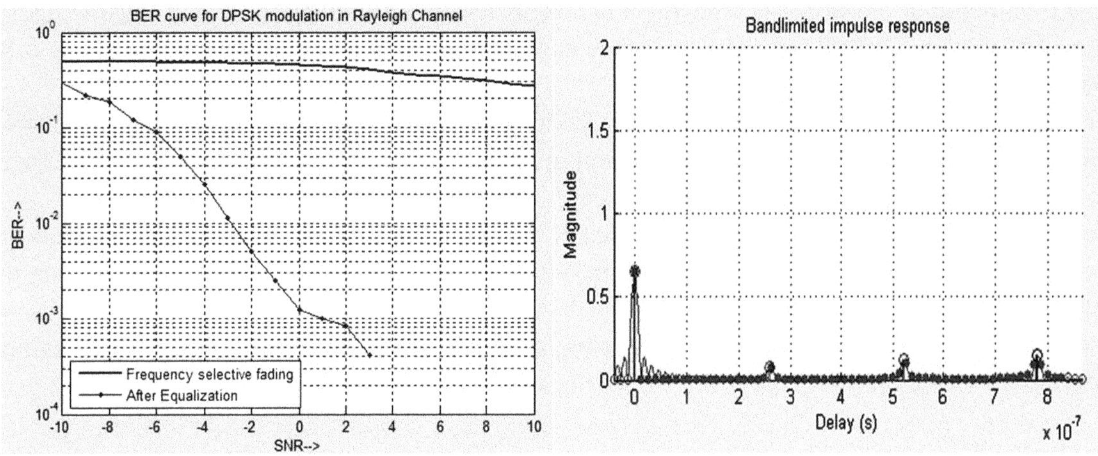

In case of frequency selective channel, the amplitudes, phases, and time delays of any one of the multipath components vary faster than the rate of change of the transmitted signal (Rappaport, 2005).

Slow and Fast Fading

In another experiment a comparative study has been done between logistic map based spreading sequence, LFSR based PN sequence and Kasami sequence under slow and fast fading channels conditions. The BER curves for slow and fast fading channels are shown Figures 28. It is clear from the figures shown below that the logistic map based spreading sequence performs considerably better than both the LFSR based PN sequence and Kasami sequence under both the fading cases.

Also observable from the figure is that the performances of the sequences were marginally better in slow fading case than in fast fading condition, which can be accounted from the dependence of the Doppler shift associated with both the fading cases.

Figure 28. BER curves in slow fading (left) and fast fading channel (right)

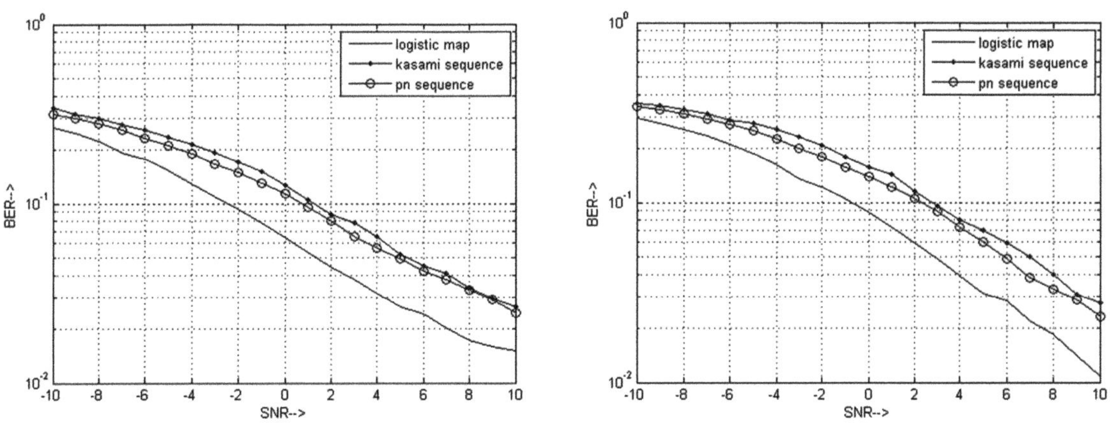

Timing Consideration

In the final phase of the work, a comparison is done between the simulation time of logistic map sequence PN sequence, LFSR based PN sequence and Kasami sequence for slow and fast fading cases, shown in Figure 29.

Here it is observed that for BPSK and DPSK modulation, the logistic map based sequence took minimum time compared to PN sequence and Kasami sequence, whereas for QPSK, logistic map sequence took more time for both slow and fast fading cases.

One possible reason for this may be because PN sequence and Kasami sequence are internally generated by Simulink, but the logistic map sequence is generated using embedded Matlab function then applied as input spreading sequence to the model. As such there is better synchronization between in built models, since BPSK and DPSK use single bits as output but QPSK uses two bits, for representing one symbol. A comparative graph is plotted for the above mentioned cases in Figure 29.

Limitations

Certain computational complexities are observed in the work. Firstly, since the spreading sequence being generated in Matlab, an 'Embedded Matlab Function' block is used to generate the spreading sequence. Sequence indexing has to be used in this block to get all the quantized bits concatenated. Secondly, a lot of additional blocks like 'real to complex' are used to negate port mismatches. Due to these additional blocks simulation process became slower. Generation of the spreading sequence is also an issue because the iterated values from the logistic equation are fractional numbers. As such, binary conversion is difficult because decimal to binary conversion of fractional numbers directly would not have yielded the exact spreading sequence length. So, 3-bit quantization is used to convert the iterated values from the logistic equation into binary.

Figure 29. Simulation times taken by 3 sequences in slow (left) and fast (right) fading channel

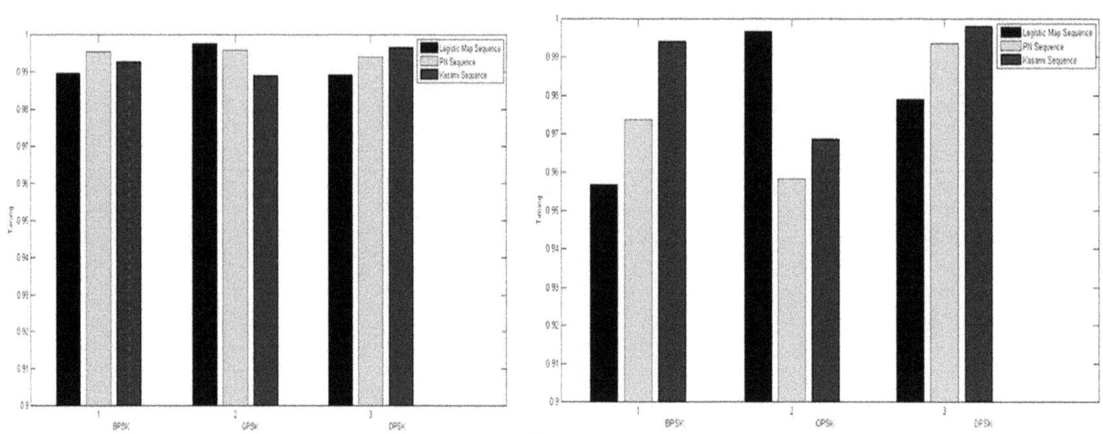

**For a more accurate representation of this figure, please see the electronic version.*

FUTURE RESEARCH DIRECTIONS

Although the results presented here has demonstrated the effectiveness of the proposed methods, there is still a lot of scope for improvement in the logistic map based PN sequence.

- The generated sequence can be developed for practical CDMA scheme where multiple user access can be employed. The generated sequence needs to have good cross correlation property without many changes in the design.
- The BER performance can be made better by improving the RAKE receiver. Diversity techniques have to be used for such improvements in the error performance.
- The performance of the generated sequence in a timing related comparative situation can be enhanced by better system level programming for the generation of the sequence.
- The generated sequence can be used in encryption algorithms for compression techniques or as cipher code.

CONCLUSION

In this chapter, an efficient and simple framework for development of a logistic map based pseudo noise sequence is proposed. In the first contribution, a system for generating a PN sequence using the one dimensional chaotic map the logistic map was developed. The sequence so generated is simple in design and has stringent requirement of digital devices for its generation from a wireless device point of view. In the second contribution, the performance of the generated sequence in a standard DSSS modulation scheme under various channel types(AWGN and Rayleigh) and conditions which are of practical consideration(multipath fading channels) was ascertained. The results so obtained are found to be satisfactory and up to common communication requirement. In the third contribution, a comparison in the performance of the generated sequence with existing spreading sequences like LFSR based PN sequence and Kasami sequence was done. Here too the performance of the logistic map sequence is found to be better than both the LFSR based PN sequence and Kasami sequence.

The other contributions are a comparison with an orthogonal spreading sequence like Hadamard code and logistic map sequence is also done. Here too the logistic map based sequence performed considerably better than the Hadamard code. The timing and synchronization aspects of the sequences are also checked and a comparative study is done on it.

REFERENCES

Ahmad, M., & Farooq, O. (2011). Chaos Based PN Sequence Generator for Cryptographic Applications. In *Proceedings of International Conference on Signal Processing and Communication Technologies*, (pp. 83-86). doi:10.1109/MSPCT.2011.6150443

Bo, W., & Chengji, P. (2012). A New Optimal Design Method of Arbitrary Limited Period Spreading Sequences Based on Logistic Mapping. In *Proceedings of Fourth International Conference on Computational and Information Sciences (ICCIS)*, (pp 1269-1272).

Cruselles, E., Soriano, M., & Melus, J. L. (1995). Uncorrelated PN sequences generator for spreading codes in CDMA systems. In *Proceedings of 6th IEEE International Symposium on Indoor and Mobile Radio Communications*, (vol. 3, pp.1335-1340). doi:10.1109/PIMRC.1995.477380

Dogaru, R. (2010). HCA101: A chaotic map based on cellular automata with binary synchronization properties. In *Proceedings of 8th International Conference on Communications,* (pp 41 – 44).

Elmirghani, J. M. H., & Cryan, R. A. (1995). Point-to-point and multi-user communication based on chaotic sequences. In *Proceedings of IEEE International Conference on Communications,* (pp 582 – 584).

Feltekh, K., & Jemaa, Z. B. J. (2011). Chaos based spread spectrum watermarking algorithm. In *Proceedings of International Conference on Communications, Computing and Control Applications (CCCA),* (pp 1 – 5).

Heidari-Bateni, G., & McGillem, C. D. (1994, March/April). A Chaotic Direct-Sequence Spread-Spectrum Communication System. *IEEE Transactions on Communications*, *42*(21314), 1524–1527. doi:10.1109/TCOMM.1994.582834

Htut, S. M., & Puthusserypady, S. (2003). A novel CDP DS/SS system with 2-dimensional Ikeda map chaotic sequence. In *Proceedings of 14th IEEE conference on Personal, Indoor and Mobile Radio Communications*, (vol. 3, pp 2734-2738). doi:10.1109/PIMRC.2003.1259238

Hwang, S. Y., Park, G. Y., Kim, D. H., & Jhang, K. S. (2010). Efficient Implementation of Pseudorandom Sequence Generator for High Speed Data Communications. *ETRI Journal*, *32*(2), 222–229. doi:10.4218/etrij.10.1409.0047

Jiang, H., & Fu, C. (2008). A Chaos-Based High Quality PN Sequence Generator. In *Proceedings of International Conference on Intelligent Computation Technology and Automation (ICICTA),* (pp 60 – 64).

Khan, A. R. (2012, January). Performance evaluation of DS-CDMA system using MATLAB. *International Journal of Advances in Engineering and Technology*, *2*(1), 269–28.

Mandayam, N. (2002). Wireless Communication Technologies. In *Lecture Notes* (2nd ed.). Rutgers University. Retrieved from http//mathworld.wolfram.com/LogisticMap.html

Martoyo, I., Susanto, A., & Wijanto, E. (2010). Chaos codes vs Orthogonal codes for CDMA. In *Proceedings of IEEE 11th International Symposium on Spread Spectrum Techniques and Applications (ISITA)*, (pp 189 – 193).

Mutagi, R. N. (April 1996). Pseudo-noise Sequences For Engineers. Electronics and Communication Engineering Journal, 79-87.

Rappaport, T. S. (2002). *Wireless Communications* (2nd ed.). Principles and Practice, Prentice- Hall India.

Tang, K. W., & Tang, W. K. S. (2005). A Chaos-based Secure Voice Communication System. In *Proceedings of IEEE International Conference on Industrial Technology*, (pp. 571-576).

Tranter, W. H., Shanmugan, K. S., Rappaport, T. S., & Kosbar, K. L. (2003). *Principles of Communication Systems Simulation with Wireless Applications* (1st ed.). Prentice- Hall.

Xiao, Lu, & Zhang. (2006). Performance Enhancement of DS/SS System Using Hybrid Chaotic Spreading Sequence. In *Proceedings of International Conference on Communication Technology*, (pp 1 - 5, 27-30). doi:10.1109/ICCT.2006.341981

Chapter 9
Sensing and Performance Issues with Reconfigurable Wireless Transceiver

Amrit Mukherjee
K. L. University, India

Sagaika Sahoo
KIIT University, India

Amlan Datta
KIIT University, India

ABSTRACT

With the wide applications of wireless communication, it becomes an important factor to keep the pace of the spectrum usage for all the new applications. To improve efficiently the spectrum usage in variety of applications, where the primary users are not active all the time, we can introduce the secondary users to exploit the radio frequency spectrum by not creating any interference to the Primary Users (PU). Unlike the pre-assigned spectrum allocation policy that has been adopted by various wireless technologies, in Cognitive Radio (CR), the whole spectrum is divided into many bands and corresponding channels where spectrum holes exist. The main challenge is thus to sense these holes and accommodate the secondary unlicensed users. In this chapter, proposing the Quantized data fusion sensing which is one of the type of cooperative sensing scheme used for unused spectrum sensing and acknowledgement to secondary signals. Simulation results with error rates are improved by the activity of PU and have been presented.

INTRODUCTION

The massive increase in Spectrum Access along with enhancing Spectrum efficiency along with the launching of Software Defined transceivers. Also the realization of dynamic machine learning is approached, with which, the transceivers will have the capability to create new areas and possibilities for wireless radio/communication researchers.

DOI: 10.4018/978-1-5225-2342-0.ch009

This chapter aims to investigate the software defined radios' next version cognitive radio, the reconfigurable wireless transceiver and also an adaptive transceiver concepts from different other perspectives (T. Yucek & H. Arslan, 2009). CR and cognitive radio networks (CRN) are studied and analyzed with respect to wireless communication system for its enhancement and simultaneously giving an extra importance on efficient use of spectrum (R.W. Broderson, et al., 2004).

Challenges

Various challenges that are linked with the issues of spectrum sensing for cognitive transceiver system are discussed in the later sections.

REQUIREMENTS FOR HARDWARE

High sampling rate is required for cognitive transceiver spectrum sensing applications, multiple analog front-end circuitry, high resolution of Analog to Digital Converters having large and dynamic range, and fast signal processors. Calculating the interference temperature or noise variance over intended narrowband (NB) transmission signals is an old development. Such calculation of noise variance techniques was accepted for different optimal receiver designs such as soft generation of information, channel estimation, also for the case of improved hand-off, channel allocation techniques and power control. For receivers being tuned to receive/collect the signals those are transmitted over desired bandwidth the noise or interference estimation issue has become easier for these purposes. In addition, receivers are able to process the narrowband (baseband) signals with sensibly low complexity and processors consuming less power. Still, terminals are required in cognitive transceivers while processing transmission for an outspreaded spectrum band to sense any other opportunity for unlicensed users.

Problem of Hidden Primary User

The problem related to the Hidden primary user in Cognitive transceivers is almost similar as like hidden node problem which mainly occurs in Carrier Sense Multiple Accessing (CSMA). These issues are caused because of many factors like shadowing or excessive multipath fading which occurs by Secondary Users (SU) observation when it's scanning the primary users' transmissions. As per the given figure shows below, hidden node problem is illustrated. Here, the unwanted interference is caused by the CR device to the PU (receiver section) as the transmitted signal of primary user will remain undetected due to device positioning in space (J. Mitola, 2000; I. F. Akyildiz, B. F. Lo & R. Balakrishnan, pp. 40-62, 2010).

Figure 1 shows the CR transceiver system having one primary band where two licensed primary users are communicating and the secondary or cognitive users are accessing the same band but in the absence of their signals.

Spread Spectrum Primary Users

The different PU's use different techniques of signaling ex. Frequency hoping or sometimes spread spectrum. In these cases, the primary user signal power is spreaded over a vast spreaded frequency ranges although using a narrower bandwidth for actual information for transmission (J. Unnikrishnan & V. V.

Figure 1. Problem of hidden primary user in cognitive radio transreceiver system

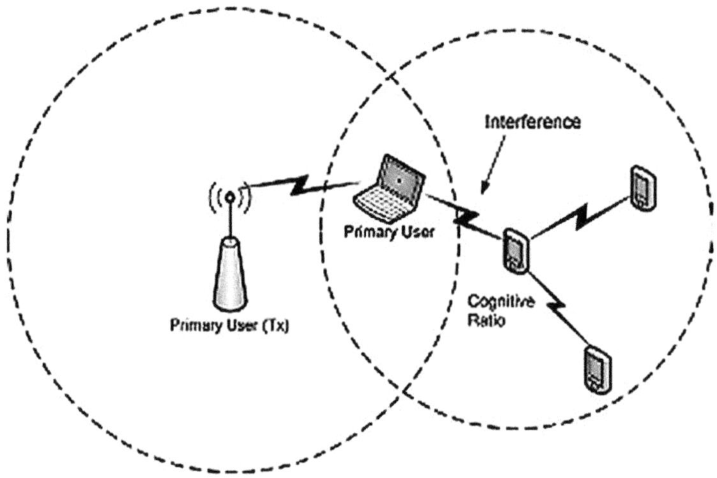

Veeravalli, 2007). Here, the detection of the presence of primary users are very difficult by analyzing their energy levels. Especially, signaling based on frequency hopping spread spectrum creates major problems for spectrum sensing. The sensing problem can be minimized by avoiding the hopping pattern which is predefined and thus making a perfect synchronization with respect to the signal (N. T. Khajavi, S. Sadeghi & S.M.S. Sadough, pp.13-17, 2010).

Sensing Time

The licensed user or the PU are authorized to use their allotted frequency bands anytime even when CR is in operation at that channel (T. Jiang & D. Qu, 2008). Thus to avoid this interference from PU's, CR must be capable of identifying the absence of PU's as fast as possible and should use the channel immediately.

Sensing methods are required to identify the absence or presence of PU within a particular time interval (Xu, G., Lu, Y., He, J. & Hu, N., pp. 777-780, 2008). These particular requirements have maximum and minimum limits for the performance of different sensing algorithms creating a challenge towards CR design.

OTHER CHALLENGES

There are few more challenges, should be considered at the time of designing effective algorithm of spectrum sensing which are implementation complexity, coherence times, the presence of multiple SU, multipath and shadowing, cooperation, competition, robustness, heterogeneous propagation losses and power consumption (Li Y. et al., 2010; T.i Khajavi, N., S. Ivrigh, S.M. S. Sadough, pp. 26002609, 2011). Some important topics are covered here; they are mainly CR applications, SDR and CR architectures, cognitive networks, soft spectrum usage, adaptive wireless system design, measurements as well as awareness of various parameters which includes geo-location information and the interference temperature. Physical layer access technologies, and cross-layer adaptation related concepts are also covered.

Need for Dynamic Spectrum Access

Wireless spectrum is a limited resource and must be spent optimally. Until recently, the electromagnetic radio spectrum has statically been allocated to licensed users by regulatory agencies. This stiff policy makes some frequency bands suffer in scarcity, while huge portions of the entire radio spectrum is left untouched disregarding time and location which adds limitation to the performance of potential users (C. Ghosh, C. Cordeiro, D. P. Agrawal & M. B. Rao, 2009). Therefore opportunistic access to these underutilized spectrum bands seems to be vital in preference to lack of availability of spectrum. As an opposition to this, CR can be seen as a revolution in flexibility, intelligence, and connectivity in the world of wireless communication (Chunli, D., & Yuning, D., W.Li, 2009; Yarkan S., Arslan H., pp. 1563-1567, 2007). One of the prominent aspects of the proposed transceiver system is about the autonomous exploitation of locally unused radio frequency band to facilitate fresh directions to spectrum access.

FREQUENCY BASED CHANNEL SENSING

The principal module of the reconfigurable radio system concept can be explained by factors such as the judicious measurement, learning and most important, to be sensible of the parameters associated with the attributes of wireless channel, operating environment, scope of pre-assigned spectrum and power, potential user requirements and other parallel limitations (Chen, Z., Qiu & R. C., pp.276-282, 2010; J. Ma & Y. Li, pp. 3139-3143, 2007). Proposed techniques can help the dynamicity of the transceiver's behavior to maximize the spectrum utility, even in an optimal licensed band.

The primary user in a cognitive radio system has the soul authority (license) to use the particularly allocated spectrum whereas the secondary users which need to have capabilities resembling that of CR opportunistically exploits the existence of underutilized spectrum such that the entire process does not disturb performance of primary user at all (Zhi Quan, Shuguang Cui, H. Vincent Poor & Ali H. Sayed, pp. 63-70, 2008). On that account, cognitive users must acquire potential for sensing the band of spectrum faithfully, so as to scan the spectrum holes that can be exploited to accommodate themselves. The typical problem rises in crafting of highly reliable spectrum sensing devices and algorithms for transacting spectrum sensing data between nodes (Zhengquan Li, Peng Shi, Wanpei Chen & Yan Yan, 2011).

The actual unutilized spectrum allocated depends on the number of Primary users. CR spectrum sensing is developing day by day and in present days there are limited techniques to detect primary signal (A. Hyvarinen, pp. 626-634, 1999; I. Akbar & W. Tranter, pp. 196-201, 2007). CR networks have important property and unique also that they can shift and also can change their access between ISM and Sensor networks (both are different radio access technologies) because different slots of frequency band arises. To achieve the criteria for channel varying quality, availability of network congestion, interference of channel to signal and the services for the channel requirements the most basic requirement of transmitter is the dynamic spectrum access. The secondary users of CR will also need to coexist with PU, as they are having the right to use spectrum and thus must have a surety not to be interfered by secondary users (S. I. Khajavi, N. S., S.M.S. Sadough, pp. 2600 – 2609, 2011). In fig.2 the spectrum sensing is shown, where main focus is on the spectrum processing in allocated channels. The system with cognitive capability able to interact with the environment by which it is surrounded. Thus proper communication parameters according to specific environment is selected. Separation of observed mixed and combined signals is done whether the PU is present or not (Arslan H & Yarkan, S., pp. 1563 – 1567, 2007). Auto

Figure 2. Spectrum sensing

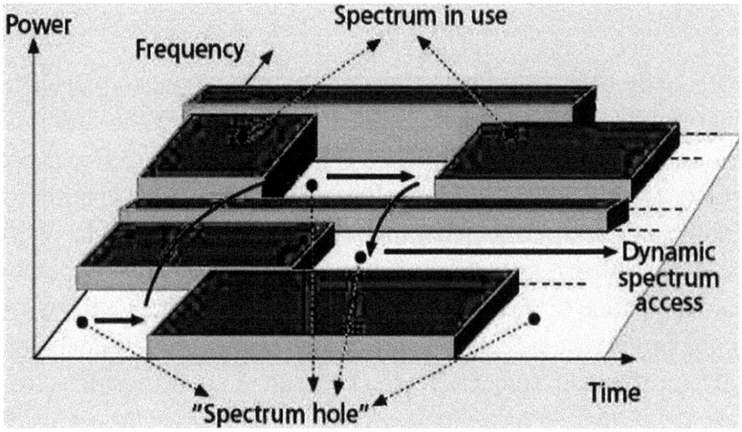

and cross correlation between two signals is the reason of this separation. The main two categories of Energy Detection Techniques are: "Transmitter Detection" and "Receiver Detection". It is assumed that at first the PU is transmitting and then PU is receiving respective to categories. This paper emphasizes on transmitter detection. Spectrum sensing along with PU's activity to assure presence of them in the channel has become the main challenge.

Figure 2 shows the spectrum is being divided into multiple bands or channels based on their frequency utilization. The secondary user is trying to sense the vacate channel and travelling in the vacant white space for its communication. The main assignment for the CR users stands the detection, whether any spectrum hole exists in the whole range under scan, so that available channel can be utilized by other unlicensed user and it is shown in the above Figure 2.

The spectrum sensing techniques so far can broadly be classified into the following types.

- Primary transmitter detection,
- Cooperative detection,
- Interference based detection.

Paying particular attention to the above mentioned schemes, the transmitter detection stands as an essential class which, in general, incorporates three main techniques: ED, detection of matched filter and the last one is cyclostationary feature detection (Qiu R. C. & Chen Z., pp. 276 – 282, 2010; Y. N. Zhang, et al. pp. 2784 – 2788, 2010). Employing simulation tools, these three schemes can be compared on basis of probability of PU detection and probability of false detection. Results say the third detection technique surpasses first two techniques in terms of low SNR and is most suited. However, the short-comings of the technique comprehend computational complexity, significantly long sensing time and essentiality of prior awareness of target signal characteristics.

Cooperative sensing, being advantageous over transmitter detection, boosts the precision as well as reliability of the licensed user sensing. Cooperation improves multipath fading, shadowing effect and receiver uncertainty eventually, mitigating the hidden node problem and minimizing required sensing time. Proposed method suggests the spectrum sensing algorithm which uses energy detection optimization technique. With the advantages of cooperative sensing schemes, the sensing results can be applied through Quantized data fusion algorithm which will improve the BER and SNR of the received signals.

Figure 3 shows the PU signal is $s(t)$ and $n(t)$ is the variance of additive zero mean white Gaussian noise(AWGN). The block diagram of Fig.3 consists of a Threshold detector and an corresponding envelope detector of it. Here it is assumed that the input noise is uncorrelated and also incoherent with the respective signal. If $n(t)$ crosses threshold value then it is considered that the signal is present and it is denoted by V_T. Now the decision hypotheses can be given as:

$s(t) + n(t) > V_T$ *Presence Detection*

$n(t) > V_T$ *False alarm Detection*

SPECTRUM SENSING CHARACTERISTICS AND SPECTRUM SENSING MODEL

The spectrum sensing process and dynamic PU activity are well designed by a finite Markov chain. To predict the nature of the PU, and Auto Regressive (AR) algorithm is been implemented. Least Square (LS) and Minimum Mean Square Error (MMSE) are the attributes of an auto regressive which are being predicted. Based on binary time series, the prediction method predicts the primary user activity. We are using the Hidden Markov Model for the primary user prediction by sensing and predicting the primary user's next sensing frame. These sensing frames are different from those of primary user's data frame. The proposed model works in the way that if the result of the prediction is showing the absence of PU for the upcoming next sensing frame, then any one of the secondary user is allowed to occupy the channel and thus can send the data frame during the sensing which is coming next. With the process taking place, the other secondary user's continuously senses the spectrum and with this the throughput of the cognitive radio network is becoming efficient since these includes the send only part of the data frame send initially.

For an optimum and fast working of spectrum sensing which tries not to interfere the primary users properly and fast white spaces detection which are present in the spectrum available for the proper and intelligent access by the unlicensed users (J. Ma & Y. Li, pp. 3139–3143, 2007; Zhi Quan, Shuguang Cui, H. Vincent Poor & Ali H. Sayed, pp. 63–70, 2008; Mukherjee A., Maiti S. & Datta A., 2014). The architecture is implemented for the cognitive network and primary network. The figure shows the CR network and the primary transmitter is using the same bandwidth available. The base station of the cognitive radio network is acting as the main source which implements the cooperative sensing of the spectrum.

As a binary hypothesis testing, the sensing is performed to ensure if any presence of the primary transmitter for a particular band of frequency and denoting them by:

Figure 3. Flow chart of the secondary receiver

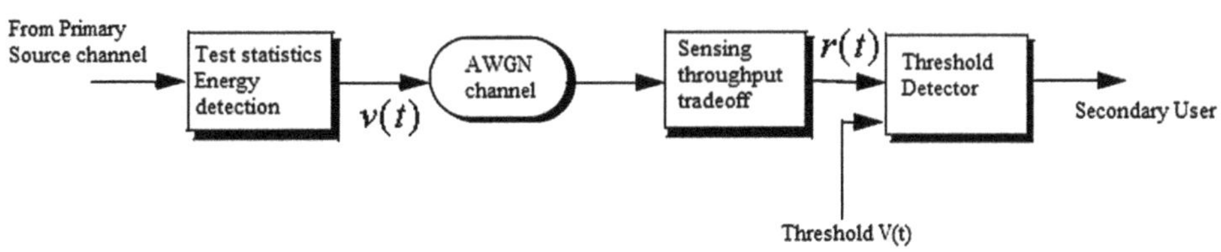

H_0: The idle primary user.
H_1: The working primary user.

The working primary user must be different with that of the idle primary user and thus from CR decision performed by spectrum sensing part (A. Hyvarinen, pp. 626–634, 1999; Teguig, Scheers & Le Nir V., 2012). Here in this case, the presence of primary signal has been given by H^{PN}_i where i=0 is the absence and i=1 shows the presence. Given *CN i H*, which is sensing the presence or absence of the primary signal and making the decision on the basis of the hypothesis. The following model structures are used which defines the below conditional probabilities:

$$P_m = P\left(H_0^{CN} \mid H_1^{PN}\right) \quad (1)$$

and

$$P_m = P\left(H_1^{CN} \mid H_0^{PN}\right) \quad (2)$$

Taking any one secondary user is then selected and allowed to send its data frame it is having when the primary user is not present and parallel the other SU acts as distributed sensors. The signal which is then received at a particular instance say j[th] is given by:

$$y_j = h_{j;1}a^{PN} + h_{j;2}a^{CN} + n_j \quad (3)$$

where,

$$a^{PN} = \left[a_1^{PN} a_2^{PN} a_3^{PN} \ldots . a_L^{PN}\right] \quad (4)$$

$$c^{AN} = \left[c_2^{PN} c_3^{PN} \ldots . c_L^{PN}\right] \quad (5)$$

$$n_j = \left[n_{j,1} n_{j,2} n_{j,3} \ldots n_{j,L}\right]$$

Here as mentioned from the above equations, primary user and secondary user channel is given by $h_{j;1}$, and channel between j-th and any other secondary user is given by $h_{j;2}$ considering Rayleigh distribution (A. Mukherjee et. al., 2016). In this case the time sensing frame takes less time than the data frame of primary user. Here the zero mean n_j is the circularly symmetric complex Gaussian (ZMCSCG) noise and distribution as $n_j \sim CN(0;\sigma_2^{nj} I_L)$. Here I_L is the L X L identity matrix. The vector a^{PN} contains the primary user symbol to be transmitted in a primary network and a^{cN} contains active symbols from any of the transmitted symbols.

Several CR sensing frames are being sensed by spectrum sensing units and the output of the sequence contains the observation which is essential by the algorithm which is predicting the next primary user state. In the Figure 4 we can see the structure of the frames which are inside the cognitive radio sensing where n-1 are the total sensing frames which can be observed and formed for prediction of the primary user state in the n-th sensing frame there. The presence of PU is indicated by the predictor, the active SU becomes idle and inactive in the n-th frame of sensing. It also indicates the absence of primary user, then the active secondary user again comes in active condition in the channel in n-th sensing frame in the structure.

Figure 4 shows that previous n-1 sensing frames are observed by the predictor unit so it can predict the next n^{th} sensing frames. The CR network is sensing all the users so as to collect as the base station of CR network. Again, on the observations based on the observations, the process of spectrum sensing takes place in the sensing period of the n-1 frames.

Practical Challenges to Sensing

The lack of radio resources being acute in the contemporary society, CR provides a great solution theoretically. Yet, CR still encounters unpreventable complications while imitating the theoretical results into practical application.

These challenges made spectrum sensing as a very active area of research for the recent years although it has long history. A very few sensing methods are proposed, those include the ED, classic likelihood ratio test (LRT), matched filtering (MF) detection, cyclostationary detection (CSD), and some recently emerging methods for example wavelet-based sensing, Eigen value-based sensing, covariance-based sensing, and blindly combined energy detection popularly known as Blind Source Separation (BSS). These techniques have different requirements for implementation and in accordance to this, it can be classified into three general categories given by:

1. Techniques which require both source signal as well information of noise power,
2. Techniques which need only noise power information is known as semi-blind detection where
3. Techniques requiring least information or no information of source signal/noise power, termed as (totally blind detection).

Figure 4. Sensing structure frame for CR

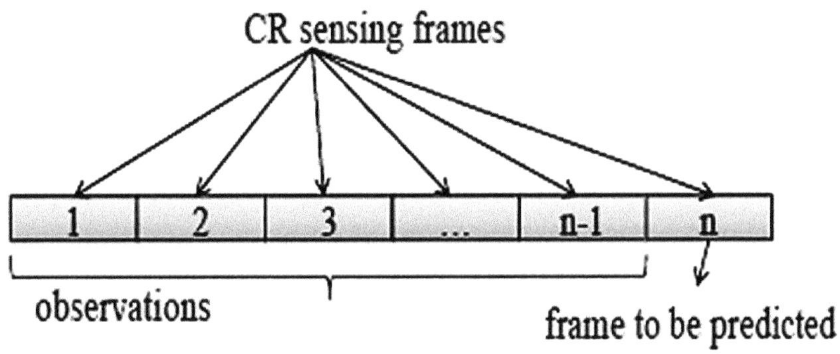

As instance, LRT, MF, and CSD belong to the first category; where the second category includes ED and wavelet-based sensing methods; Eigen value-based sensing, blindly combined energy detection (BSS) and covariance-based sensing belong to category C. Here in this chapter, main focus is on the methods of second and third categories, though in first category some of other methods are discussed to make it complete. Multi transceiver systems was broadly deployed to gain the high channel capacity and/or improve the reliability of transmission in case of wireless communications and in addition to this, for the enhancement of the performance of direction, range and velocity estimations formation of array radar or MIMO radar is required and for this purpose multiple transceivers are more often used. As a result, MIMO techniques also can be implemented for the improvement of the spectrum sensing performance. Hence, in this chapter, in general a multi transceiver system model is assumed, while the single-transceiver system is considered as special case.

When multiple secondary receivers are distributed at separate locations then there is a possibility that they can cooperate to achieve improved sensing reliability. In the present scenario there are several sensing cooperation schemes. These schemes generally are of two categories:

1. Data fusion, where raw/unprocessed data or processed data of each user are sent by its own to a particular user destination and it processes the collected data finally to take the decision.
2. Decision fusion, where data of multiple users are processed independently and then the individually obtained decisions send by them to a predefined user destination and finally the decision is taken.

Where CR needs highly dynamic sensing capabilities for identification of spectrum holes to allow secondary users to access the same, which is desired. However, with energy detection, the cognitive radio user stands incapable of distinguishing the transmission of PU and SU separately, it can only sense presence of transmission leading to false alarm detection and thus reduction of spectrum opportunities. Cyclostationary feature detection is superior to transmitter detection when ability of noise rejection is considered, yet it adds complexity in terms of design aspect. Although cooperative sensing improves detection accuracy, certain issues like packet retransmission, optimization, higher latency in accumulating detection information of the same and network traffic still remains a challenge.

Figure 5 shows the graph of probability of false alarm versus threshold. As the level of threshold increase, the values of probability of false alarm decreases. The theoretical and practical values are simulated considering the Rayleigh distribution of the channels. One of the most important aspect that any communication system must be vigilant over is Information security, which in case of CR, is risked by various factors like Dynamic spectrum access and tampering of spectrum sensing data, PU emulation, interfering PUs and similar considerations.

Proposed Methods for the Spectrum Sensing Challenges and Allocation

As mentioned above, the challenges are mainly related to the detection and sensing of the spectrum and implementation of secondary users efficiently. Considering the BSS algorithm which is applied to the channels by considering the unknown nature of the signal and predicting the output based on HMM (Hidden Markov Model) which calculates the presence of holes.

Assuming that the primary signal statistics are known, in the presence of stationary Gaussian noise the optimal detector acting as a matched filter method sensing which then compares it with threshold. For the primitive type CR, this coherent type of detection may be a viable but implementing the second-

Figure 5. Probability of false alarm versus threshold (theoretical and practical)

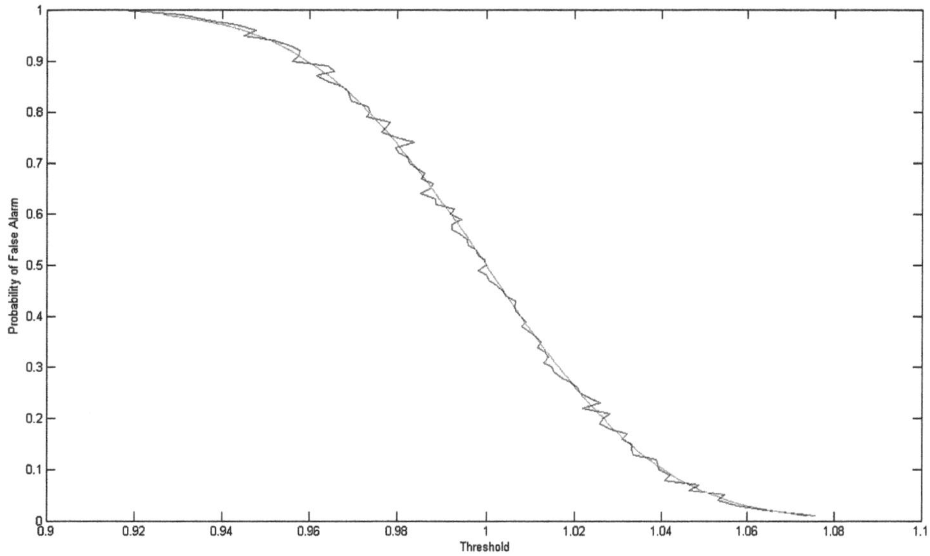

ary users should be restricted to only a limited number of channels. As the number of primary channels allowed for open access, the complexity and the cost effectiveness with respect to the system started increasing. Here it concludes the necessity of a dedicated system to obtain balance in each available licensed user (PU) for implementing coherent detection.

In the ED technique helps in calculation of the present energy in the PU channel for a particular instant and spectrum holes are present if the level of that energy is less than that of threshold. In case of matched filter as discussed above, it takes the processing time more for sensing operation fin a particular interval but the implementation cost and simplicity in device circuitry makes it a good option for spectrum sensing in CR. The disadvantage associated with the energy sensing method is all about the differentiating the primary signal and noise of received source and thus the probability of changes in noise power in background for low SNR condition occurs. AS the spectral correlation of different PU''s are different features which permits CR for specific PU signal to get detected in presence of interference and noise.

As per the application point of view, different combinations of the sensing techniques are required for different requirements. For example, in case of quick and coarse scan of broad frequency range, the sensing method can of ED type. The present white spaces can further be discovered accurately using feature detection method. In spite of having underlying and overlying concepts, sensing performance can be increased scanning the available channel for along interval and thus increase in gain due to signal processing. One should take into account the regulatory constraints that might have been imposed for a particular sensing time.

In case of Hard decision fusion scheme, where each user decides the presence or absence of the PU and the data fusion center receives a decision of one bit. The requirement of limited bandwidth made this method easy and this is the main advantage. If it is reported that the binary decisions are in the common node then three decision taking rules are there, the "and", "or" and another one is majority rule. The individual statistics Δk, quantized to one bit where $\Delta k=0, 1$; is assumed as the kth CR user's hard decision. Where 1 and 0 represents presence and absence of signal respectively. The AND rule decides that a signal is present if all users have detected a signal. In the other hand if a signal is detected by any one of the users then the presence of signal is decided by The OR rule. Last rule is the voting rule which determines the presence of signal if at least a signal is detected by M of the K users with

$$K = M = 1 \tag{6}$$

In case of soft data fusion, CR users sensing result E_k is forwarded to their center fusion. In this case there is no performance of any local decisions in their respective nodes. At Fusion Center (FC) the combined decision is being made with the help of different combining rules which includes Selection Combining, Maximal ratio combining and square law combining techniques. Comparing the performance of these two above fusion techniques, soft combination yields a better response although it requires larger bandwidth for working of its control channel. Also more overheads are being generated in soft fusion as compared with hard fusion. In this Scenario the Quantized data fusion holds great advantage. Details of the Quantized data fusion are described later in this chapter.

The Hidden Markov model (HMM) (Simmons) is one of the Markov model which is having a hidden i.e the state which is unobservable. This model is a five tuple < S,A,O,T,Z,p >, in which S, A, and T with the Markov model are same and p is given as the distribution for initial state. Also O (a finite set), Z denotes the observation function given by: O×S×A→R, and Z(o, s, a) =p(o| s, a), that is given as the probability of O (receiving observation) when the system is ended in S state which will give action a after execution. In various applications, Z remains unchanged for all the values of a Z(o, si, aj)=Z(o, si, ak)). Implementing Z(o, s) as the substitute for the Z(o, s, a), given that Z remains constant for each "a". Considering the Markov model, while predicting the state of actions, the state of the system remains uncertain although the actions are executed. Thus with this drawback, we have selected HMM where the observations and the transitions can be used to infer the distribution of the next state. In the Markov models, equations given are to predict state distribution when the time t+1 occurs and providing state distribution for a particular time t. Here the action is denoted by a^{t+1}. The quantization can be of any type which includes vector and scalar quantization. Here we can process many number of sub-frequency bands simultaneously using HMM approach. Proposed model can be implemented to almost all type of wideband applications.

Figure 6 shows, the actual states, unknown states and the known value from observations are marked. As it can be seen in the figure below where a sub-frequency band an observation value is obtained at the end of phase of spectrum sensing in each slot which is o^t as per our proposed model. Certain channel is associated in selected and each slot for time for a sub frequency band along with channel is denoted active or passive. Practically it is very hard to find the state of the channel. Pin the proposed model and

Figure 6. Illustration of channel state prediction

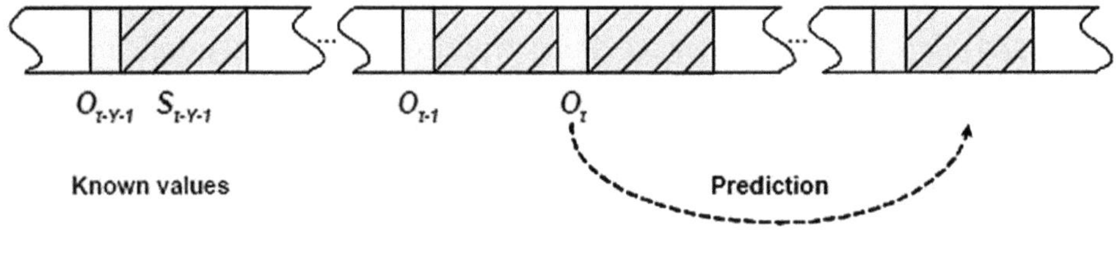

it is assumed that the states of the channel are known and fixed by such verifications. The maximal mean of verification delay denoted in time slots as "Y". Channel state are predicted by using known values after observations and the states of original channel for finding and predicting the future states of channel. Here in this case, HMM is improved from its standard given form.

The following are several spectrums sensing algorithm which may be used in cognitive transceiver system:

1. Blind Source Separation (BSS) model,
2. Hidden Markov Model (HMM) technique,
3. Quantized Data Fusion model.

Blind Source Separation

There are different spectrum sensing techniques based on energy detection, likelihood ratio test, cyclostationary detection, matched filter method and many more. Requirements of these methods are different with additive advantages and disadvantages simultaneously. Especially when the above mentioned techniques are implemented in case of cooperative spectrum sensing, the complexity increases. In case of LRT when being applied, the presence of signal and the noise distribution across the channel is required which is tedious job to find in practice. Matched filter based method seems to require precise knowledge for the corresponding channel responses from different primary users towards their respective receivers. Synchronization to a large extent is necessary aid in case of MF based sensing method which is again a pitfall. These sensitive parameters are to be taken care of in proper manner or else, the performance may be reduced expectedly low. At crucial point of instance, it might not be applicable if there exists differences in cooperation between the primary and secondary users in the network. In CSD method, cyclic frequencies of the primary signal are needed to be known, which, for many applications, may not be realistic. Moreover, high computing complexity defects the method. ED, unlike the three other techniques, never requires any knowledge about the source signal. It is considered to be robust to dispersed unknown channel and fading. To detect independent and identically distributed (IID) signals this is optimal, but for detecting correlated signals it is not optimal. Again ED has to estimate power of noise to implement statistic decision. Practically in the sensing of spectrum the uncertainty of noise limits ED. So, proposed blind spectrum sensing algorithm is here to overcome the limitation of ED. BSS algorithm is implemented for spectrum sensing. There is no requirement of any information related to the source signal or the channel. Thus, for cooperative spectrum sensing it becomes suitable. Also, for correlated signals its performance is well and the effect of noise uncertainty is overcome. Further, for the improvement of sensing reliability, cooperative spectrum sensing can be exploited and it is considered as the advantage of blind spectrum sensing. The system equations mentioned in system model can be shown by this matrix:

$$Y = H \cdot A + N \tag{7}$$

where, our observation denoted by Y relates with one secondary user containing L different symbols. In case of blind source separation method, the observation matrix Y values are known and then calculating H for the estimation of A matrix as an independent source. Calculating H can be done by different other methods ex. TALS, Fast ICA, COMFAC MUK. Here the BSS uses fast ICS method for recovering the original signal. These original signals are then use the Kurtosis metric which is used to find attributes of the separated signals. Recently, learning and predictions methods are adopted by some spectrum sensing techniques. In these methods there is no randomness in PU transmission activity but some particular patterns and correlations is followed. More precisely, the behavior of the PU is like a real and physical user, and not like a jammer. So, now an argument may come about the PU activity that it is predictable. In the literature there are many proposed different algorithms to predict the future state of a primary user which is on the basis of training sequences and the observed values. To predict the activity of PU an Auto Regressive (AR) model is used. These parameters are then calculated by LS and MMSE methods. Based on HMM some algorithms are there which are proposed in the next section for predicting the primary user's activity.

The optimization is performed by calculating the best value of **W**. The overall cost function is maximized by this value. Since Kurtosis of the Gaussian noise is always zero Kurtosis matrix is used to measure the non-Gaussianity. As given below the cost function:

$$F(G) = \sum_{j=1}^{2} \left|K(Z_j)\right| = \left|K(Z_1)\right| + \left|K(Z_2)\right| \tag{8}$$

Increasing the value of $F(G)$, every term is individually increased and comparing the absolute value with the value of threshold, if the value is greater than threshold, it shows the presence of an independent signal. If the transmitter of the transceiver is operation:

1. For any random two independent components with their absolute values more than the threshold implies the primary user transmitter is in operation.
2. For any single independent component with single absolute is having more value as compared with the other, which concludes PU transmitter in non operational.

If transmitter side of the transceiver is not in operation:

1. For any single independent component having only a single of the absolutes is having more value compared with the other, implies primary user transmitter is in operation.
2. If for no independent component where both the absolute values are small as compared with the threshold, implies primary user transmitter is not in a mode of operation.

The above rule is also applicable for sensing the channel in Blind source detection.

Hidden Markov Method

Basically a Hidden Markov method (HMM) refers to a doubly type and with stochastic process where the symbol observations when generated relies on nature of the states. Thus, known states generate unknown value of observation although both are not linked directly.

An HMM can be illustrated by denoting as a pair $\{X, Y ; t \in N\}$. This stochastic processes pair is explained on the set of probability space with components (Ω, F, P). Here X_t refers to the hidden state sequence along with Y_t which denotes the observation sequence of each of the components. Here the stationary finite state system (SFSS) is assumed to be finite by fulfilling the criteria as:

$$(X_t, Y_t) \text{ are jointly stationary} \tag{9}$$

$$\Pr\left(Y_{t+1} = y_{t+1}, X_{t+1} \mid Y_1^t = y_1^t, X_1^t = x_1^t\right) = \Pr\left(Y_{t+1} = y_{t+1}, X_{t+1} = x_{t+1} \mid X_t = x_t\right) \tag{10}$$

With the above equations X_s and Y_t here are referred to as the state and the output of stationary finite state model respectively. In this case, the mathematical model which can generate this type Hidden Markov Process is known as HMM. An HMM can be stated as a finite state machine where sequence of observation is a probabilistic function of state.

Application of HMM is specifically been used to predict the channel usage by the PUs and corresponding behavior of the users in frequency band identified. The whole discussion lets into the conclusions whether the SU needs to alter to another band or remain in the same. Channel state predictor can be utilized using HMM based pattern recognition. In the proposed chapter, overall gist to an approach regarding BSS and implemented modified HMM through it. As shown below the modified HMM can be represented by:

$$a_{i,j} = \Pr\left(q_{t+(x+y)} = \theta_j \mid q_t = \theta_i\right) \tag{11}$$

$$i, j = 1, \ldots., N$$

$$b_{j,k} = \Pr\left(o_t = v_k \mid q_t = \theta_j\right) \tag{12}$$

$$j = 1, \ldots\ldots, N$$
$$k = 1, \ldots\ldots M$$

As shown above, X is pre-assumed as the period of prediction that works with the maximal response delay possibility with respect to time slots. Other parameters which are associated with HMM (modified) '{π,A,B}', are statistically calculated. The below equations shows extraction of parameters of HMM (modified) by implementing the training sequence:

$$\pi_i = \left\{1 / (L - X)\right\} \sum_{t=1}^{(L-X)} Eval(q_t = \theta_i) \tag{13}$$

$$i = 1, \ldots\ldots, N$$

$$a_{i,j} = \left(\sum_{t=1+Y}^{L-X} Eval\left(q_{(t+X)} = \theta_j \mid q_{(t-Y)} = \theta_i\right)\right) / \left(\sum_{t=1+Y}^{L-X} Eval\left(q_{(t+X)} = \theta_j\right)\right) \quad (14)$$

$$i, j = 1, \ldots.., N$$

$$b_{j,k} = \left(\sum_{(t=1)}^{(L-X)} Eval\left(o_t = v_k \mid q_{(t+X)} = \theta_j\right)\right) / \left(\sum_{(t=1)}^{(L-X)} Eval\left(q_{(t+X)} = \theta_j\right)\right) \quad (15)$$

$$j = 1, \ldots\ldots, N$$
$$k = 1, \ldots\ldots M$$

Here, length of the above training sequence is denoted by L which is presented in the time slots, and

$$Eval\left(\bullet\right) = \begin{cases} 1(\bullet) & established \\ 0 & otherwise \end{cases} \quad (16)$$

Now implementing this with one step ahead as per the trained parameters '{π,a,b}' which uses two methods they are "π b" which uses π and 'b' and another method is "ab" which uses 'a' and 'b' respectively for prediction, the channel state X-Slot can be predicted. Following equations define these with T = 1

1. Initialization:

$$\delta_1\left(i\right) = \pi_i b_i\left(o_1\right) \quad (17)$$

$$i = 1, \ldots.., N$$

2. Iteration:

$$\delta_t\left(j\right) = \max_{1 \le i \le N} \left[\delta_{t-1}\left(i\right) a_{ij} b_j\left(o_t\right)\right] \quad (18)$$

$$j = 1, \ldots.., N$$
$$t = 2, \ldots.., T$$

3. Termination:

$$P^* = \max_{1 \le i \le N} \left[\delta_T \left(i \right) \right] \tag{19}$$

$$q_T^* = \arg \max_{1 \le i \le N} \left[\delta_T \left(i \right) \right] \tag{20}$$

The method, termed as "ab" which uses 'a' and 'b' to predict, which is defined earlier with T = 1.

$$\delta_1 \left(i \right) = a_{ji} b_i \left(o_1 \right) \tag{21}$$

$$q_{1-Y} = \theta_j$$
$$i = 1, \ldots, N$$

If we use one of above equations, the channel state which is predicted q*T and their respective likelihood probabilities, indicated by P* can be calculated. Now, using this, the performance can be predicted by using two different metrics precisely probability of detection (PD) and probability of false alarm (PFA). The prediction of probability of detection indicates the rate at which the prediction model estimates the state of the channel precisely when the channel state is "busy" or the actual channel state is "idle", PFA describes the rate of failing to predict the state of the channel correctly. The combined lower PFA and higher PD is prescribed for better prediction.

Taking the two above methods discussed "πB" and "AB" as the proposed prediction model into account, now, for Y=0, in the second method "AB" which can probably perform have a high prediction in the multi band which is measured spectrum applications. In few of special cases, the method "AB"'s performance may degrade where Y>0. The first method "πB" turns out to be a little less that is independent of "Y" and so it is better to use "πB" method as for Y>0, the HMM based prediction model is concerned.

Quantized Fusion Method

In this scheme, a tradeoff has been tried to be realized between the overhead and the detection performance of the network. Instead of using one bit hard combining, where only one threshold is appointed for dividing the whole range of detected energy into two regional sections, a better performance of detection can be obtained when the number of thresholds is increased in order to get more clear regions of observed energy. A two-bit hard combining scheme can be implemented to divide the whole range of the energy that has been detected into four distinct regions. At the fusion center, the presence of the signal of interest is

$$\sum_{i=0}^{4} w_i n_i \geq L \tag{22}$$

Here L is taken as the threshold which equals the upper region weight, whereas the number of different ascertained energies are denoted by n_j, which is in region i. The weight of the corresponding regions of energy is represented by "w_i" with: w_0=0, w_1=1, w_2= 2 and w_3= 4 respectively.

In the next part we will be discussing the different Data fusion techniques that are to be applied in cooperative sensing of CRN corresponding to 4-bit fusion combining technique. In three-bit scheme, seven threshold λ_1, λ_2... and λ_7, divides the complete range of the statistic into 8 different regions. But in four-bit scheme, fifteen distinct thresholds denoted as λ_1, λ_2... and λ_{15}, divides the range into 16 different regions. Each user in the CR network forwards 4 bit of information to identify the region of the intended and observed energy (A. Mukherjee & A. Datta, pp. 85-87, 2016; S. Sahoo, et. al., pp. 179-187, 2015). Threshold having upper regions are considered for higher energies through the nodes and the value from the individual nodes are forwarded and compared with nodes having less energy corresponding to the lower threshold regions.

The four-bit fusion combining technique is calculated as below:

1. A quantization threshold λ_i (i=1 . . . 15) is defined for each quantized region according to the maximum received energy of the signal being examined.
2. A local decision is taken by each individual user by making comparison between the received energy and the thresholds predefined, and sends 4-bits information to the Fusion Centre (FC).
3. The FC then sums the obtained local decisions calculated with weights w_i (i=0.., 15).
4. The final decision is then made by comparing the total sum with a threshold L such as:

$$\sum_{k=1}^{K} w_{i,k} \geq L \tag{23}$$

Square law combination (SLC) can be one of the simplest linear soft combining schemes. In SLC, the fusion center is fed with the estimated energy in each node to be added together (A. Mukherjee, et. al, 2015). This summation is then compared with a fixed threshold to decide on the existence or absence of the PU and a decision statistic is given by:

$$E_{slc} = \sum_{k=1}^{K} E_k \tag{24}$$

where, E_k denotes the statistic from the kth CR user.

Figure 7 shows the four bit hard combination, where the numbers of thresholds are 1- the number of regions. The detection probability and false alarm probability are formulated as follow:

$$Q_{d,SLC} = Q_{mK}\left(\sqrt{2\gamma_{slc}}, \sqrt{\lambda}\right) \tag{25}$$

Figure 7. Four-bit hard combination scheme

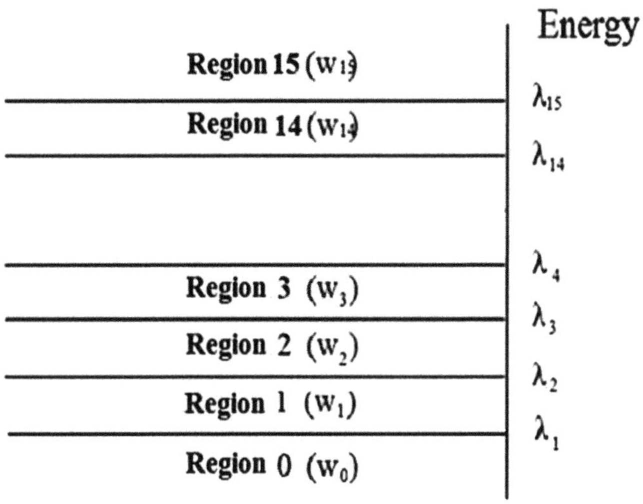

$$Q_{d,SLC} = \frac{\Gamma\left(mK, \frac{\lambda}{2}\right)}{\Gamma\left(mK\right)} \tag{26}$$

$$\gamma_{slc} = \sum_{k=1}^{K} \gamma_k \tag{27}$$

Here the received SNR is γ_k at k[th] CR user.

Diverse measurements are utilized to settle on a choice like correlation metric, negentropy metric. They are acclimated to measure the non gaussianity of the separated signal. The non-gaussianity measured is then compared with the summed threshold so that the decision regarding the primary signal in the frame is within the sensing period.

Figure 8 shows the theoretical comparison of the probability of Detection versus probability of false alarm.

Figure 9 shows the graph between probability of detection and probability of false alarm the simulation result (practical) in smoother as compared with the practical case where we have considered noise and interference. The above mentioned method does not any kind of synchronization with the primary signal. The BER (Bit Error Rate) and SNR (Signal to Noise Ratio) is further improved by quantized data fusion scheme. In this case, we divide the threshold level into 16 levels and allotting corresponding fixed weights to the channel as shown below.

Figure 8. Graph of probability of detection vs. probability of false alarm (theoretical)

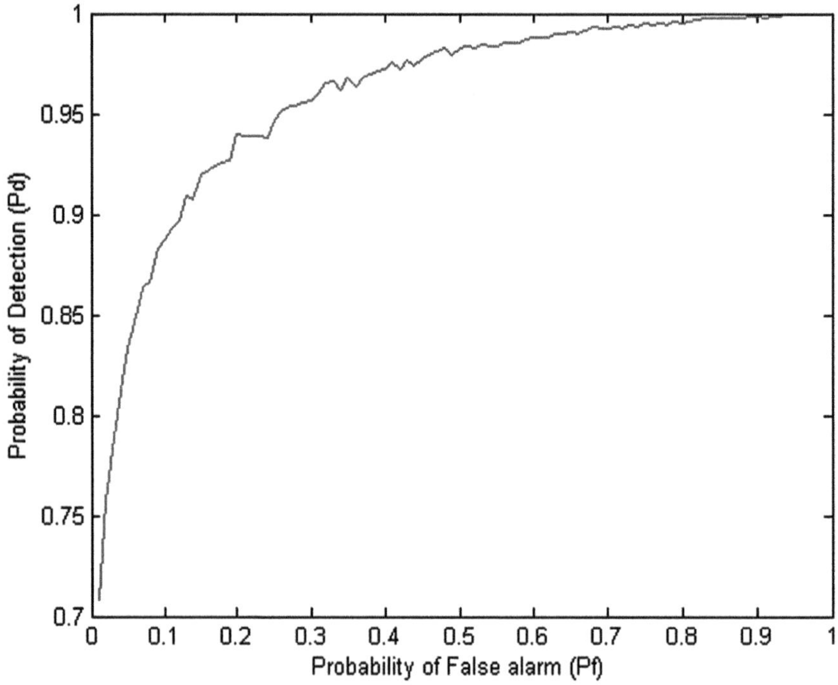

Figure 9. Graph of probability of detection vs. probability of false alarm (practical)

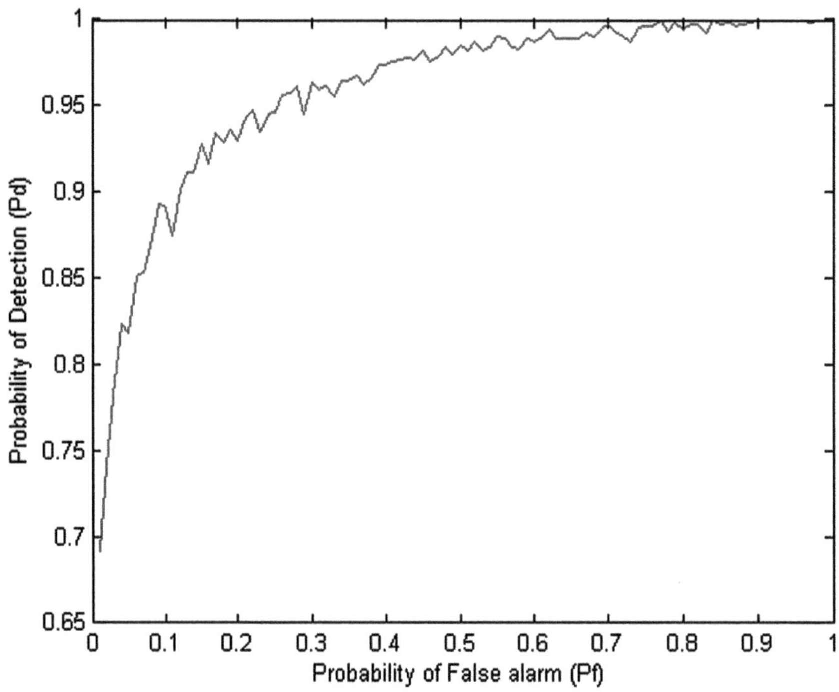

CONCLUSION

The proposed method illustrates clearly over the demerits of the primitive spectrum sensing method and the optimization where the spectrum was not being used efficiently. Also the primary user's state was not known which was then improved using HMM. Depending on the requirements and sensing environment considering secondary user's location (fixed or variable) the choosing of algorithm depends. Some of the disadvantage that are associated with BSS (Blind detection) that includes beam forming demerit, that is it can only distinguish signals that have well separated angle of arrivals. Thus this disadvantage is taken care of by HMM and quantized hard fusion scheme. The HMM can generate alignments where every known state corresponding to a particular observed value. Here the same state can be used repeatedly in a given sequence but the observed value can be once in that sequence. The sensing performance of the Transceiver and their spectrum sensing errors on the performance corresponding to average BER is improved while performing multi node decision stream of two different channel using optimization of energy based sensing scheme as shown below.

Though being modified the conventional spectrum we have seen there are still some disadvantages. Thus both of these disadvantages is taken care of by HMM and quantized hard fusion scheme. BER is improved while performing multi node decision stream of two different channel using optimization of energy based sensing scheme as shown below:

Though there is a tradeoff between overhead and detection performance the proposed quantized four-bit combination scheme gains the advantage of the soft and the hard decisions schemes. Whereas, in practical application, according to the requirement of detection performance a perfect method of data fusion and decision algorithms can be selected and also the requirement of the available bandwidth for the reporting channel.

REFERENCES

Akbar, I., & Tranter, W. (2007). Dynamic spectrum allocation in cognitive radio using hidden Markov models: Poisson distributed case. *Proceedings of IEEE Southeast Conference*, 196–201. doi:10.1109/SECON.2007.342884

Akyildiz, I. F., Lo, B. F., & Balakrishnan, R. (2010, March). Cooperative Spectrum Sensing in Cognitive Radio Networks: A Survey. *Physical Communication*, *4*(1), 40–62. doi:10.1016/j.phycom.2010.12.003

Arslan, H., & Yarkan, S. (2007). Binary time series approach to spectrum prediction for cognitive radio. *IEEE 66th Vehicular Technology Conference,* 1563 – 1567.

Broderson, Wolisz, Cabric, Mishra, & Willkomm. (2004). *White paper: Corvus: A cognitive radio approach for usage of virtual unlicensed spectrum.* Academic Press.

Chen, Z., & Qiu, R. C. (2010). Prediction of channel state for cognitive radio using higher-order hidden Markov model. *Proceedings of the IEEE Southeast Conference Concord*, 276 – 282. doi:10.1109/SECON.2010.5453870

Chunli, D., Yuning, D., & Li, W. (2009). Autoregressive channel prediction model for cognitive radio. *5th International Conference on Wireless Communications, Networking and Mobile Computing*, 1 – 4.

Ghosh, C., Cordeiro, C., Agrawal, D. P., & Rao, M. B. (2009). Markov chain existence and hidden Markov models in spectrum sensing. *Proceedings of IEEE International Conference on Pervasive Computing and Communications*. doi:10.1109/PERCOM.2009.4912868

Hyvarinen, A. (1999). Fast and robust fixed-point algorithms for independent component analysis. *IEEE Transactions on Neural Networks*, *10*(3), 626–634. doi:10.1109/72.761722 PMID:18252563

Hyvarinen, A. (1999). Fast and robust fixed-point algorithms for independent component analysis. *IEEE Transactions on Neural Networks*, *10*(3), 626–634. doi:10.1109/72.761722 PMID:18252563

Ivrigh, S. (2011). A novel framework for spectrum sensing in cognitive radio networks. *IEICE Transactions on Communications*, *E9-B*(9), 2600–2609.

Jiang, T., & Qu, D. (2008). On minimum sensing error with spectrum sensing using counting rule in cognitive radio networks. *Proceedings on 4th Annual International Conference on Wireless Internet*, 1–9. doi:10.4108/ICST.WICON2008.4954

Joseph Mitola, I. I. I. (n.d.). *An Integrated Agent Architecture for Software Defined Radio* (Dissertation). KTH/IT/AVH—00/01—SE.

Khajavi, N. T., Sadeghi, S., & Sadough, S. M.-S. (2010). An improved blind spectrum sensing technique for cognitive radio systems. *5th International Symposium on Telecommunications*, 13-17.

Li, Shi, Chen, & Yan. (2011). Square Law Combining Double threshold Energy Detection in Nakagami Channel. *International Journal of Digital Content Technology and its Application*, *5*(12).

Li, Y., Dong, Y. N., Zhang, H., Zhao, H. T., Shi, H. X., & Zhao, X. X. (2010). Spectrum usage prediction based on high-order markov model for cognitive radio networks. *IEEE 10th International Conference on Computer and Information Technology*, 2784 – 2788.

Ma, J., & Li, Y. (2007). Soft combination and detection for cooperative spectrum sensing in cognitive radio networks. *Proceeding of IEEE Global Telecommunication Conference*, 3139–3143. doi:10.1109/GLOCOM.2007.594

Ma, J., & Li, Y. (2007). Soft combination and detection for cooperative spectrum sensing in cognitive radio networks. *Proceedings of IEEE Global Telecommunication Conference*, 3139–3143. doi:10.1109/GLOCOM.2007.594

Mukherjee, et al. (2015). Source Number detection and Realization of ROC for Spectrum sensing in Cognitive Radio. *Indian Journal of Science and Technology*, *27*.

Mukherjee, A. (2016). HML based Smart positioning of Fusion Center for Cooperative Communication in Cognitive Radio Networks. *IEEE Communications Letters*, *20*(11). doi:10.1109/LCOMM.2016.2602266

Mukherjee, A., & Datta, A. (2016). Vector Quantization based Power Allocation for Non-Ergodic Cognitive Radio Systems. *Journal of Engineering Science and Technology Review*, *9*(5), 85–87.

Mukherjee, A., Maiti, S., & Datta, A. (2014). Spectrum Sensing for Cognitive Radio Using Blind Source Separation and Hidden Markov Model. *IEEE Proceedings of Advanced Computing & Communication Technologies*, 409 – 414. doi:10.1109/ACCT.2014.63

Qiu, R. C., & Chen, Z. (2010). Prediction of channel state for cognitive radio using higher-order hidden Markov model. *Proceedings of the IEEE Southeast Conference*, 276 – 282.

Quan, Z., Cui, S., Poor, H., & Sayed, A. (2008). Collaborative wideband sensing for cognitive radios. *IEEE Signal Processing Magazine*, *25*(6), 63–70. doi:10.1109/MSP.2008.929296

Quan, Z., Cui, S., Poor, H., & Sayed, A. (2008). Collaborative wideband sensing for cognitive radios. *IEEE Signal Processing Magazine*, *25*(6), 63–70. doi:10.1109/MSP.2008.929296

Sahoo, S. (2015). Advanced Energy Sensing Techniques Implemented Through Source Number Detection for Spectrum Sensing in Cognitive Radio. In *Computational Vision and Robotics* (pp. 179–187). Springer. doi:10.1007/978-81-322-2196-8_21

Tafaghodi Khajavi, N., Sadeghi Ivrigh, S., & Sadough, S. M. S. (2011). A novel framework for spectrum sensing in cognitive radio networks. *IEICE Transactions on Communications*, *E9-B*(9), 2600–2609. doi:10.1587/transcom.E94.B.2600

Teguig, S., & Le Nir, V. (2012). Data fusion schemes for cooperative spectrum sensing in cognitive radio networks. *Communications and Information Systems Conference*, 1–7.

Unnikrishnan & Veeravalli. (2007). *Cooperative Spectrum Sensing and Detection for Cognitive Radio*. IEEE GLOBCOM.

Xu, G., Lu, Y., He, J., & Hu, N. (2008). Primary users detect for multiple antenna cognitive radio based on blind source separation. *8th International Symposium on Intelligent Information Technology workshops*, 777 – 780. doi:10.1109/IITA.Workshops.2008.112

Yarkan, S., & Arslan, H. (2007). Binary time series approach to spectrum prediction for cognitive radio. *IEEE 66th Vehicular Technology Conference*, 1563 - 1567. doi:10.1109/VETECF.2007.332

Yucek, T., & Arslan, H. (2009). A Survey of Spectrum Sensing Algorithms for Cognitive Radio Applications. *IEEE Communications Surveys and Tutorials*, *11*(1), 116–130. doi:10.1109/SURV.2009.090109

Zhang, Zhao Li, Shi, Zhao, & Dong. (2010). Spectrum usage prediction based on high-order markov model for cognitive radio networks. *IEEE 10th International Conference on Computer and Information Technology (CIT)*, 2784 – 2788.

Chapter 10
Wireless Sensor Networks Localization

Ahmed Elsayed Abo-Elhassab
Ain-Shams University, Egypt & TEDATA Company, Egypt

Sherine Mohamed Abd El-Kader
Electronics Research Institute, Egypt

Salwa Elramly
Ain-Shams University, Egypt

ABSTRACT

Localization is a fundamental problem in wireless sensor networks. Localization means the determination of geographical locations of sensor nodes, consequently detecting the event location and to initiate a prompt action whenever necessary. The localization process passes with three phases distance and/or angle estimation phase, position phase and algorithm phase. There are many techniques can be used in each phase, some of these techniques may add additional devices, cost, power consumption, or delay to the network. This chapter includes a classification and overview of the most popular localization techniques in each phase, also high lighting the problems of these techniques and their solutions in such a way that helps in understanding localization concepts especially with respect to WSN localization.

INTRODUCTION

Wireless sensor network (WSN) consists of many sensors deployed in a certain area to monitor or detect an event or some events depending on the application and sensor specifications; sensors can detect temperature, sound, pressure, etc. Wireless sensor networks (WSNs) are composed of hundreds, possibly thousands, of tiny low-cost and smart devices called sensor nodes that are capable of measuring various physical values, this sensors are deployed in large numbers, and they are communicating with each other and organizing themselves in order to cooperatively achieve a desired task It can provide opportunities for monitoring and controlling homes, cities, and the environment. Wireless sensor network can be used in many applications such smart home, precision agriculture, industrial monitoring, environmental/earth

DOI: 10.4018/978-1-5225-2342-0.ch010

monitoring. In addition, WSNs have a broad spectrum of applications in the defense area, generating new capabilities for reconnaissance and surveillance as well as other applications. The main challenges of a WSN design and implementation are power consumption constrain for nodes using batteries, scalability to large scale of deployment, Ability to withstand harsh environmental conditions and Ease of use. An important aspect in most of the sensor networks application is the localization of the individual nodes, localization capability is a highly desirable characteristic of wireless sensor networks in environmental monitoring applications such as bush fire surveillance, water quality monitoring and precision agriculture, and other applications. The measurement data are meaningless without knowing the location from where the data are obtained. Moreover, location estimation may enable a myriad of applications such as intrusion detection, road traffic monitoring, health monitoring, reconnaissance and surveillance. Localization enables the efficient routing: A typical sensor network has large number of nodes which communicate at very short distance (a few meters). The data sensed by a node has to be communicated to the central unit through several other nodes. Thus multi-hop routing is a must. In order to impellent multi-hop routing it is necessary that nodes are of their locality, namely, they know their relative position with respect to their neighbors. Consequently localization becomes very important. Localization provides the power saving: Suppose we have deployed the sensor network for pollution monitoring. Now the neighbor sensor nodes will have data which will not be dramatically different from each other. Thus to save power it makes sense to combine the data from neighboring nodes and then communicate the combined, reduced data set, thereby conserving power (since communication takes lot more power than local processing). In order to do this local data fusion, we will need the location information. Once again localization becomes vital. Localization assists in the applications like target tracking: In this application, we typically need to determine the range, speed and the direction of the target. Sensors are deployed in region which sense the moving target, using which the range, speed and direction of the target may monitored. In order to calculate the global orientation of the target it is necessary that we know the location of the sensor nodes. Once again localization becomes necessary. Localization useful in locating the source of the data: In many applications, an event based sensor network is used. Here, the nodes are normally in sleep mode and when an event occurs (say sudden vibrations take place) the nodes are awakened. The nodes than sense and transmit the data. Such data requires a location stamp and therefore localization becomes necessary. From the previous examples we see that localization is indeed a necessity for sensor networks. Moreover, localization techniques vary depending on applications. Patil, Shaha,Desai and Merchant(2011) showed that in their work .Localization systems can be divided into three distinct components as shown in Figure.1: Distance/angle estimation: This component is responsible for estimating information about the distances and/or angles between two nodes. This information will be used by the other components of the localization system. Position computation: This component is responsible for computing a node's position based on available information concerning distances/angles and positions of reference nodes, reference nodes are nodes which it is positions is known. Localization algorithm: This is the main component of a localization system. It determines how the available information will be manipulated in order to allow most or all of the nodes of a WSN to estimate their positions.

The chapter is divided into three sections; First section includes angle and distance estimation phase and it consists of two parts; part one Angle Estimation, part two Distance estimation. Angle Estimation part consists of; technique that uses the receiver antenna's amplitude response for angle estimation and technique that use the receiver antenna's phase response for angle estimation. Distance estimation part consists of four sub-sections, first sub-section includes one-way propagation time measurements, second sub-section includes round-trip propagation time measurements and Maximum Likelihood Method,

Figure 1. Localization systems components

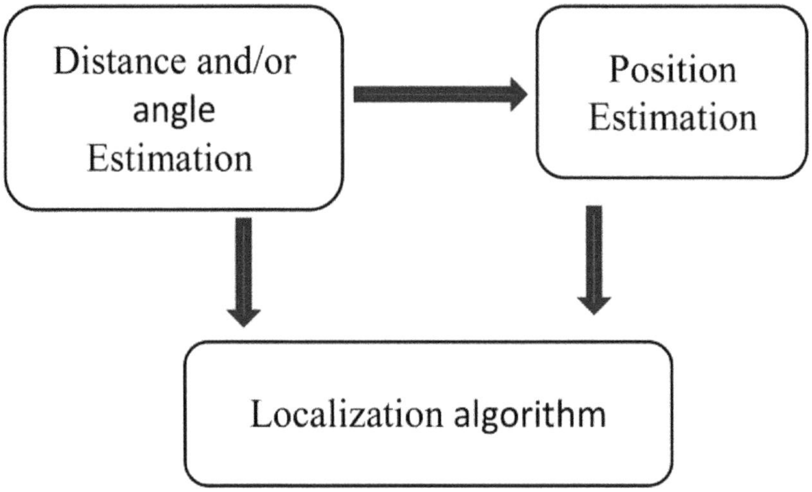

third sub-section includes time-difference-of-arrival measurements and the forth sub-section includes distance estimation using received signal strength measurements. Second Section includes Position estimation which consists of trilateration subsection, probabilistic approaches and spring-relaxation technique subsection and. Third section includes localization based on RSS profiling algorithm, Using bearing lines algorithm which is based on AOA and hyperbola principle algorithm which is based on TDOA in localization.

ANGLE ESTIMATION

Angle estimation based on Angle-Of-Arrival (AOA) techniques estimates the angle at which signals are received depending on either amplitude or phase. It can be divided into two subclasses; technique that uses the receiver antenna's amplitude response and technique that uses the receiver antenna's phase response.

Technique that Uses the Receiver Antenna's Amplitude Response

It considers that the beam of the receiver antenna is rotated, and the direction corresponding to the maximum signal strength is taken as the direction of the transmitter, from Figure.2 if the arrow is referring to the transmitter direction which is in direction with maximum signal strength. Mao, Fidan and Anderson(2007) showed that in their work.

Problems

A technical problem to be faced and overcome arises when the transmitted signal has varying signal strength, the receiver cannot differentiate the signal strength variation due to the varying amplitude of the transmitted signal and the signal strength variation caused by the anisotropy in the reception pattern,

Figure 2. Receiver horizontal antenna pattern with varying amplitude

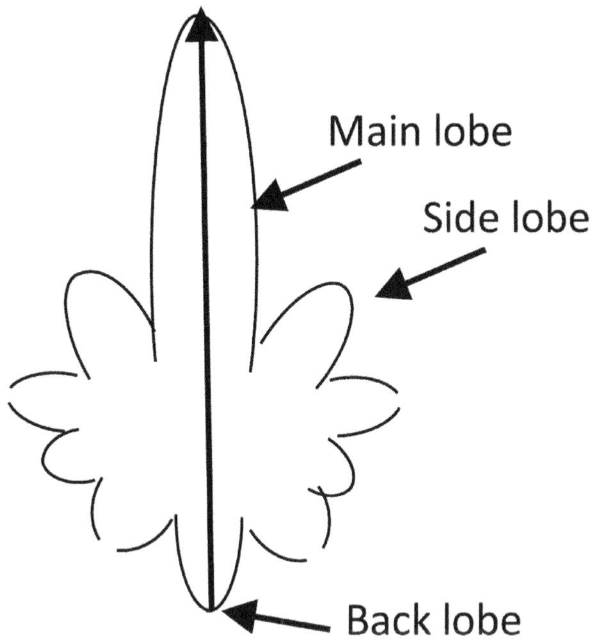

from Figure.2 if the side lobe strength is almost equal to the main lobe strength this problem will occur. Boudhir, Mohamed and Mohamed (2010) showed that in their work.

Solutions and Recommendations

To deal with the problem we can use a second non-rotating and Omni directional antenna at the receiver. By normalizing the signal strength received by the rotating anisotropic antenna with respect to the signal strength received by the non-rotating Omni directional antenna, the impact of varying signal strength can be largely removed.

Technique that Uses the Receiver Antenna's Phase Response

In this technique AOA can be measured from the measurements of the phase differences in the arrival of the wave front; it typically requires a large receiver antenna relative to the wavelength of the transmitter signal or an antenna array. Figure 3 shows an antenna array of N antenna elements, the adjacent Antenna elements are separated by a uniform distance d.

The distance between a transmitter and the ith antenna element can be approximated by:

$$R_i \approx R_0 - id\cos\theta . \quad (1)$$

where θ is the AOA and it is the bearing of the transmitter with respect to the antenna array. The transmitter signals received by adjacent antenna elements have a phase difference of: $2\pi \frac{d\cos\theta}{\lambda}$. This allows

Figure 3. Antenna array used in AOA measurement

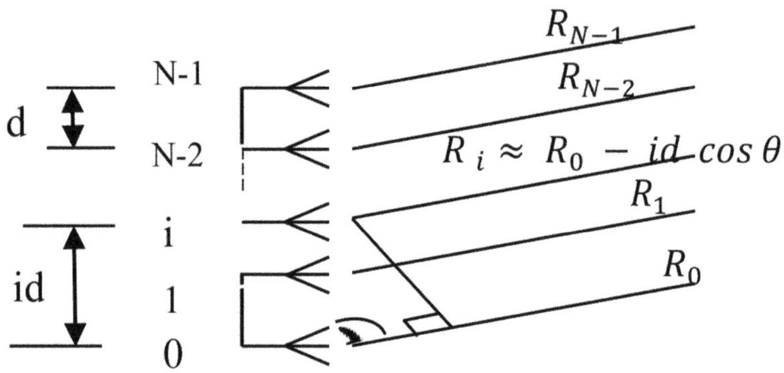

obtaining the bearing of the transmitter from the measurement of the phase difference. Boudhir, Mohamed and Mohamed (2010)showed that in their work.

Problems

Angle-Of-Arrival measurement using the receiver antenna's phase response needs a large size antenna that increases it cost and restrict its applicability. Angle-Of-Arrival measurement either by using the receiver antenna's amplitude response or by using the receiver antenna's phase response works quite well for high SNR but may fail in the presence of strong co-channel interference and/or multipath signals. The accuracy of AOA measurements is limited by the directivity of the antenna, by shadowing and by multipath reflections. The multipath component may appear as a signal arriving from an entirely different direction and can lead to very large errors in AOA measurements. Mao, Fidan and Anderson (2007), and Gemayel, Koslowski and Jondral (2009) showed that in their work.

Solutions and Recommendations

Multipath problem in AOA can be solved by using the maximum likelihood (ML) method this method is explained in details later. ML method can be used to estimate the AOA of each separate path in a multipath environment. The path that its parameters maximize the likelihood function is chosen Mao, Fidan and Anderson (2007) showed that in their work.

DISTANCE ESTIMATION

Distance estimation techniques used to estimate the distance between nodes, it can be divided into four subclasses; distance estimation based on one-way propagation time measurements, distance estimation based on round-trip propagation time measurements, distance estimation based on time-difference-of-arrival measurements, and distance estimation using received signal strength measurements.

One-Way Propagation Time Measurements

One-way propagation time measurements measure the difference between the sending time of a signal at the transmitter and the receiving time of the signal at the receiver to determine the distance between the transmitter and the receiver, the distance between the transmitter and the receiver regardless the processing delay time can be given by $d = c \times \| t_T - t_R \|$, where t_T is the transmitting time and t_R is the receiving time, this processing happens at the receiver.

Problems

This technique requires the local time at the transmitter and the local time at the receiver to be accurately synchronized. This requirement may add to the cost of sensors by demanding a highly accurate clock and/or increase the complexity of the sensor network by demanding a sophisticated synchronization mechanism. This disadvantage makes one-way propagation time measurements a less attractive option than measuring round-trip time in WSNs

Solutions and Recommendations

By using a combination of RF and ultrasound hardware on each transmission, a transmitter sends an RF signal and an ultrasonic pulse at the same time. The time difference between the receipt of the RF signal and the receipt of the ultrasonic signal is used as an estimate of the one way acoustic propagation time. This can be used to adjust clock on both the transmitter and the receiver

Round-Trip Propagation Time Measurements

It also known as time of arrival (TOA). Ou, C.-H. (2011) showed that in their work. it measures the difference between the time when a signal is sent by a sensor and the time when the signal returned by a second sensor is received at the original sensor. Since the same clock is used to compute the round-trip propagation time, there is no synchronization problem. The major error source in round-trip propagation time measurements is the delay required for handling the signal in the second sensor. This internal delay is either known via a priori calibration, or measured and sent to the first sensor to be subtracted. The performance of TOA based ranging depends on the availability of the direct path (DP) signal. In its presence, for example, short distance line - of - sight (LOS) conditions, accurate estimates are feasible. The challenge, however, is ranging in non - LOS (NLOS) conditions, which can be characterized as site - specific and dense multipath environments. These environments introduce several challenges. The first corrupts the TOA estimates due to the multipath components (MPCs), which are delayed and attenuated replicas of the original signal, arriving and combining at the receiver shifting the estimate. The second is the propagation delay caused by the signal traveling through obstacles, which adds a positive bias to the TOA estimates. The third is the absence of the DP due to blockage, also known as undetected direct path (UDP). The bias imposed by this type of error is usually much larger than the first two and has a significant probability of occurrence due to cabinets, elevator shafts, or doors that are usually cluttering the indoor environment. Mao, Fidan and Anderson (2007)showed that in their work.

Maximum Likelihood Method

ML estimation is a very general method used to measure the true value of some quantity, applicable to wide range of problems. It allows estimation of models nonlinear in parameters. ML estimator requires a distributional assumption about the error term. ML estimation, parameters should be chosen so as to maximize probability of the observed sample of data is generated by a hypothesized process. B. C. (2003) showed that in his work.

ML estimation can be used as follows:

- Assume that it is needed to measure the true value of some quantity (x_T). Repeating the measurements of this quantity $\{x_1, x_2, \dots x_n\}$. The standard way to estimate x_T from the measurements is to calculate the mean value:

$$\mu_x = \frac{1}{N}\sum_{i=1}^{N} x_i \tag{2}$$

- Set $X_T = \mu_x$
- Assume there is N measurements of x $\{x_1, x_2, \dots x_n\}$.
- Assume that the probability distribution function that describes x: f(x, α) is known.
- Assume that α is the parameter which is needed to be determined. Where α will maximize the probability of getting the measurements (the x_i's)

Let:

- The probability of measuring x_1 is $f(x_1, \alpha)dx$
- The probability of measuring x_2 is $f(x_2, \alpha)dx$
- The probability of measuring x_n is $f(x_n,\alpha)dx$

If the measurements are independent, the probability of getting these measurements is:

$$L = f(x_1, \alpha)dx.f(x_2, \alpha)dx \dots f(x_n, \alpha)dx = f(x_1, \alpha) . f(x_2, \alpha) \dots f(x_n, \alpha)dx^n$$

We can drop the dx^n term as it is only proportionality constant

Likelihood Function is: $L = \prod_{i=1}^{N} f(x_i, \propto)$

We want to estimate α that maximizes L: $\left.\frac{\partial L}{\partial \propto}\right|_{\propto=\propto^*} = 0$

Both L and $\ln L$ have maximum at the same location.

Maximize $\ln L$ rather than **L** itself because $\ln L$ converts the product into a summation

$$\ln L = \sum_{i=1}^{N} \ln f\left(x_i, \alpha\right)$$

New maximization condition:

$$\left.\frac{\partial \ln L}{\partial \propto}\right|_{\propto=\propto^*} = \sum_{i=1}^{N} \frac{\partial}{\partial \alpha} \ln f\left(x_i, \alpha\right)\Big|_{\propto=\propto^*} = 0$$

α could be an array of parameters (e.g. slope and intercept) or just a single variable. L. (n.d.) (2009) showed that in his work.

Example

- Let $f\left(x, \propto\right)$ be given by a Poisson distribution.
- Let $\propto = \mu$ be the mean of the Poisson.
- We want the best estimate of a from our set of n measurements $\left\{x_1, x_2, \ldots. x_n\right\}$.
- The likelihood function for this problem is:

$$L = \prod_{i=1}^{n} f(x_i, \propto) = \prod_{i=1}^{n} \frac{e^{-\alpha}\alpha^{x_i}}{x_i!} = \frac{e^{-\alpha}\alpha^{x_1}}{x_1!} * \frac{e^{-\alpha}\alpha^{x_2}}{x_2!} * \ldots\ldots\ldots\ldots \frac{e^{-\alpha}\alpha^{x_n}}{x_n!} = \frac{e^{-n\alpha}\alpha^{\sum_{i=1}^{n} x_i}}{x_1!x_2!\ldots x_n!}$$

- Find a that maximizes the log likelihood function:

$$\frac{d \ln L}{d \propto} = \frac{d}{d \propto}\left(-n \propto + \ln \propto \cdot \sum_{i=1}^{n} x_i - \ln\left(x_1!x_2!\ldots.x_n!\right)\right) = -n + \frac{1}{\propto}\sum_{i=1}^{n} x_i = 0$$

$$\propto\Big|_{average} = \frac{1}{n}\sum_{i=1}^{n} x_i$$

Some general properties of the Maximum Likelihood Method:

- For large data samples (large n) the likelihood function (L) approaches a Gaussian distribution.
- Maximum likelihood estimates are usually consistent.

For large n the estimates converge to the true value of the parameters we wish to determine.

- Maximum likelihood estimates are usually unbiased.

For all sample sizes the parameter of interest is calculated correctly.

- Maximum likelihood estimate is efficient: the estimate has the smallest variance.

- Maximum likelihood estimate is sufficient: it uses all the information in the observations (the x_i's).
- The solution from MLM is unique. Bulusu, Heidemann, and Estrin (2000) showed that in their work.

Time-Difference-Of-Arrival Measurements (TDOA)

It measures the TDOA of a transmitter signal at a number of receivers with known location information to estimate the location of the transmitter. Mao, Fidan and Anderson (2007) showed in their work that the TDOA between a pair of receivers *i* and *j* is given by:

$$\Delta t_{ij} \triangleq t_i - t_j = 1/c\left(\left\|r_i - r_t\right\| - \left\|r_j - r_t\right\|\right), i \neq j \tag{4}$$

From equation (4) if Δt_{ij} is measured so r_t can be estimated. Figure.4 shows the localization using time-difference-of-arrival measurements where t_i and t_j are the time when a signal is received at receivers *i* and *j* respectively, C is the light speed .The most widely used method in measuring the TDOA of a signal at two receivers at separate locations is the generalized cross correlation method, where the cross correlation function is given by:

$$\rho i, j\left(\tau\right) = \frac{1}{T}\int_0^T s_i\left(t\right) s_j\left(t-\tau\right) dt \tag{5}$$

Cross correlation method steps is shown in Figure.5. Gemayel, Koslowski and Jondral (2009) showed in their work that the TDOA is given by:

$$\Delta tij = \arg\max\left(\rho i, j\left(\tau\right)\right) \tag{6}$$

TDOA system requires sensors to be accurately synchronized among each other. Cross-correlating signals of a pair of sensors (r_i, r_j) yields the requested TDOA by detecting the peak position. For an accurate estimation of the cross-correlation, the used time window per time difference estimation should

Figure 4. Localization using time-difference-of-arrival measurements

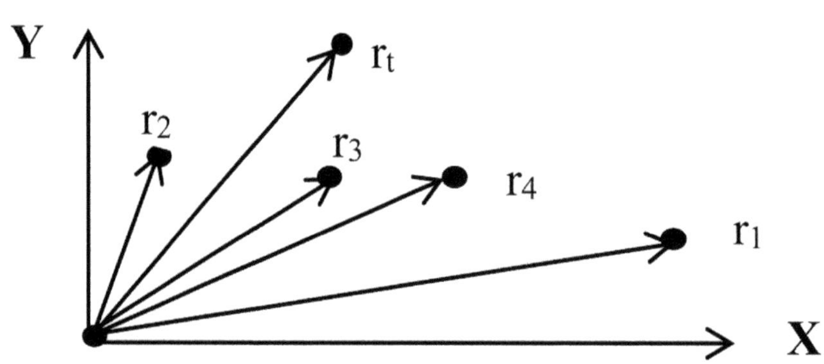

be much larger than the maximum delay: T >> τ_{max}. also Gemayel, Koslowski and Jondral (2009) showed in their work that the cross-correlation function can also be obtained from an inverse Fourier transform of the estimated frequency domain cross-spectral density function. Frequency domain processing is often preferred because the signals can be filtered prior to computation of the cross-correlation function. The cross correlation approach requires very accurate synchronization among receivers but does not impose any requirement on the signal transmitted by the transmitter. The accuracy and temporal resolution capabilities of TDOA measurements will improve when the separation between receivers increases because this increases differences between time-of-arrival. Closely spaced multiple receivers may give rise to multiple received signals that cannot be separated. For example, TDOA of multiple signals that are not separated by more than the width of their cross-correlation peaks (whose location on the time-delay axis corresponds to TDOA) usually cannot be resolved by conventional TDOA measurement techniques. Yet another factor affecting the accuracy of TDOA measurement is multipath problem. Boudhir, Mohamed and Mohamed (2010) showed that in their work.

Problems

Overlapping cross-correlation peaks due to multipath often cannot be resolved. Even if distinct peaks can be resolved, a method must be designed for selecting the correct peak value, such as choosing the largest or the first peak. Mao, Fidan and Anderson(2007) showed that in their work.

Solutions and Recommendations

Multipath problem can be solved by using an ultra-wide band (UWB) signals for accurate distance estimation. A UWB signal is a signal whose bandwidth to center frequency ratio is larger than 0.2 or a signal with a total bandwidth of more than 500 MHz, UWB can achieve higher accuracy because its bandwidth is very large and therefore its pulse has a very short duration. This feature makes fine time resolution of UWB signals and easy separation of multipath signals possible. Mao, Fidan and Anderson(2007) showed that in their work.

Distance Estimation Using Received Signal Strength Measurements

Mao, Fidan and Anderson (2007), and Ou, C.-H. (2011) showed in their work that the distances between neighboring sensors from the received signal strength measurements. These techniques are based on a

Figure 5. Block diagram of generalized cross correlation method

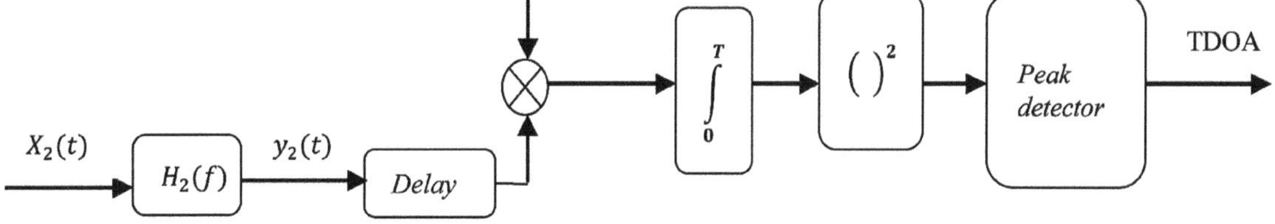

standard feature found in most wireless devices, a received signal strength indicator (RSSI). They are attractive because they require no additional hardware, and are unlikely to significantly impact local power consumption, sensor size and thus cost. In free space, the RSS varies with the inverse square of the distance d between the transmitter and the receiver:

$$\mathrm{RSSI} = \Pr\left(d\right) = \frac{P_t G_t G_r \lambda^2}{\left(4\pi\right)^2 d^2}, d = \sqrt{\frac{P_t G_t G_r \lambda^2}{\left(4\pi\right)^2 P_r\left(d\right)}} \tag{7}$$

where P_t is the transmitted power, G_t is the transmitter antenna gain, G_r is the receiver antenna gain and λ is the wavelength of the transmitter signal in meters. The free-space model however is an over-idealization, and the propagation of a signal is affected by reflection, diffraction and scattering. Of course, these effects are environment (indoors, outdoors, rain, buildings, etc.) dependent. However, it is accepted on the basis of empirical evidence that it is reasonable to model the RSS= $P_r\left(d\right)$ at any value of d at a particular location as a random and log-normally distributed random variable with a distance-dependent mean value, That is,

$$RSSI\left[dBm\right] = P_r\left(d\right)\left[dBm\right] = P_t - PL\left(d_0\right) - 10n_p \log_{10}\left(\frac{d}{d_0}\right) + X_\sigma \tag{8}$$

Let

$$P_0\left(d_0\right)\left[dBm\right] = P\_t - PL\left(d_0\right) \tag{9}$$

Then

$$RSSI\left[dBm\right] = P_r\left(d\right)\left[dBm\right] = P_0\left(d_0\right)\left[dBm\right] - 10n_p \log_{10}\left(\frac{d}{d_0}\right) + X_\sigma \tag{10}$$

where P_t is the transmit power of the sender in dBm, $PL\left(d_0\right)$ is the power loss in dBm at a reference distance d_0. $P_0\left(d_0\right)\left[dBm\right]$ is a known reference power value in $dBmilliwatts$ at a reference distance d_0 from the transmitter, n_p is the path loss exponent that measures the rate at which the RSS decreases with distance and the value of n_p depends on the specific propagation environment, X_σ is a zero mean Gaussian distributed random variable with standard deviation σ and it accounts for the random effect of shadowing. It is possible to conclude from equation (10) that given the RSS measurement P_{ij} between a transmitter i and a receiver j, a maximum likelihood estimate of the distance, d_{ij}, between the transmitter and the receiver is:

$$\hat{d}_{ij} = d_0 \left(\frac{P_{ij}}{P_0 d_0} \right)^{-1/n_p} \tag{11}$$

Using equation (10) and equation (11), the estimated distance $\hat{d}_{ij}$ can be related to the true distance:

$$\hat{d}_{ij} = d_{ij} 10^{-\frac{X_\sigma}{10 n_p}} = d_{ij} 10^{-\frac{\ln(10) X_\sigma}{10 n_p \ln(10)}} = d_{ij} e^{-\frac{X_\sigma}{\eta n_p}} \tag{12}$$

where: $\eta = \frac{10}{\ln(10)}$

The expected value of $\hat{d}_{ij}$ is:

$$E\left(\hat{d}_{ij}\right) = \frac{1}{\sqrt{2\pi\sigma}} \int_{-\infty}^{\infty} d_{ij} e^{-\frac{X_\sigma}{\eta n_p}} e^{-\frac{X_\sigma}{2\sigma^2}} dX_\sigma = d_{ij} e^{\frac{\sigma^2}{2\eta^2 n_p^2}} \tag{13}$$

Thus the maximum likelihood estimate in equation (11) is a biased estimate of the true distance and an unbiased estimate is given by:

$$\hat{d}_{ij} = d_0 \left(\frac{P_{ij}}{P_0 d_0} \right)^{-1/n_p} e^{-\frac{\sigma^2}{2\eta^2 n_p^2}} \tag{14}$$

POSITIONE ESTIMATION

Position estimation techniques use angle or distance estimation techniques to estimate nodes positions. Several methods can be used to compute the position of a node, trilateration and probabilistic approaches methods are included in this section. By knowing the distance to three anchors, a node can find its 2D position using

Trilateration

Trilateration is the process of finding the position of a node in space based on its distance to three anchors as shown Figure.6. Niculescu and Nath (2001) showed in their work that assuming the positions of the three fixed anchors be defined by vectors: $\vec{n}_0$, $\vec{n}_1$, and $\vec{n}_2 \in \mathcal{R}^2$. Further, let $\vec{p} \in \mathcal{R}^2$ be the position vector to be determined. Consider three circles, centered at each anchor, having radii of d_i meters, equal to the distances between $\vec{p}$ and each anchor $\vec{n}_i$. These geometric constraints can be expressed by the following system of equations:

$$/\vec{p} - \vec{n}_0 /^2 = d_0^2 \tag{15}$$

$$/\vec{p} - \vec{n}1 /^2 = d_1^2 \tag{16}$$

$$/\vec{p} - \vec{n}_2 /^2 = d_2^2 \tag{17}$$

Since

$$/\vec{p} - \vec{n}_i /^2 = \left(\vec{p} - \vec{n}_i\right) \cdot \left(\vec{p} - \vec{n}_i\right) = /\vec{p} /^2 - 2\vec{n}_i \cdot \vec{p} + /\vec{n}_i /^2,$$

the equations (15), (16), (17) can be rewritten as follows:

$$/\vec{p} /^2 - 2\vec{n}_0 \cdot \vec{p} + /\vec{n}_0 /^2 = d_0^2,$$
$$/\vec{p} /^2 - 2\vec{n}1 \cdot \vec{p} + /\vec{n}_1 /^2 = d_1^2,$$
$$/\vec{p} /^2 - 2\vec{n}2 \cdot \vec{p} + /\vec{n}_2 /^2 = d_2^2.$$

Subtracting the second and third equations from the first, results in the following two equations:

$$2\left(\vec{n}_1 - \vec{n}_0\right) \cdot \vec{p} = d_0^2 - d_1^2 - /\vec{n}0 /^2 + / \vec{n}1 /^2$$
$$2\left(\vec{n}_2 - \vec{n}_0\right) \cdot \vec{p} = d_0^2 - d_2^2 - /\vec{n}0 /^2 + / \vec{n}2 /^2$$

By solving the following linear system, $\vec{p}$ (expressed as a column vector) can be determined:

Figure 6. Trilateration

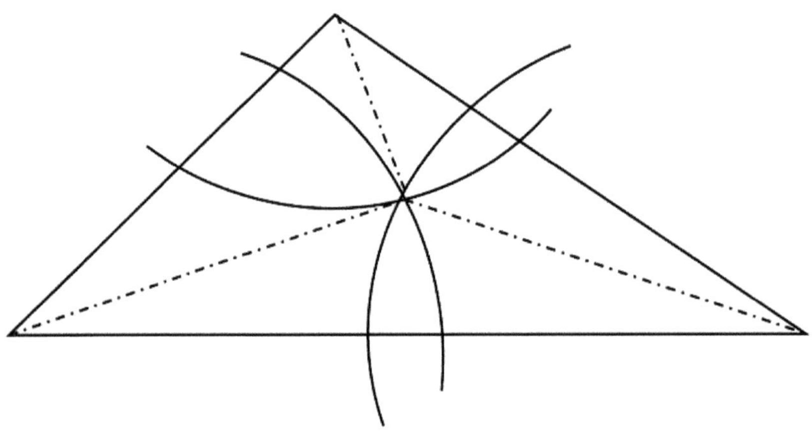

$$A = \begin{pmatrix} \vec{n}1 & \vec{n}0 \\ \vec{n}2 & \vec{n}0 \end{pmatrix}$$

$$\vec{b} = \begin{pmatrix} d_0^2 - d_1^2 - /\vec{n}0\,/^2 + /\,\vec{n}1\,/^2 \\ d_0^2 - d_2^2 - /\vec{n}0\,/^2 + /\,\vec{n}2\,/^2 \end{pmatrix}$$

$$2A \cdot \vec{p} = \vec{b}$$

More generally, to determine $\vec{p} \in \mathcal{R}^N, N+1$ fixed anchors are required: $\vec{n}_0, \ldots, \vec{n}_N \in \mathcal{R}^N$ Additionally, the distances between $\vec{p}$ and the N + 1 fixed anchors need to be known. These geometric constraints may be expressed by the following set of equations:

$$\left\|\vec{p} - \vec{n}_i\right\|^2 = d_i^2, 0 \leq \mathrm{i} \leq \mathrm{N} \tag{66}$$

Using a similar reasoning as before, subtracting equations 2, . . ., N + 1 from the first, results in the following system of N linear equations:

$$A = \begin{pmatrix} \vec{n}_1 - \vec{n}_0 \\ . \\ . \\ . \\ . \\ . \\ . \\ . \\ \vec{n}_N - \vec{n}_0 \end{pmatrix}, \vec{b} = \begin{pmatrix} d_0^2 - d_1^2 - \left|\vec{n}_0\right|^2 + \left|\vec{n}_1\right|^2 \\ d_0^2 - d_N^2 - \left|\vec{n}_0\right|^2 + \left|\vec{n}_N\right|^2 \end{pmatrix}$$

$$2A \cdot \vec{p} = \vec{b}$$

Assuming that the distances between the fixed anchors are known, but the anchors absolute positions are unknown, it is possible to construct a generic coordinate system for locating $\vec{p}$. Consider three fixed anchors: $\vec{n}0$, $\vec{n}1$, and $\vec{n}2 \in R2$, such that only the distances between each pair of anchors are known. One possible way to define these vectors is described next:

$$\vec{n}_1 = \begin{pmatrix} q \\ 0 \end{pmatrix}$$

$$\vec{n}_2 = \begin{pmatrix} \vec{n}_{2x} \\ \vec{n}_{2y} \end{pmatrix}$$

By expanding ‖ 〖n ⃗〗 _2^ -n ⃗1‖2, the following is obtained:.

$$/\vec{n}_2 - \vec{n}1\,/^2 = \left(\vec{n}_2 - \vec{n}_1\right)\cdot\left(\vec{n}_2 - \vec{n}_1\right) = /\vec{n}_2\,/^2 - 2\vec{n}_1 \cdot \vec{n}_2 + /\vec{n}1\,/^2 \tag{18}$$

Rearranging equation (18) in terms of $\vec{n}_1 \cdot \vec{n}_2$, yields the following:

$$\vec{n}_1 \cdot \vec{n}_2 = \frac{1}{2}\left(\left\|\vec{n}_1\right\|^2 + \left\|\vec{n}_2\right\|^2 - \left\|\vec{n}_2 - \vec{n}_1\right\|^2\right),$$

$$= ½\left(q^2 + r^2 - s^2\right)$$

The components of vector $\vec{n}_2$ can now be determined as follows:

$$\begin{pmatrix} q \\ 0 \end{pmatrix} \cdot \begin{pmatrix} \vec{n}_{2x} \\ \vec{n}_{2y} \end{pmatrix} = ½\left(q^2 + r^2 - s^2\right)$$

$$\begin{pmatrix} q \\ 0 \end{pmatrix} \cdot \begin{pmatrix} \vec{n}_{2x} \\ \vec{n}_{2y} \end{pmatrix} = ½\left(q^2 + r^2 - s^2\right)$$

$$q \cdot \vec{n}_{2x} = ½\left(q^2 + r^2 - s^2\right)$$

$$\vec{n}_{2x} = \frac{1}{2q}\left(q^2 + r^2 - s^2\right)$$

$$\vec{n}_{2x}^2 + \vec{n}_{2y}^2 = r^2$$

$$\vec{n}_{2y} = \pm\sqrt{r^2 - \vec{n}_{2x}^2}$$

Assuming that $n_{2y} \geq 0$, then $\vec{n}_2$ is given by:

$$\vec{n}_2 = \begin{pmatrix} \frac{1}{2q}\left(q^2 + r^2 - s^2\right) \\ \sqrt{r^2 - \vec{n}_{2x}^2} \end{pmatrix}$$

Provided vectors $\vec{n}_1$ and $\vec{n}_2$ are not collinear, i.e. $n_{1x} \neq 0$ and $n_{2y} \neq 0$, they form a basis for this two-dimensional vector space, and $\vec{p}$ can be determined by substituting $\vec{n}_1$ and $\vec{n}_2$ into linear system.

Problems

The main problem with this technique is that it relies on exact measurements to determine a position without taking into consideration error in distance measurement

Solutions and Recommendations

To solved this problem a good path loss model should be used.

Probabilistic Approaches

The uncertainty in distance estimations has motivated the appearance of probabilistic approaches for computing a node's position. An example of a probabilistic approach is proposed in [13], where the errors in distance estimations are modeled as normal random variables. When an unknown node receives a packet from a reference node, it can be in any place around the reference node with equal probabilities. When another packet is received from another reference node, the unknown node computes its position again as depicted in Figure 7. When new position information is received from other nodes, it becomes possible to identify the probable location of the unknown node, as depicted in Figure 8. The main drawbacks of this approach are the high computational cost and the space required to store the information. Patwari, Hero, Perkins, Correal and Dea (2003) showed that in their work.

Figure 7. Location probability when receiving packets from two reference nodes

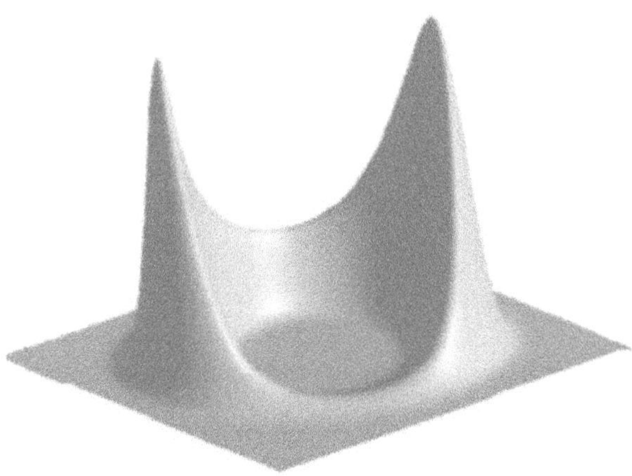

Figure 8. Location probability when receiving packets from two reference nodes

Every unknown node estimates its own position using a probability density function (pdf) $f_{X,Y}\left(x,y\right)$ of the two-dimensional coordinate variable (X,Y). Therefore, the probability of an unknown being placed at the coordinate $\left(x_{\varsigma},y_{\varsigma}\right)$ is:

$$Prob\left(x_{\varsigma},y_{\varsigma}\right) \approx \int_{y_{\varsigma-\Delta y}}^{y_{\varsigma+\Delta y}} \int_{x_{\varsigma-\Delta x}}^{x_{\varsigma+\Delta x}} f_{X,Y}\left(x,y\right) dxdy \tag{19}$$

where

$$x_{\min} \leq x \leq x_{\max}$$

and

$$y_{\min} \leq x \leq y_{\max}$$

The probability

$$prob\left(x_{\varsigma},y_{\varsigma}\right) = prob\left(X = x_{\varsigma}, Y = y_{\varsigma}\right),$$

the constants $x_{\min}$, $x_{\max}$, $y_{\min}$ and $y_{\max}$ are the bounding coordinates of the network, and both Δx and Δy are arbitrarily small values. The higher the probability $prob\left(x_{\varsigma},y_{\varsigma}\right)$, the more likely that the unknown node is placed at $\left(x_{\varsigma},y_{\varsigma}\right)$.

By using distribution functions from distance or angle Estimation techniques $prob\left(x_{\varsigma},y_{\varsigma}\right)$ can be estimated. The next procedure is considering RSS technique used in distance estimation.

Initially Calibration Process is needed. During the calibration phase the RSS is measured at different distances between a transmitter and a receiver pair. Let P and D denote the random variables of RSS (in dB) and distance (in meters), respectively. The mean of the RSS measurements *P*(*d*) and the standard deviation σ*P*(*d*) at each distance D = d can be calculated from the calibration measurements and using equation (20)

$$RSS\left[dBm\right] = P_t\left[dBm\right] - PL\left(d_0\right)\left[dBm\right] - 10\alpha\log_{10}\left(d / d_0\right) + X_{\sigma RSS}\left[dBm\right] \tag{20}$$

P(d) and $\sigma p(d)$ can be used to generate a mapping from any RSS to a pdf of the distance random variable *D*. Using this mapping which is a log-normal mapping of the RSS measurements, each unknown estimates its position pdf. Initially, each unknown sets its initial estimation to an even distribution over the entire network area,

$f_{X,Y}\left(x,y\right) = \dfrac{1}{A}$; Where *A* is the total area of the network. Nodes with position information, including both beacons (anchors) and unknowns with updated pdf estimations, send out beacon packets to their neighbors. Upon receiving a beacon packet, an unknown node executes the following algorithm:

- It measures the RSS of the received beacon packet;
- It maps the RSS to a one dimensional pdf using equation (20) and generate a pdf constraint $\Psi_{X_C,Y_C}\left(x,y\right)$, which is a function of the coordinate random variable $\left(X_C, Y_C\right)$;
- It updates the old pdf estimation by intersecting it with the generated constraint;
- Finally, the unknown with the updated pdf estimation will broadcast to all its neighbors.

There are two classes of beacon packets:

- Beacon packets from a beacon, and
- Beacon packets from an unknown.

If the beacon packet is from a beacon b placed at $\left(X_b, Y_b\right)$, we assume the position of the beacon is accurate (although the scheme may take into account any inaccuracies of the beacon's position). An unknown j within b 's transmission range receiving the beacon packet with $RSS_{p_{b,j}}\ dB$ maps $p_{b,j}$ to a pdf of the distance with mean $\mu_D\left(p_{b,j}\right)$ and standard deviation $\sigma_D\left(p_{b,j}\right)$; both $\mu_D\left(p_{b,j}\right)$ and $\sigma_D\left(p_{b,j}\right)$ are calculated during a calibration stage that should be at first. The unknown j then calculates a pdf constraint as:

$$\Psi_{X_C,Y_C}\left(x,y \mid p_{b,j}\right) = \frac{\varphi\left(x,y,x_b,y_b\right)}{\int_{y_{\min}}^{y_{\max}}\int_{x_{\min}}^{x_{\max}} \varphi\left(x,y,x_b,y_b\right) dx dy} \tag{21}$$

$$\varphi\left(x,y,X_b,Y_b\right)=\frac{1}{\sqrt{2\pi\sigma_D^2\left(p_{b,j}\right)}d_{b,j}}e^{-\frac{\left(\log d_{b,j}-\mu_D\left(p_{b,j}\right)\right)^2}{2\sigma_D^2\left(p_{b,j}\right)}} \tag{22}$$

$$d_{b,j}=\sqrt{\left(x-x_b\right)^2+\left(y-y_b\right)^2} \tag{23}$$

If the beacon packet is from an unknown node i with pdf estimation $f_{X_i,Y_i}\left(x,y\right)$, the unknown j, receiving a beacon packet from i with $RSS_{p_{b,j}}\ dB$, estimates the pdf constraints as:

$$\Psi_{X_C,Y_C}\left(x,y\mid p_{i,j}\right)=\frac{\varphi\left(x,y\right)}{\int_{y_{\min}}^{y_{\max}}\int_{x_{\min}}^{x_{\max}}\varphi\left(x,y\right)dxdy} \tag{24}$$

$$\varphi\left(x,y\right)=\int_{y_{\min}}^{y_{\max}}\int_{x_{\min}}^{x_{\max}}\varphi\left(x,y,x_i,x_i\right)f_{X_i,Y_i}\left(X_i,Y_i\right)dx_idy_i \tag{25}$$

where

$$\varphi\left(x,y,x_i,x_i\right)=\frac{1}{\sqrt{2\pi\sigma_D^2\left(p_{i,j}\right)}d_{i,j}}e^{-\frac{\left(\log d_{i,j}-\mu_D\left(p_{i,j}\right)\right)^2}{2\sigma_D^2\left(p_{i,j}\right)}} \tag{26}$$

Assume the original pdf estimation for unknown j is $f_{(X_i,Y_i)}^{old}\left(x,y\right)$; a new pdf estimation for unknown j can be calculated by intersecting the pdf constraint described above, either from a beacon or an unknown, with the original estimation, if vector random variables $\left(X_i,Y_i\right)$ and $\left(X_C,Y_C\right)$ are mutually independent.

$$f_{X_i,Y_i}\left(x,y\right)=\frac{f_{(X_i,Y_i)}^{old}\left(x,y\right)\Psi_{X_C,Y_C}\left(x,y\mid p\right)}{\int_{y_{\min}}^{y_{\max}}\int_{x_{\min}}^{x_{\max}}f_{X_i,Y_i}^{old}\left(x,y\right)\Psi_{X_C,Y_C}\left(x,y\mid p\right)dxdy} \tag{27}$$

Figure 9 depicts a network with three beacons (b_1,b_2,b_3) and two unknowns (u_1,u_2).Assume the following sequence of events:

Figure 9. Network topology with three beacons (b_1, b_2, b_3) and two unknowns (u_1, u_2)

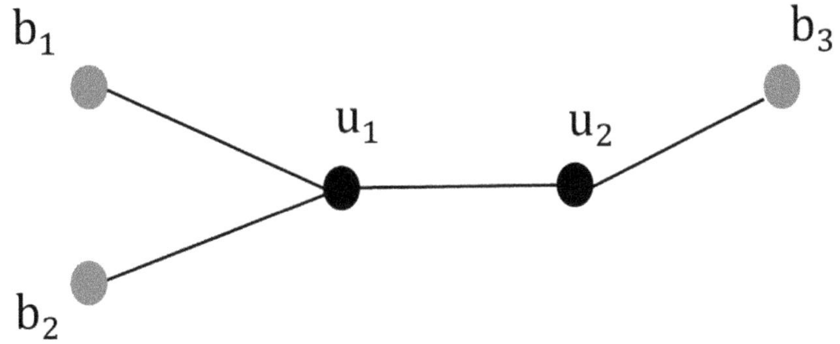

- Unknown u_1 uses the packets received from b_1 and b_2 to compute two constraints, which are intersected with each other to generate a new constraint. The new constraint is then intersected with the initial estimation (uniform distribution) to obtain a new pdf estimation as shown in Figure 10(a).
- Unknown u_2 computes a pdf estimate of its position as shown in Figure 10 (b) using the beacon packet from b_3.
- Both u_1 and u_2 send out beacon packets to their neighbors. When receiving the beacon packet from u_2, u_1 calculates a pdf constraint as shown in Figure 10 (c).
- Node u_1 intersects the pdf constraint in Figure 10 (c) with its old pdf estimation in Figure 10 (a) to update its estimation. The results of the intersection are shown in Figure 10 (d).
- The same sequence is carried for u_2 (i.e., it computes the constraint and intersects it with the old pdf estimate). Using the beacon packet sent by u_1, u_2 computes the pdf constraint shown in Figure 10 (e).
- Finally, u_2 intersects the constraint in Figure.10(e) with its old pdf estimation in Figure.10(b); the new pdf estimation of u_2 is shown in Figure.10(f). Intersecting the current position pdf estimation and the new constraint is only possible if the two vector random variables (X_j,Y_j) and (X_C,Y_C) are mutually independent. The position of the unknown is determined at the place with the largest probability. The probability estimations for unknown u_1 and u_2 have the same shapes as their pdf estimations shown in Figure.10(d) and Figure.10(f), respectively. While the probability estimation for u_1 has only one peak which is at the true position of u_1, however, the probability estimation for u_2 has two peaks, with the true position on lower peak. To improve the estimation accuracy and reduce the uncertainty, negative constraints are introduced in the next section.

Spring-Relaxation Technique

The concept of spring-relaxation technique is explained be considering the following example. The example consists of five beacons and a sensor whose location is to be determined. In the concept of spring-relaxation technique, the considered example is equivalent to having a moving particle (i.e., sensor) attaching with five springs. For each spring, while its one end attaches to the particle, its other end is nailed by a pin (i.e., beacon) at a fixed location. Figure.11 depicts the described example. In the illustration, the black rings are the beacons or the pins in fixed locations, and white ring is the sensor or

Figure 10. Localization evolution for the network topology in Figure.10 (a) pdf estimation of u1 using beacon packets from b1 and b2 (b) pdf estimation of u2 using beacon packet from b3 (c) pdf constraint on u1 calculated by using u2's beacon packet (d) Final position estimate of u1 computed as the intersection of the old pdf estimation and the pdf constraint for u1 (e) pdf constraint on u2 calculated by using u1's beacon packet (f) Final position estimate of u2 computed as the intersection of the old pdf estimation and the pdf constraint for u2.

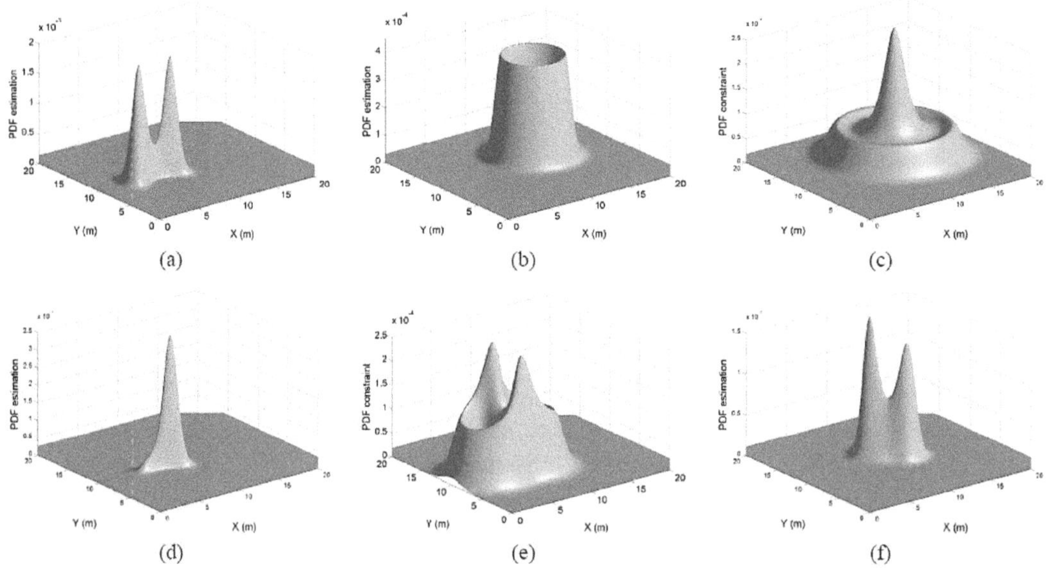

the particle in its initial guessed location. The natural length of a spring is the length where the spring is in the rest state. When the length of a spring becomes shorter (resp. longer) than its natural length, the spring is compressed (resp. stretched) and forces are produced at each end of the spring. The particle attached to a set of springs receives forces from them when they are compressed or stretched. The net force applies on the particle is the vector sum of all received forces. When the particle begins at a particular location with nonzero net force, the net force moves the particle to a new location and the net force changes accordingly. The particle continues to move until the net force becomes zero, and the particle comes to rest. This resting location, indicated by the grey ring in Figure 11, is also the final stopping location of the particle. Localization using spring-relaxation technique does not have real springs connecting the particles and pins. It uses the concept to simulate the movements of the particle under the spring forces computed, and find the final stopping location, which is the estimated location of the particle. From Hooke's law, the magnitude of the force F from each spring is

$$F = k\left(L_0 - L\right) \quad (1)$$

where L_0 is the natural length of the spring, L is the current length of the spring, and k is the spring constant. The difference which $L_0 - L$ describes is the stretch or compression of the spring. Let $\vec{F}$ be the force vector applied on a particle by a spring, and $\overrightarrow{F_{net}} = \sum \vec{F}$ be the net force applying on the particle by all the attached springs. By Newton's first and second law, we have the relationship

$$\overrightarrow{F_{net}} = \frac{d(m\vec{v})}{dt}.$$

where m is the mass of the particle and $\vec{v}$ is the instantaneous velocity of the particle due to the net force. The instantaneous displacement of the particle can be determined by

$$\vec{D} = \int \vec{v}dt = \frac{\overrightarrow{F_{net}}}{m}dt^2.$$

The instantaneous displacement will cause a change in force applied by each spring on the particle. This change in force leads to a new net force on the particle which then changes the instantaneous displacement of the particle again. As this process continues, the particle moves and will eventually rest at an equilibrium location where the net force is zero Zhang, Foh, Seet and Fong (2010) showed that in their work.

Figure 11. Spring-relaxation of five beacons and a sensor

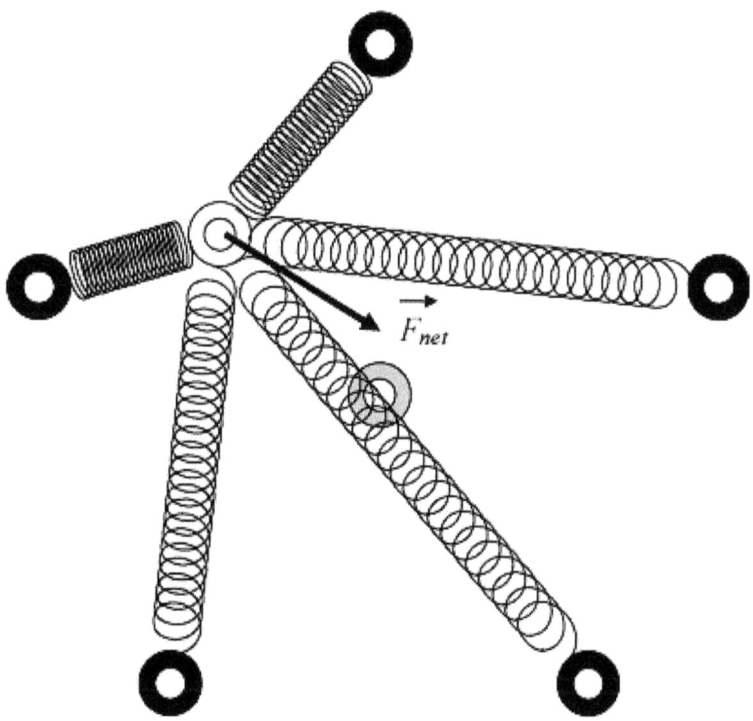

LOCALIZATION ALGORITHM

It determines how the available information will be manipulated in order to allow most or all of the nodes of a WSN to estimate their positions.

Localization Based on RSS Profiling Algorithm

Santos (2008) Showed in his work that the distance estimation phase is the initial step performed when locating a node's position in space using RSS profiling techniques. By estimating distances to neighbors with known coordinates, a node can determine its own position using positioning techniques. Once a node determines its position, it becomes an anchor node and can then help other neighbors to find their positions. Alternatively, nodes attached to GPS-devices can rely on this instrument for obtaining coordinates, without needing to estimate distances to their neighbors. Stoep and Vander (2009) showed in their work that the received signal strength (RSS) algorithm does not require any hardware in addition to the radio transceiver. Knowing the transmitted signal's power and path-loss model, the inter-node separation can be calculated. RSS based localization systems do not require hardware components in addition to the radio transceiver. Moreover, no dedicated packets need to be sent over the network for such systems to function. However, RSS measurements are very unreliable, even when both sender and receiver are stationary. Ranging errors of $\pm 50\%$ have been observed, leading to inaccurate distance estimates. Hence, it is important to understand the sources of error before relying on this technique for locating nodes. Multipath and shadowing are two major phenomena affecting the reliability of RSS measurements. Different magnitude signals arriving out of phase at the receiver cause constructive and destructive interference. Spread spectrum radios effectively mitigate this problem by averaging the received power over multiple frequencies. Shadowing effects are caused by obstructions (e.g. thick vegetation, walls, furniture) that attenuate the signal's strength. Additionally, not all RSSI circuits are factory calibrated, resulting in device-dependent RSS measurements for the same signal strength. The actual signal power can be different from the transceiver's intended transmission power, causing further discrepancies in RSS measurements. Various techniques have been proposed for mapping RSS measurements to distance estimates. The basic formula used in RSS localization is given by the formula

$$RSS\left[dBm\right] = P_t\left[dBm\right] - PL\left(d_0\right)\left[dBm\right] - 10\alpha\log_{10}\left(d / d_0\right) + X_{\sigma_{RSS}}\left[dBm\right], \quad (20)$$

where RSS is the received signal strength in units of decibels with respect to milli watts (dBm). In equation (20), d is the true distance from the sender to the receiver, $\alpha = n_p$ is the path-loss exponent, P_t is the transmit power of the sender in dBm, $PL(d_0)$ is the power loss in dBm at a reference distance d_0. The quantity $X_{\sigma RSS}$ in dBm is a random variable representing the noise in the measured RSS and is often assumed to be a zero-mean Gaussian random variable with variance σ_{RSS}. The source of the noise $X_{\sigma RSS}$ in measured RSS can come from both time varying and time-invariant sources. Time varying errors, such as interference, can be averaged out by taking multiple measurements corresponding to the same distance. Time-invariant errors, Such as shadowing due to heterogeneity in the medium resulting from objects such as walls or buildings can cause the signal to degrade contrary to the path-loss model. These errors cannot be averaged out by taking multiple measurements, as the path loss model cannot be specifically designed for each wireless channel in each deployed network. It is observed that the random effects of shadowing are appropriately modeled by assuming the error $X_{\sigma_{RSS}}$ is Gaussian. In a WSN performing RSS-based localization according to traditional techniques, each node measures an RSS value corresponding to each neighboring beacon node. This measured value RSS is mapped to a distance estimate $\hat{d}$. This mapping from RSS to $\hat{d}$ requires an expression for the distance $\hat{d}$ as a function of RSS and can be obtained by solving (20), yielding

$$\hat{d} = k10^{\left(P_t - RSS + X_{\sigma_{RSS}}\right)/(10\alpha)} \tag{28}$$

where k is a constant incorporating both $PL\,(d_0)$ and $\alpha \log_{10}\left(d_0\right)$. To obtain a practical fit for the model parameters, Stoep and Vander (2009)showed in their work that the RSS measurements is taken at a set of reference distances d_{ri}. For each distance d_{ri}, the sample mean $\mu_{RSS}\left(d_{ri}\right)$ and sample standard deviation $\sigma_{RSS}\left(d_{ri}\right)$ are computed and plotted in Figure.12. The model parameters are computed from the set of sample means using a best-fit curve using equation (28) and averaged as $\mu_{RSS}\left(d_{ri}\right)$ Patwari, Hero, Perkins, Correal and Dea (2003) showed that in their work. A curve is fit to this sample mean using equation (28). An example of this model-fitting process is illustrated in Figure.12. From Figure.12, the resulting sample standard deviation $\sigma_{RSS}\left(d_{ri}\right)$ is approximately constant as a function of the reference distance d_{ri}. To clarify this point, Figure.13 illustrates the sample standard deviation as a function of d_{ri} and the corresponding linear approximation. The variance in the Gaussian random variable $X_{\sigma_{RSS}}$ is thus independent of the distance d between the transmitter and receiver so given the empirical values of k and α, the resulting model in (28) can be used to compute estimated distance $\hat{d}$ from the measured RSS as $\hat{d} = k10^{\left(P_t - RSS\right)/(10\alpha)}$ (29)

Figure 12. The sample mean and standard deviation are plotted for a set of reference distances d_{ri}. The model parameters k and α are computed by fitting a curve to the reference model using the samples

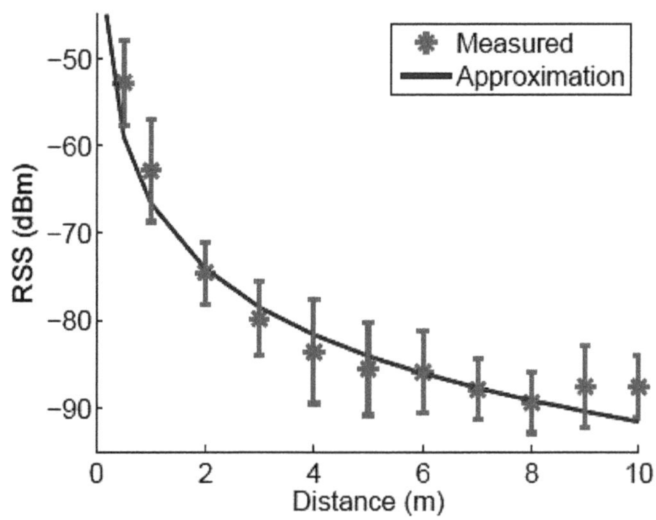

Figure 13. The sample standard deviation $\sigma_{RSS}\left(d_{ri}\right)$ over the set of reference distances is well-approximated by a constant function in d_{ri}

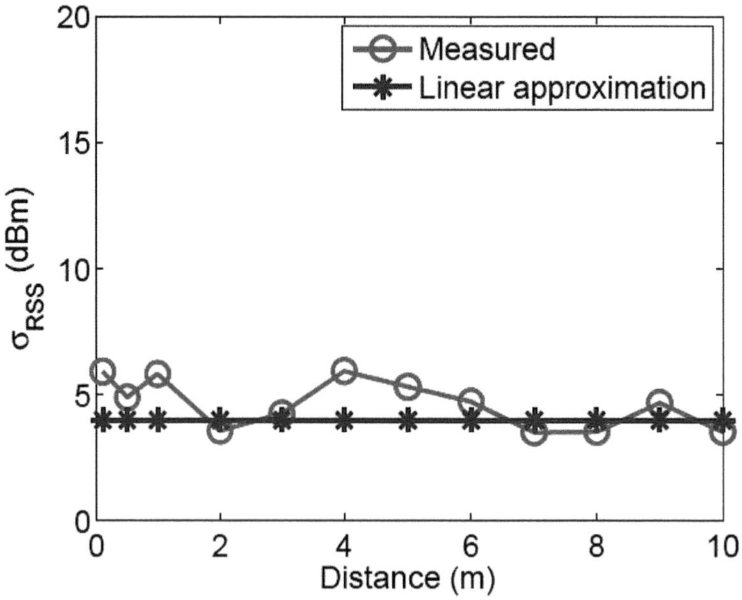

This estimation can be solved by using the Maximum Likelihood Method which is discussed in a previous section for the distance *d* given the measured RSS. Patwari, Hero, Perkins, Correal and Dea (2003) showed that in their work.

Rice (1995) showed in his work that due to the non-linearity of the mapping from the measured RSS to the distance estimate $\hat{d}$ in (29), the constant standard deviation σ_{RSS} in the RSS measurement maps to a standard deviation σ_d in the distance estimate that increases linearly with distance. This result can be derived analytically using a delta approximation based on the Taylor series expansion for dependent random variables X and Y such that $Y = f\left(X\right)$. The delta approximation states that the standard deviation σ_Y is approximated by $\sigma_Y \approx \sigma_X \cdot f'\left(\mu_X\right)$ (30) where μ_X and σ_X are respectively the mean and standard deviation of X and $f'\left(x\right)$ is the derivative of $f\left(x\right)$. Applying this result to the function $\hat{d}(RSS)$ mapping a measured RSS to a distance estimate $\hat{d}$ yields:

$$\sigma_d \approx \sigma \hat{d}'\left(\mu_{RSS}\right)$$

$$= \sigma_{RSS} K \log\left(10^{1/10\alpha}\right) 10^{\left(P_t - \mu_{RSS}\right)/10\alpha}$$

$$= \sigma_{RSS} \log\left(10^{1/10\alpha}\right) d \qquad (31)$$

Hence, the variance in the distance estimate $\hat{d}$ is directly proportional to the true distance **d** at which the measurement was taken. Bulusu,Heidemann and Estrin (2000), and Capkun and Hubaux(2005), and He, Huang, Blum, Stankovic, and Abdelzaher (2003), and Kwon, Mechitov, Sundresh, Kim and Agha (2005) suggested in their work to discard the measurements from beacons beyond a threshold distance due to the high degree of uncertainty in such measurements. However, the ability to discard measurements while obtaining a sufficient number of measurements from nearby beacons requires a relatively high beacon density. In particular, when beacons with radio range **r** are randomly deployed with density ρ**,** the probability that a node is within range of at least **k** beacons, denoted by $p(k,r,\rho)$ is given by:

$$p(k,r,\rho) = 1 - e^{-\rho\pi r^2}\sum_{i=0}^{k-1}\frac{(\rho\pi r^2)^i}{i!} \tag{32}$$

which follows by use of spatial statistics for a Poisson point process of intensity ρ B. C. (2003) Showed that in his work. If RSS measurements beyond a threshold distance $r_T < r$ are to be discarded then the beacon density must increase to $\rho_T = \rho\left(\frac{r}{r_T}\right)^2$,

In order to maintain the coverage probability $p(k, r_T, \rho_T)$ equal to $p(k,r,\rho)$ with the modified parameters. Furthermore, Stoep and Vander (2009) showed in their work that the when beacons are randomly deployed, the probability that each beacon measurement is within the threshold distance r_T decays rapidly as the threshold becomes stricter, as the coverage area of a beacon scales quadratically in r_T. This problem is depicted in Figure.14 for random deployment of beacons. As illustrated in Figure.14 if measurements from beacons in the shaded region beyond a threshold distance $r_T < r$ are to be discarded, the beacon density ρ must increase by a factor $(r / r_T)^2$ to maintain the same average beacon

Figure 14. area of deployment

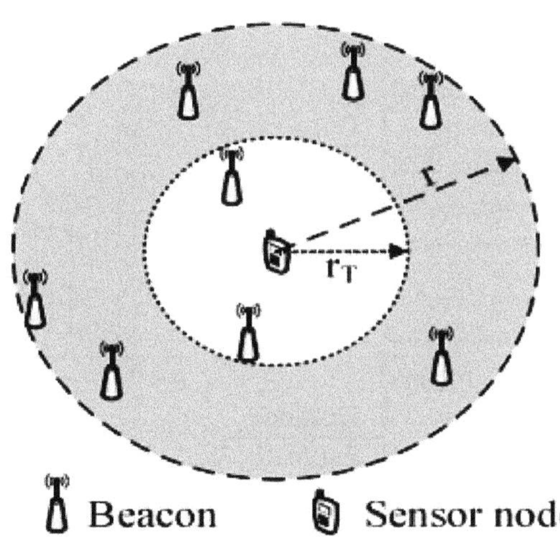

coverage at the localizing node. Hence, Stoep and Vander (2009) take in their work the advantage of the estimation error in the distance estimates $\hat{d}_i$ from distant neighboring beacons instead of discarding this information.

As illustrated in Figure.15 traditional RSS-based localization algorithms consist of two independent estimation problems. The first problem is to estimate the distance $\hat{d}_i$ to each beacon i from a measurement RSS_i, and the second problem is to estimate the location of the node using the distance estimates $\hat{d}_i$ and the beacon coordinates (x_i, y_i). In traditional RSS-based localization algorithms, the typical operation is to estimate a distance $\hat{d}_i$ from the measured RSS_i corresponding to each neighboring beacon *i*. These distance estimates $\hat{d}_i$ are then passed to the localization algorithm along with the beacon coordinates (x_i, y_i) in the plane of the network. This two-step protocol operation is illustrated in the schematic in Figure.15. Since the standard deviation in each distance estimate $\hat{d}_i$ increases proportional to the true distance d_i to the beacon *i*, as given in (31), the accuracy of the resulting location estimate depends indirectly on the standard deviation σ_{d_i}. However, most RSS-based localization algorithms operate only with the estimates $\hat{d}_i$ and corresponding coordinates, without considering the inherent uncertainty in the estimates. Given that the mapping from a measurement RSS to a distance estimate $\hat{d}$ has standard deviation proportional to the distance *d* between the transmitter and receiver, the ability to take advantage of the distance estimates from distant beacons requires more information about the measurement RSS to be passed to the localization algorithm. It is good to mention that Stoep and Vander (2009) in their work that the following framework for RSS-based distance estimation is proposed. As illustrated in Figure.16, the Gaussian distribution of noise in RSS measurements with constant variance σ^2_{RSS} can be mapped to a skewed distribution on the beacon distance d_i as also shown in Figure.16 while measurements from distant beacons have high estimation variance, they still provide information to the localization algorithm. Instead of constructing an estimate $\hat{d}_i$ for each distance d_i, corresponding to keeping only a single point in the RSS noise distribution, the entire distribution $f_i\left(d_i \mid RSS_i\right)$ which characterizes the relationship between the RSS measurement RSS_i and the true beacon distance d_i is

Figure 15. Two-step protocol operation

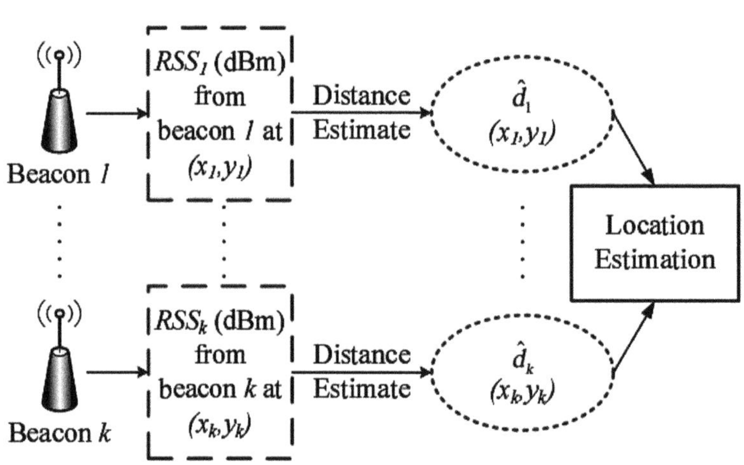

Figure 16. Mapping the Gaussian distribution of RSS measurement error to a distribution of the beacon distance d_i

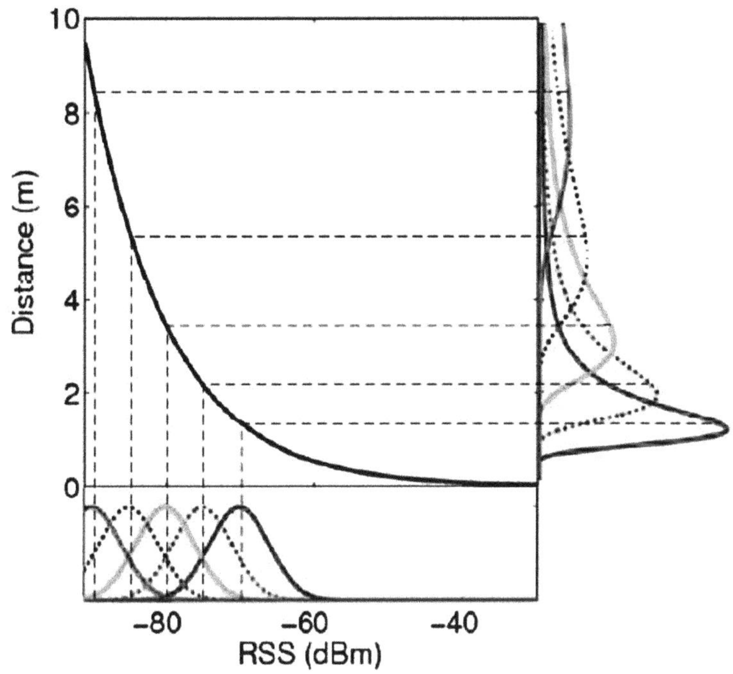

constructed. This distribution can be passed to the localization algorithm which can leverage the uncertainty in the set of distributions f_i to estimate the node location.

The RSS profiling-based localization techniques, work by constructing a form of map of the signal strength behavior in the coverage area. The map is obtained either offline by a priori measurements or online using sniffing devices deployed at known locations. They have been mainly used for location estimation in WLANs, but they would appear to be attractive also for wireless sensor networks Mao, Fidan and Anderson (2007), and Wang, Ghosh and Das (2010)showed that in their work. In this technique, in addition to there being anchor nodes and non-anchor nodes, a large number of sample points (e.g., sniffing devices) are distributed throughout the coverage area of the sensor network. At each sample point, a vector of signal strengths is obtained, with the jth entry corresponding to the jth anchor's transmitted signal. Many entries of the signal strength vector may be zero or very small, corresponding to anchor nodes at larger distances (relative to the transmission range or sensing radius) from the sample point. The collection of all these vectors provides (by extrapolation in the vicinity of the sample points) a map of the whole region. The collection constitutes the RSS model, and it is unique with respect to the anchor locations and the environment. The model is stored in a central location. By referring to the RSS model, a non-anchor node can estimate its location using the RSS measurements from anchors. Each non-anchor node unaware of its location uses the signal strength measurements it collects, stemming from the anchor nodes within its sensing region, and thus creates its own RSS finger print, which is then transmitted to the central station. Then the central station matches the presented signal strength vector to the RSS model, using probabilistic techniques or some kind of nearest neighbor based method, which chooses the location of a sample point whose RSS vector, is the closest match to that of the non-anchor node to be the estimated location of the non-anchor node. In this way, an estimate of the location of the non-anchor node can be obtained. The estimate is transmitted to the non-anchor node from the central station. Obviously, a non-anchor node could also obtain the full RSS model from the central station and perform its own location estimation. The accuracy of this technique depends on two distinct factors: the particular technique used to build the RSS model, with the resultant quality of that model, and the technique used to fit the measured signal strength vector from a non-anchor node into the appropriate part of the model. Mao, Fidan and Anderson (2007), and Wang, Ghosh and Das (2010) showed that in their work. The major practical obstacle in the RSS-profiling based localization is the extensive amount of profiling data required. Substantial and possibly costly initial experiments are needed to establish the model. Subsequent changes in the environment (e.g. inside a building, office occupancy can change) can affect the model, and so a static model derived from a single-shot experiment may be inadequate in some applications. Recently, there has been proposed a method of online profiling, which would reduce or possibly eliminate the amount of profiling required before deployment, but at the expense of deploying a large number of additional devices (termed "sniffers") at known locations. Together with a large number of stationary emitters (anchor nodes) deployed at known locations, the "sniffers" can be used to construct and update the RSS model online.

Using Bearing Lines Algorithm Which Is Based on AOA in Localization

In the absence of noise and interference, bearing lines from two or more receivers will intersect to determine a unique location, i.e., the location estimate of the transmitter. In the presence of noise, more than two bearing lines will not intersect at a single point and statistical algorithms, sometimes called triangulation or fixing methods, are required in order to obtain the location estimate of the transmitter this is shown in Figure.17.

Mao, Fidan and Anderson (2007) showed in their work that the 2D localization problem using bearing measurements can be formulated as follows.

Let $x_t = [x_t, y_t]^T$ be the true coordinate vector of the transmitter to be estimated from bearing measurements $\beta=[\beta_1,\ldots, \beta_N]^T$ where N is the total number of receivers. Let $X_i = [x_i, y_i]^T$ be the known location of the receiver associated with the ith bearing measurement.

Figure 17. Bearing lines from multiple receivers can be used to localization

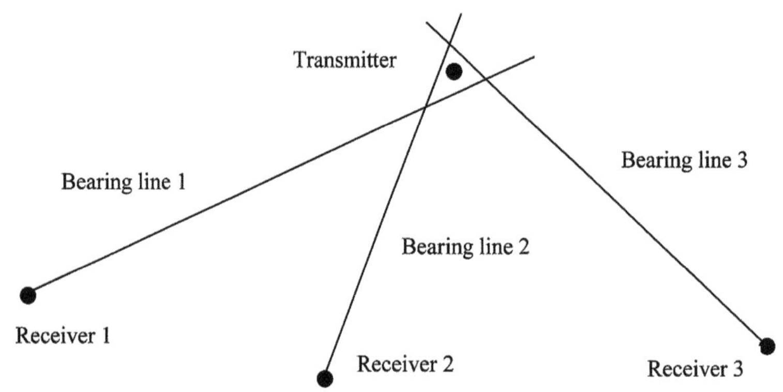

Denote by $\theta(x) = [\theta_1(x), \ldots\ldots\ldots, \theta_N(x)]^T$ the bearings of a transmitter located at $X = [x, y]^T$ from the receiver locations, where $\theta_i(x)$; $1 \leq i \leq N$ is related to x by: $\tan\theta_i\left(X\right) = \frac{y - y_i}{x - x_i}$. This is shown in Figure 18.

The measured bearings of the transmitter consist of the true bearings corrupted by additive noises $\varepsilon = [\varepsilon_1, \ldots\ldots, \varepsilon_N]^T$, which are assumed to be zero-mean Gaussian noises with N * N Covariance matrices

$$S = diag\left\{\sigma_1^2, \ldots\ldots\ldots, \sigma_N^2\right\},\ \beta = \theta\left(X_t\right) + \varepsilon \tag{33}$$

When the receivers are identical and much closer to each other than to the transmitter, the variances of bearing measurement errors are equal:

$$\sigma_1^2 = \ldots\sigma_N^2 = \sigma^2$$

The ML estimator of the transmitter location x_t is given by:

$$\hat{X}_t = \arg\min \frac{1}{2}\left[\theta\left(\hat{X}_t\right) - \beta\right]^T S^{-1}\left[\theta\left(\hat{X}_t\right) - \beta\right]$$

$$\arg\min \frac{1}{2}\sum_{i=1}^{N} \frac{\left(\theta_i\left(\hat{X}_t\right) - \beta_i\right)^2}{\sigma_i^2} \tag{34}$$

where arg min stands for the argument of the minimum and it is the set of points of the given argument for which the given function attains its minimum value.

The nonlinear minimization problem in equation (34) can be solved by Newton-Gauss iteration:

Figure 18. Bearings of a transmitter T_x from the receiver R_i

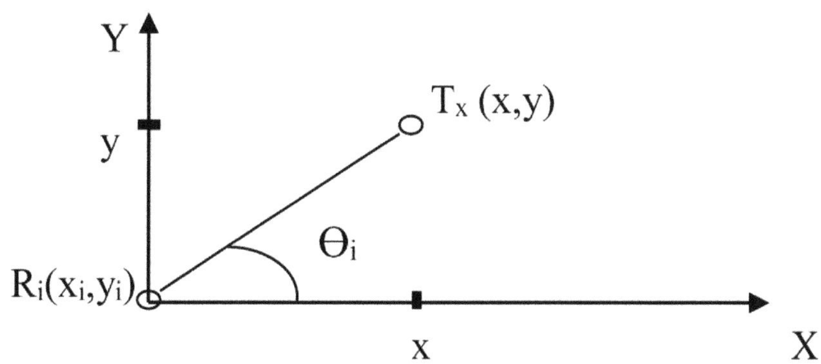

$$\hat{X}_{t,k+1} = \hat{X}_{t,k} + \left(\theta_X\left(\hat{X}_{t,k}\right)^T S^{-1}\theta_X\left(\hat{X}_{t,k}\right)\right)^{-1} \theta_X\left(\hat{X}_{t,k}\right)^T S^{-1}\left[\beta - \theta_X\left(\hat{X}_{t,k}\right)\right] \tag{35}$$

where $\theta_X\left(\hat{X}_{t,k}\right)$ denotes the partial derivative of θ with respect to x evaluated at $\left(\hat{X}_{t,k}\right)$. The use of equation (34) requires an initial estimate close enough to the true minimum of the cost function. Such an initial estimate may be obtained from prior information, or using a suboptimal procedure. There is an approach called Stansfield approach which assumes that the measurement error is small enough such that

$\varepsilon_i \approx \sin \varepsilon_i$, $1 \leq i \leq N$.

In that case, the cost function in equation (34) becomes:

$$\frac{1}{2}\sum_{i=1}^{N} \frac{\sin^2\left(\theta_i\left(\hat{X}_t\right) - \beta_i\right)}{\sigma_i^2}$$

Using the relation

$$\sin\left(\theta_i(X_t) - \beta_i\right) = \sin\theta_i(X_t)\cos\beta_i - \cos\theta_i\left(X_t\right)\sin\beta_i$$

$$= \frac{\left(y_t - y_i\right)\cos\beta_i - \left(x_t - x_i\right)\sin\beta_i}{r_i} \tag{36}$$

where:

$$r_i = \sqrt{\left(x_t - x_i\right)^2 + \left(y_t - y_i\right)^2}$$

Equation (3) becomes:

$$\frac{1}{2}\sum_{i=1}^{N} \frac{\left[\left(y_t - y_i\right)\cos\beta_i - \left(x_t - x_i\right)\sin\beta_i\right]^2}{\sigma_i^2 r_i^2}$$

$$= \frac{1}{2}\left(AX_t - b\right)^T R^{-1} S^{-1} \left(Ax_t - b\right) \tag{37}$$

where:

$$A = \begin{bmatrix} \sin\beta_1 & -\cos\beta_1 \\ \vdots & \vdots \\ \sin\beta_N & -\cos\beta_N \end{bmatrix}$$

$$b = \begin{bmatrix} x_1 \sin\beta_1 & -y_1 \cos\beta_1 \\ \vdots & \vdots \\ x_N \sin\beta_N & -y_1 \cos\beta_N \end{bmatrix}$$

$$R = diag\left\{r_1^2, \ldots.., r_N^2\right\}$$

Stansfield implicitly assumes that even though R is not perfectly known, a rough estimate of R can be obtained. Since the cost function weakly depends on R, the fact that the estimate is rough will not significantly affect the solution. Under these assumptions, the minimization of equation (37) with respect to x_t is a well-known problem and the solution is given by:

$\hat{x}_t = (A^T R^{-1} S^{-1} A)^{-1} A^T R^{-1} S^{-1} b$

Note that the closed form solution in the Stansfield approach depends on two assumptions: first, the measurement error is small such that $\varepsilon_i \approx \sin \varepsilon_i$, $1 \leq i \leq N$, second, R is known. One may choose to accept the first assumption but reject the second assumption. In that case an iterative procedure can be used to obtain the solution to the minimization problem, which has no advantage over the ML (Maximum likelihood) technique. It was shown that the Stansfield approach provides biased estimates even for a large number of bearing measurements and the ML approach is asymptotically unbiased at a large number of measurements. However the RMS (root mean square) error of Stansfield approach is not necessarily larger than that of the ML approach. Mao, Fidan and Anderson (2007) showed that in their work.

Using Hyperbola Principle Algorithm Which Is Based on TDOA in Localization

Given the TDOA measurement Δt_{ij} and the coordinates of receivers i and j, equation (4) defines one branch of a hyperbola whose foci are at the locations of receivers i and j and on which the transmitter r_t must lie. In 2D measurements from a minimum of three receivers are required to uniquely determine the location of the transmitter. In a system of N receivers, there are N - 1 linearly independent TDOA equations, which can be written compactly as:

$$\begin{bmatrix} \|r_1 - r_t\| - \|r_N - r_t\| - c\Delta t_1 N \\ . \\ . \\ . \\ \|r_{N-1} - r_t\| - \|r_N - r_t\| - c\Delta t_{N-1} N \end{bmatrix} = 0 \tag{38}$$

In practice, Δt_{ij} is not available; instead we have the noisy TDOA measurement $\Delta \hat{t}_{ij}$ given by:

$\Delta \hat{t}_{ij} = \Delta t_{ij} + n_{ij}$,

where n_{ij} denotes an additive noise, which is usually assumed to be an independent zero-mean Gaussian distributed random variable. Equation (38) is a nonlinear equation that is difficult to solve, especially when the receivers are arranged in an arbitrary fashion. Moreover, in the presence of noise Equation (38) may not have a solution. A noisy version of Equation (38) can be written as:

$$\begin{bmatrix} \Delta \hat{t}_{1N} \\ . \\ . \\ . \\ \Delta \hat{t}_{N-1N} \end{bmatrix} = \begin{bmatrix} \frac{\|r_1 - r_t\| - \|r_N - r_t\|}{c} \\ . \\ . \\ . \\ \frac{\|r_{N-1} - r_t\| - \|r_N - r_t\|}{c} \end{bmatrix} + \begin{bmatrix} \varepsilon_{1N} \\ . \\ . \\ . \\ \varepsilon_{N-1N} \end{bmatrix}$$

Denote by $\Delta \hat{t}$ the TDOA measurement vector

$$\Delta \hat{t} \left[\Delta \hat{t}_{1N}, \ldots\ldots, \Delta \hat{t}_{N-1N}\right]^T.$$

Denote by $f(r_t)$ the vector:

$$\frac{1}{c}\left(\|r_1 - r_t\| - \|r_N - r_t\|\right)\left[\ldots\ldots\ldots 1/c\left(\|r_{N-1} - r_t\| - \|r_N - r_t\|\right)\right]^T$$

and denote by S the covariance matrix of the TDOA measurement errors. The ML estimator minimizes the following quadratic function:

$$Q(\hat{r}_t) = \left[\Delta \hat{t} - f(\hat{r}_t)\right] TS^{-1}\left[\Delta \hat{t} - f(\hat{r}_t)\right].$$

In which $f(r_t)$ is a nonlinear vector function. In order to obtain a reasonably simple estimator, $f(r_t)$ can be linearized using Taylor series around a reference point r_0:

$$f(r_t) \approx f(r_0) + f(r_0)(r_t - r_0)$$

where $f(r_0)$ is a matrix of dimensions $(N-1)\times 2$ of partial derivative of f with respect to r evaluated at r_0. ML method can be used to guess the transmitter location. This method relies on a good initial guess of the transmitter location. Moreover, in some situations this method can result insignificant location estimation errors due to geometric dilution of precision (GDOP) effects. GDOP describes a situation in which a relatively small ranging error can result in a large location estimation error because the transmitter is located on a portion of the hyperbola far away from both receivers

FUTURE RESEARCH DIRECTIONS

In future work, more localization algorithms including mobility of sensor nodes will be studied.

CONCLUSION

TDOA methods are impressively accurate under line-of-sight conditions. But this line-of-sight condition is difficult to meet in some environments. Furthermore, the speed of sound in air varies with air temperature and humidity, which introduce inaccuracy into distance estimation. Acoustic signals also show multi-path propagation effects that may impact the accuracy of signal detection TDOA also needs synchronization. AOA techniques provide more accurate localization result than RSSI based techniques but the cost of hardware of very high in AOA. However Angle-Of-Arrival measurements techniques and distance related measurements techniques can be used to determine the location of sensor nodes, RSS can be considered the most suitable localization technique for WSN because no additional hardware need to be added (rotating antenna, antenna array, synchronization equipment's) and is unlikely to significantly impact local power consumption which is a very important issue in WSN, sensor size and thus cost.

REFERENCES

Boudhir, A. A., Mohamed, B., & Mohamed, B. A. (2010). New Technique of Wireless Sensor Networks Localization based on Energy Consumption. *International Journal of Computers and Applications*, *9*(11), 25–28. doi:10.5120/1436-1935

Bulusu, N., Heidemann, J., & Estrin, D. (2000). GPS-less low-cost outdoor localization for very small devices. *IEEE Personal Communications*, *7*(5), 28–34. doi:10.1109/98.878533

Capkun, S., & Hubaux, J.-P. (2005). Secure positioning of wireless devices with application to sensor networks. *Proc. 24th Annual Joint Conference of the IEEE Computer and Communications Societies (INFOCOM'05)*. doi:10.1109/INFCOM.2005.1498470

Function, L. (2009). *Lecture 5 Maximum Likelihood Method*. Academic Press.

Gemayel, N., El Koslowski, S., & Jondral, F. K. (2009). *A low cost TDOA Localization System : Setup, Challenges and Results*. Academic Press.

He, T., Huang, C., Blum, B. M., Stankovic, J. A., & Abdelzaher, T. (2003). Range free localization schemes for large scale sensor networks. In *Proc. 9th Annual International Conference on Mobile Computing and Networking, (MobiCom'03)*, (pp. 81–95). doi:10.1145/938985.938995

Kwon, Y. M., Mechitov, K., Sundresh, S., Kim, W., & Agha, G. (2005). Resilient localization for sensor networks in outdoor environments. In *Proc. 25th IEEE International Conference on Distributed Computing Systems (ICDCS'05)*, (pp. 643–652). doi:10.1109/ICDCS.2005.68

Mao, G., Fidan, B., & Anderson, B. D. O. (2007). Wireless sensor network localization techniques. *Computer Networks*, *51*(10), 2529–2553. doi:10.1016/j.comnet.2006.11.018

Models, B. C. (2003). *ML Estimation, Binary Choice Models, (Ml)*. Academic Press.

Ou, C.-H. (2011). Wireless Sensor Networks A Networking Perspective. *IEEE Sensors Journal*, *11*(7), 1607–1616. doi:10.1109/JSEN.2010.2102748

Oureiro, A. N. A. F. L., Niversity, F. E. U., & Inas, O. F. M. (2007). *Localization systems for wireless sensor networks*. Academic Press.

Pal, A. (2010). Localization Algorithms in Wireless Sensor Networks: Current Approaches and Future Challenges. *Network Protocols and Algorithms*, *2*(1), 1607–1616. doi:10.5296/npa.v2i1.279

Patil, M. M., Shaha, U., Desai, U. R., & Merchant, S. N. (2011). Localization in wireless sensor networks using three masters. In *2005 IEEE International Conference on Personal Wireless Communications, 2005. ICPWC 2005.* (Vol. 11, pp. 384–388). IEEE. doi:10.1109/ICPWC.2005.1431371

Patwari, N., Hero, A. O., Perkins, M., Correal, N. S., & Dea, R. J. O. (2003). Article. *Wireless Sensor Networks*, *51*(8), 2137–2148.

Peng, R., & Sichitiu, M. L. (n.d.). Robust, Probabilistic. *Constraint-Based Localization for Wireless Sensor Networks.*

Rice, J. A. (1995). Mathematical Statistics and Data Analysis (2nd ed.). Duxbury Press.

Santos, F. (2008). Localization in Wireless Sensor Networks. *ACM Journal,* 1–19.

Vander Stoep, J. (2009). Design and Implementation of Reliable Localization Algorithms using Received Signal Strength. *IEEE Sensors Journal*, 1–47.

Wang, J., Ghosh, R. K., & Das, S. K. (2010). A survey on sensor localization. *Journal of Control Theory and Applications*, *8*(1), 2–11. doi:10.1007/s11768-010-9187-7

Zhang, Q., Foh, C. H., Seet, B.-C., & Fong, A. C. M. (2010). Location estimation in wireless sensor networks using spring-relaxation technique. *Sensors (Basel, Switzerland), 10*(5), 5171–5192. doi:10.3390/s100505171 PMID:22363204

ADDITIONAL READING

Capkun, S., Hamdi, M., & Hubaux, J.-P. (2001). GPS-free positioning in mobile adhoc networks. In *Proc. 34th Annual Hawaii International Conference on System Sciences (HICSS'01)*, pages 1–15, Maui, Hawaii, USA.

B. M. M. El-Basioni, S. M. A. El-kader, and M. Abdelmonim,, (2013)"Smart home design using wireless sensor network and biometric technologies," information technology, vol. 1, p. 2.

El-kader, S. M. A., & El-Basioni, B. M. M. (2013). Precision farming solution in Egypt using the wireless sensor network technology. *Egyptian Informatics Journal, 14*(3), 221–233. doi:10.1016/j.eij.2013.06.004

Ferris, B., Hahnel, D., & Fox, D. (2006). Gaussian processes for signal strength-based location estimation. In Proc. Robotics: Science and Systems, Philidelphia, PA, USA. doi:10.15607/RSS.2006.II.039

He, T., Huang, C., Blum, B. M., Stankovic, J. A., & Abdelzaher, T. (2003). Range free localization schemes for large scale sensor networks. In *Proc. 9th Annual International Conference on Mobile Computing and Networking, (MobiCom'03)*, pages 81–95, San Diego, CA, USA. doi:10.1145/938985.938995

Holland, M., Aures, R., & Heinzelman, W. (2006). Experimental investigation of radio performance in wireless sensor networks. In *Proc. 2nd IEEE Workshop on Wireless Mesh Networks (WiMesh'06)*, pages 140–150, Reston, VA, USA. doi:10.1109/WIMESH.2006.288630

LaMarca, A., Hightower, J., Smith, I., & Consolvo, S. (2010) Self-mapping in 802.11 location systems. In *Proc. 7th International Conference on Ubiquitous Computing (Ubicomp'05)*, pages 87–104, Tokyo, Japan.

Lazos, L., & Poovendran, R. (2005). SeRLoc: Robust localization for wireless sensor networks. *ACM Transactions on Sensor Networks, 1*(1), 73–100. doi:10.1145/1077391.1077395

Li, Z., Trappe, W., Zhang, Y., & Nath, B. Robust statistical methods for securing wireless localization in sensor networks. In *Proc. 4th International Symposium on Information Processing in Sensor Networks (IPSN'05)*, pages 91–98, Los Angeles, CA, USA, April 2005.

Liu, D., Ning, P., & Du, W. K. (2005) Attack-resistant location estimation in sensor networks. In *Proc. 4th International Symposium on Information Processing in Sensor Networks (IPSN'05)*, pages 99–106, Los Angeles, CA, USA.

Lymberopoulos, D., Lindsey, Q., & Savvides, A. (2006) An empirical characterization of radio signal strength variability in 3-d IEEE 802.15.4 networks using monopole antennas. In *Proc. European Workshop on Wireless Sensor Networks (EWSN'06)*, pages 326–341, Zurich, Switzerland. doi:10.1007/11669463_24

Nabeel, M. M., el Deen, M. F., & El-Kader, S. (2013). Intelligent Vehicle Recognition based on Wireless Sensor Network. *International Journal of Computer Science Issues*, *10*(4), 164–174.

Nagpal, R., Shrobe, H., & Bachrach, J. (2003). Organizing a global coordinate system from local information on an ad hoc sensor network. In *Proc. 2nd International Workshop on Information Processing in Sensor Networks (IPSN'03)*, Palo Alto, CA, USA. doi:10.1007/3-540-36978-3_22

Niculescu, D., & Nath, B. (2001). Ad hoc positioning system (APS). In *Proc. IEEE Global Telecommunications Conference (GLOBECOM'01)*, San Antonio, TX, USA. doi:10.1109/GLOCOM.2001.965964

Chapter 11
Self-Organizing Network Solutions:
A Principal Step Towards Real 4G and Beyond

Mohammadreza Behjati
Loughborough University, UK

John Cosmas
Brunel University London, UK

ABSTRACT

Quality of service (QoS) and network capacity are being insisted as the two dominant factors for the utmost network satisfaction within any mobile network contracts. On the other hand, the heterogeneous network (HetNets), which are constructed based on sub-network layers' cooperation between macrocell and shorter-range applications like micro, femto and relay nodes, are also introduced as an open door to the recent researches towards the desired network satisfaction for the recently upgraded networks like LTE-Advanced (LTE-A). Nevertheless, since any network cooperation is expected to include a number of challenges; this cooperation is not excluded of dealing with degrading effects, such as interference, among the sub-network elements. This chapter presents a detailed discussion in self-organizing network (SON) methodology, as a novel solution to deal with network challenges, e.g. inter-cell interference coordination (ICIC), mobility, power control, etc. to improve the network quality and capacity.

INTRODUCTION

Today's mobile media and communications are being affected, more than any other period of time, by communal needs and requirements within the geographical areas. The huge number of clients demands for a smooth and reliable communications from one side, and the financial and technical strategies of the telecommunication companies and network operators from the other side, propose a new challenge to deal with mobile networks' clients. Therefore, designing up to dated network solutions to alleviate

DOI: 10.4018/978-1-5225-2342-0.ch011

the growing challenges of clients and network designers must be considered as a big step towards the public satisfactions, as well as an open door to the future networking plans.

Long Term Evolution (LTE) has been proposed by network operators to satisfy their commercial and domestic clients, who are permanently looking for more upgrades in mobile communication services. In continue with this strategy, and as a dominant enhancement to the existing LTE systems, LTE-Advanced (LTE-A) is then known as real 4G evolution step, that has recently been standardized in 3GPP Rel-10 and approved by ITU and IMT-Advanced (Yang, Hu, Xu, & Mao, 2009). The decisive factors of data rate, delay and capacity, as well as energy consumption and implementation costs are the main evolving points on LTE-A, which have been targeted by network operators to fulfill their responsibilities and promises to their clients (Mahmoud, 2007).

As a supplementary step to the existing LTE systems, LTE-Advanced is recently presented to the network subscribers by its dominant characteristics, ubiquitous supports and quality-based evolutions, which could create the center of attentions to the anticipated 4G enhancements in various points of views, such as the social aspects.

In recent years, most of the technology followers initially look for the improvements of telecommunication technologies to upgrade their subscribed services, and therefore the heterogeneity of LTE-Advanced networks has been a superior motivation for them to receive more reliable coverage and supports.

This chapter argues the novel self-organizing paradigms, applications and solutions in upgrading towards real 4G mobile communications, based on LTE-Advanced networks. The self-organizing solutions in this part are mainly considered to face the foreseen challenges with the minimum of user responsibilities and costs for the end users. Most of the focuses in this work have been spent on the added values of the advanced versions of LTE, comparing to the existing conventional networks, proposed by its new self-organizing features and strategies. In addition, the effects of the proposed elucidations on ubiquity and quality of service in new releases of LTE-Advanced are discussed in details, and the real-work implementation capabilities are implied at the end.

CELLULAR NETWORKS EVOLUTION TOWARDS 4G

The LTE-Advanced network is already unified by adopting the transmission technologies such as multi-input multi-output (MIMO) and multiple access schemes to obtain maximum possible throughput in minimum possible time.

Figure 1. LTE-Advanced known as the real 4G evolution

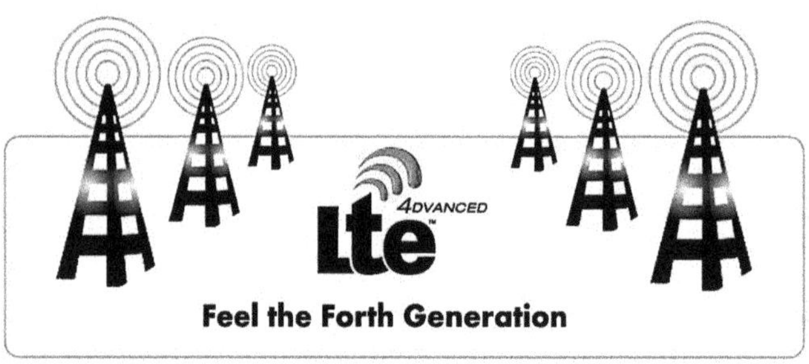

Figure 2. Cellular network evolution towards the real 4G systems

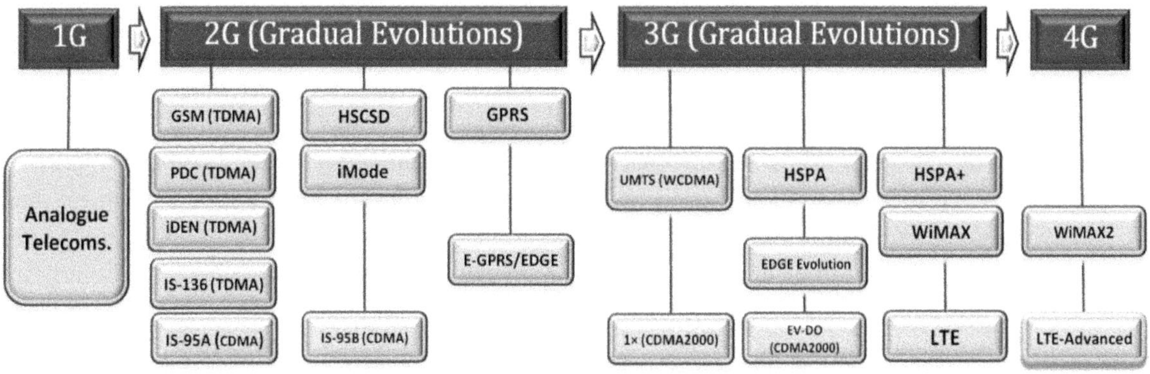

The peak data rate of 1 Gbit/sec with 100 Mbit/sec data rate in downlink using orthogonal frequency division multiple access (OFDMA) and 50 Mbit/sec data rate in uplink using single carrier frequency division multiple access (SC-FDMA) highly motivate network users to employ this quality technology.

Furthermore, intra-cell interference is already avoided due to the use of OFDMA technique in downlink, but inter-cell interference needs to be considerably mitigated.

By having a backward look to the existing systems, the added enhancements over the prior generations of cellular network confirm this fact that the foremost center of attention in LTE and LTE-Advanced enhancements has been regarded to the capacity and quality of service. As shown on picture 3, carrier aggregation process takes up to 5 component carriers (up to 100 MHz) from the same or different carrier bands, multi-antenna transmission improves spectral efficiency and channel capacity by expanding transmission dimension.

Figure 3. Dominant improved techniques within the LTE-Advanced networks

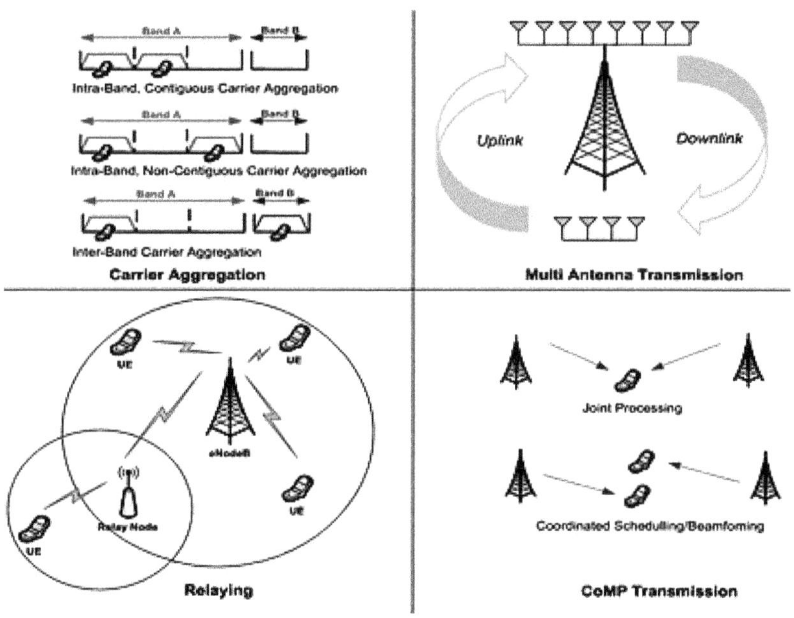

In addition to this, the relay station (RS) also helps to forward user information from the neighboring mobile station (MS) to a local eNodeB and vice versa, and coordinated multipoint (CoMP) transmission provide accountable support for mobile users by joining more than a single base station within the network (Behjati & Cosmas, 2012).

In case of spectrum allocation, the strategies for the intelligent use of spectrum have also been urgently required, since the currently available unlicensed spectrum is reaching its limits as well as the growing number of demands for the wireless access and applications. To alleviate this problem, spectrum sensing and dynamic spectrum allocation (DSA) technologies are being proposed to be implemented (Attar, Krishnamurthy, & Gharehshiran, 2011; Attar, Krishnamurthy, & Gharehshiran, 2011). Furthermore, using of short-range indoor applications should also be taken into account, which could also open new doors to the emerging technologies, like LTE-A, as well as alleviating the existing difficulties (Behjati & Cosmas, 2012).

In terms of the complimentary ideas over the existing LTE, number of novel ideas has caused undeniable influences on customer satisfactions, taking into account the existing coverage and quality difficulties of the prior systems. The strategies such as radio over fiber (RoF), cognitive radio (CR) and machine to machine (M2M) communications are assumed as supplementary plans over the mentioned baseline techniques to overcome the satisfaction challenges of network operators, caused by the growing number of customer demands, especially in urban and suburban areas. Figure 4 depicts spectrum allocation for existing and cognitive LTE base stations, to support the primary and secondary groups of mobile users. Many of the recent research works have been focused on investigations about cognitive radio and its applications within the existing LTE networks, which result notable spectrum efficiency on the distributed spectrums for LTE-Advanced.

Figure 4. Cognitive radio spectrum allocation over the existing macrocells

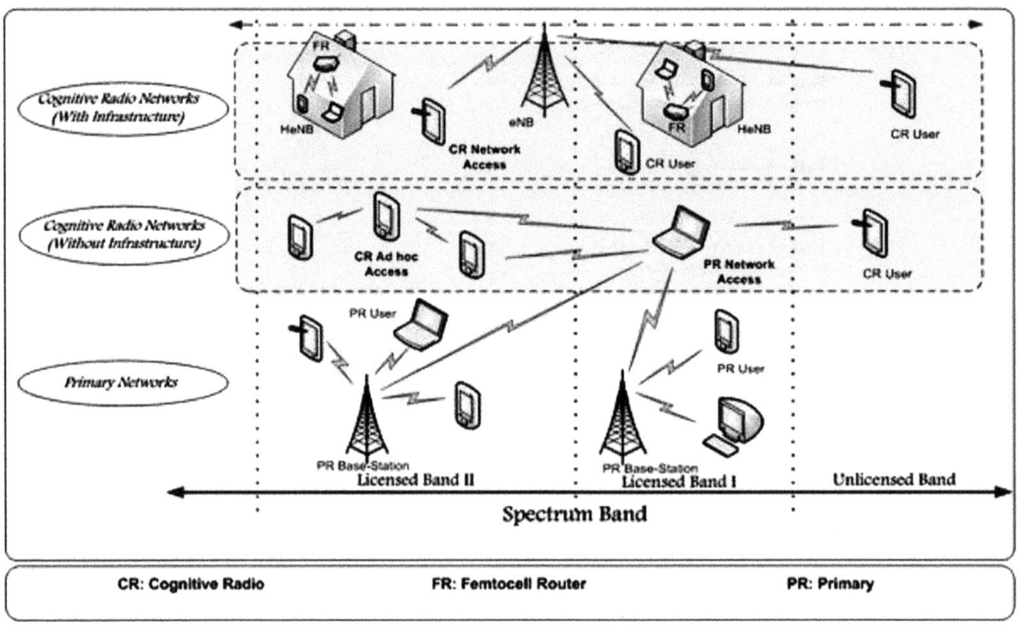

HETEROGENEOUS NETWORKS WITH SELF-ORGANIZING SOLUTIONS

Frameworks for Network Heterogeneity in New Releases

As already mentioned at the beginning of this chapter, the upgraded, enhanced and advanced version of LTE, known as LTE-Advanced, is presented as a compatible infrastructure with the current equipments, more focusing on coverage and capacity issues. Heterogeneity in telecommunication networks is considered as a decisive factor for the network expertise to advantage from merging of various types of technologies, from short-range to long-range, and from low-power to high-power applications, within a unique network platform. Therefore design and planning of a heterogeneous network (HetNet) constructed by a multi-layer network structure and different ranges of base stations seems to be a suitable solution for this shortage (Behjati & Cosmas, 2013). As shown in Figure 5, short-range applications like picocells, relay nodes and femtocells are planned under the HetNet infrastructure, which is linked to the core network through the designated backhaul.

The pico base stations have substantially lower transmission power comparing to macro base stations, as well as mostly ad-hoc placement in the network. On the other hand, the networks which use low-power base stations are being expected to have lower signal to interference conditions, because of their unplanned deployment on the network, which resulting in a challenging RF situation for control channel transmissions to the cell-edge users. Beside the mentioned facts, an important tip regarding macro-pico deployment is the large difference of the transmit power between macro and pico base-stations (e.g. 20 dB) in the network, which causes the smaller downlink coverage of a pico base-station comparing to the macro. But, this is not the case for uplink, which is proposing the same transmit power strength from user terminal to all the base stations, because it only depends on the user terminal's transmit power itself (Khandekar, Bhushan, Tingfang, & Vanghi, 2010).

Relay stations (RS) also provide wide range of applications within HetNets. There are two types of RSs defined in 3GPP LTE-Advanced and 802.16j standards: type 1 (non-transparency), which could

Figure 5. Heterogeneous architecture opens new doors to LTE-Advanced trend

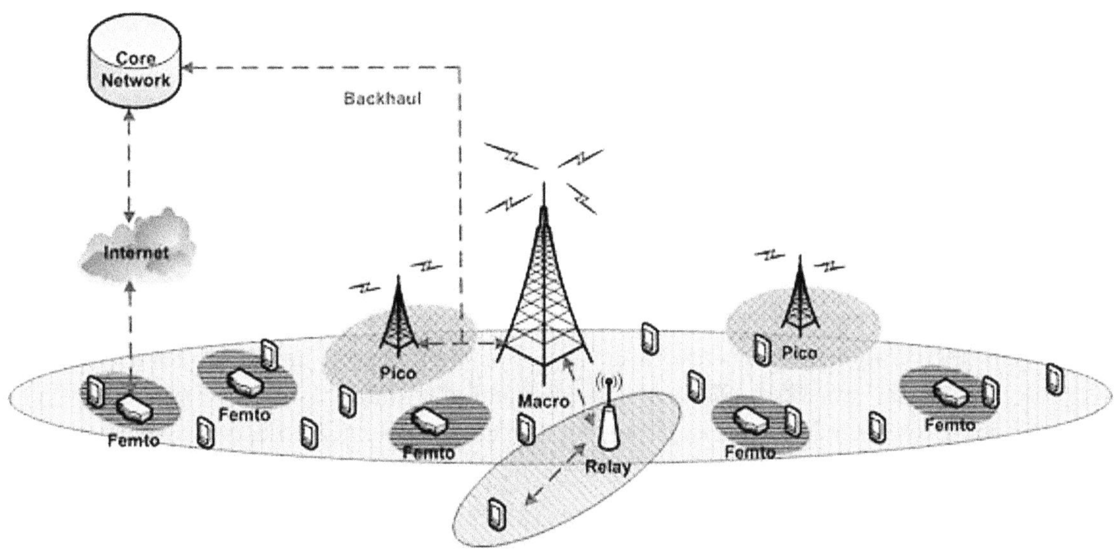

help a remote UE located far away from eNB to access to the eNB. This type of relays needs to transmit the common reference signal and the control information for the eNB, and its main aim is to extend the signal coverage and services. On the other hand, type 2 (transparency) could help a local UE, which is located within the coverage of eNB and has a direct communication link with eNB to improve the link capacity and service quality. So it does not transmit the common reference signal and the control information and its main aim is just to increasing of the overall system capacity, by achieving the transmission gain and multipath diversity for the local UEs (Yang, Hu, Xu, & Mao, 2009).

SON strategy deployment is a new strategy to escalation the network satisfaction, by using the internal communicative coordination amongst the existing and neighboring cellular base stations, inside the cellular LTE-Advanced network. This could be practically reached by gaining a smooth transmission platform via mitigating the inter-cell interference in current releases of LTE Apart from the resource allocation strategies on shared or separated radio channels, advanced interference management schemes are nevertheless required to be designed within the femtocells, to avoid either the interference among femtocells, or between a femto and a macro, known as co-channel interference (Behjati & Cosmas, 2013).

Therefore, as the first step towards network optimization, two-tier deployment should be followed by required techniques to deal with the interference. Before applying the interference coordination scheme, the power control interference avoidance techniques which contain a novel interference management for the macro users in heterogeneous networks, are applicable to both macro and femto users. Femtocells are being considered as low-power, user-deployed base stations, which provide high-quality cellular service in indoor environments via operating in licensed spectrum. The trade-off between the improvement of macrocell users' throughput and the degradation of femtocell user's throughput is the key benefit of this scheme compared to the smart power control. At the initial stage of work, a multi-layer heterogeneous LTE system is planned to intend a SON solution for the interference coordination, including both macro stations and femto indoor station, to be realized as the first and second layers of networks respectively (Behjati, Cosmas, & Nilavalan, 2013). The enhanced channel estimation and interference mitigation will result in less inter-cell interference and higher signal to noise ratio (SNR) and therefore lesser number of packet drops within the network which totally leads to get higher Quality of Service (QoS). Furthermore, reduce in packet drops also results in obtaining higher data traffic throughput, and higher SNR in transmission aims to obtain better quality of received signal within the network (Behjati & Cosmas, 2013; Khandekar, Bhushan, Tingfang, & Vanghi, 2010).

Femtocells as a New Trend in LTE-A Recent Releases

Femtocell are known as intelligent access points which support any 3G mobile device which use cellular air interface, e.g. CDMA2000, UMTS and LTE. The femto base stations are tightly integrated with the existing macro networks, and so their using and switching between macro and femto are seamless for the users in particular. Femtocells also use their Radio Frequency (RF) awareness to control interference between themselves and between femtocells and macrocells within the heterogeneous networks, with no need to do any modification in current mobile devices. The femto network architecture and its specifications allow the ordinary users to install them with plug-and-play simplicity. Femto applications in LTE network are being considered a fundamental step for the network operators to satisfy a great part of their customers' expectations. The major mobile operators are planning aggressive LTE service roll-outs.

There are three basic options for the mobile operators to allocate the available frequencies in femtocell deployments. On the first scenario, femto and macro deployments work on a separate radio channels. The

Figure 6. Femtocell technology focus on indoor applications

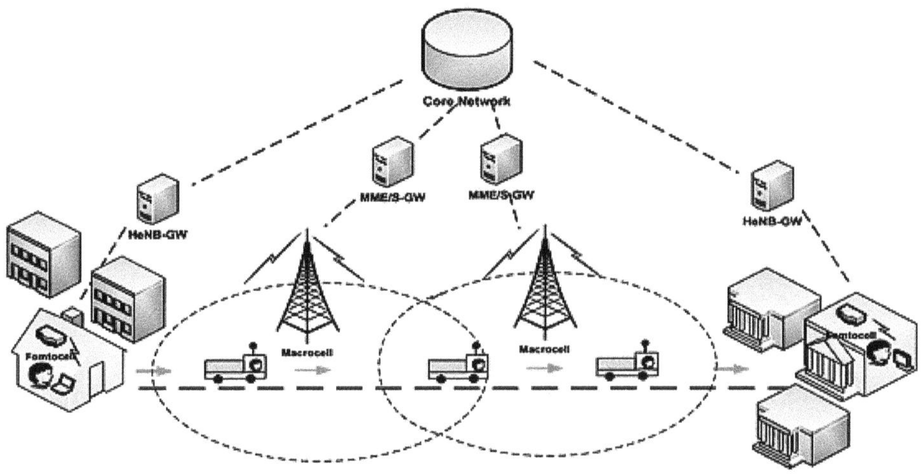

advantage of this implementation is minimizing interference between the two networks and simplifying the initial deployment of the femto application through the network. This is more suitable in rural areas and more feasible in CDMA2000 systems which have got narrower bandwidth and therefore the radio channels are more numerous.

In the third scenario, all the available radio channels are shared between the macro and femto networks. The advantage of this strategy is providing more freedom to manage interference among the femtocellls, especially in dense urban deployments, but requiring greatest degree of the interference management to minimize the co-channel interference with macrocells. But in the second scenario of his case, as a compromise between the scenarios one and three, some of the available radio channels are shared between the macro and femto networks, and the other available radio channels are fully reserved for the macro. In this scenario, the macrocell can redirect the mobile devices which are being served on the shared radio channel to a dedicated macrocell radio channel when they approach a femtocell. One example of deploying this strategy is using the mentioned shared radio channel for macrocell data services (HSxPA), and for femtocell voice and data services, and leaving the dedicated radio channels for macrocell voice services as a separated application.

Apart from these strategies regarding the shared or separated radio channels, the advanced interference schemes required to be designed within the femrocells anyway, to avoid either the interference among femtocells, or between a femto and a macro, known as co-channel interference (Airvana Mobile Broadband, 2010).

Interference Consideration Through Macro and Femto Cooperation

Heterogeneous network studies need to consider interference while applying of different applications within the same frequency bands. As a critical point, the intra-cell interference, which is regarding to the interference between eNBs inside the cell, is basically avoided in LTE network due to the orthogonality between subcarriers in OFDMA technique (Behjati & Cosmas, 2012; Tabia, Gondran, Baala, & Caminada, 2011). LTE standard also supports the CoMP transmissions in which a given UE will be served simultaneously by more than one eNB. But it is not feasible in practical networks, because of the

Figure 7. Co-channel interference in multi-cellular platforms

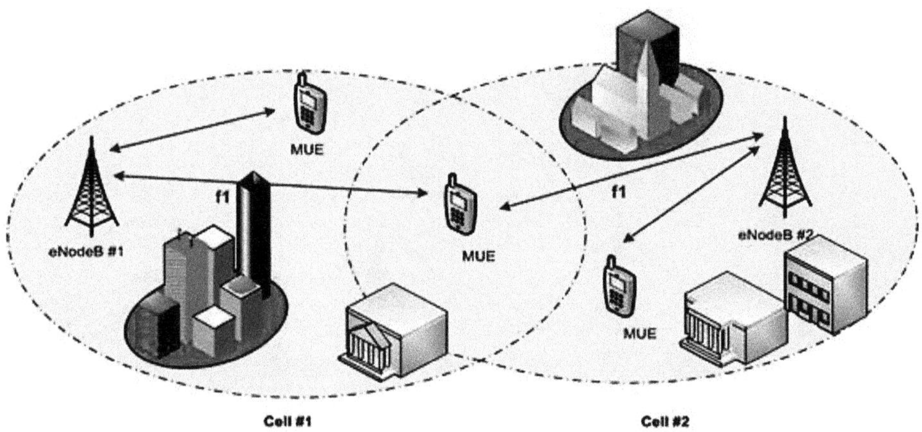

information exchanges as well as resource allocation is changing quickly in time. This problem could be solved by forming a self organizing plan by each eNB and information of the environment is fed back to the cognitive eNB by UEs in periodic intervals. Therefore, the efficient inter-cell dynamic resource allocation can be developed (Attar, Krishnamurthy, & Gharehshiran, 2011).

There would be also some problems due to deployment of a larger number of low-power base stations. Especially in co-channel deployment scenarios in heterogeneous networks, the interference may become serious due to use of low-power nodes, which could be leading to high outage of the macro users.

The smart power control is generally planned to mitigate the substantial interference caused by macro base station (MeNB) on a Home UE (HUE) and also the interference caused by home base station (HeNB) on a macro UE (MUE) in the scenarios that MUE and HUE are far from their own stations (Z. Wang, Xiong, Dong, J. Wang, & Li, 2011).

There are two different cases of interference could happen with uplink transmission in heterogeneous networks. First is the very high interference levels exist at the edge of a macrocell, as is shown on Figure 8. It is because the macro-UEs transmit with high powers to overcome high path-loss towards their serving macro-eNBs located in a farther distance. The femto-UEs are the main victims in this case. The second case is the interference caused by a femto-UE on a close macro-eNB. Comparing of these two different cases of interference in uplink, the level of the interference at the edge of a macrocell generated by macro-UE is more intensive, and therefore the femto-UEs in such locations should use higher power to overcome this interference. On the other hand, the femto-UEs close to the macro-eNB should

Figure 8. Co-channel interference in an isolated cell

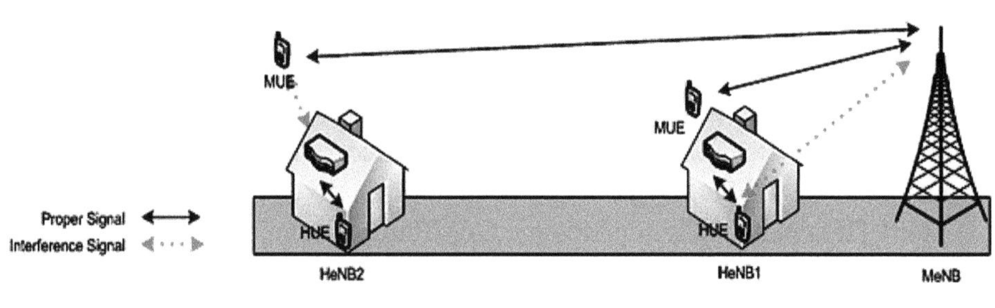

take lower transmit power to avoid causing interference on the macrocell (Góra, Pedersen, Szufarska, & Strzyz, 2010).

The proposed HetNet LTE-A plan contains two scenarios. Firstly, the conventional macro-only network is initially proposed and simulated without using of indoor applications (e.g. femtos), in which a single macro base-station has the duty of serving all the users. The presented multilayer configuration could be assumed as the main phase to employ a ubiquitous LTE-A network, which supports the increasingly higher demand within urban and sub-urban access networks. Therefore, as the second scenario, the multilayer macro+femto HetNet architecture is demonstrated to evaluate the effects of the femtocell indoor application as a vast enhancement to the existing macro networks within the system. The simulations are planned and executed on OPNET modeler software to be able to enhance the network layers such as network layer, node model, processing model, etc.

In case of the quality of service (QoS) studies, the value of signal to noise ratio (SNR) contains the interference noise (Pi), as well as the background noise (Pb) and the received power (Pr) for packet transmissions through the physical layer. The value of SNR is measured from the user point of view, which generates the applications, e.g. FTP, voice and video conferencing traffics, based on Equation 1.

$$SNR = 10\log_{10}\left[\frac{P_r}{P_b + P_i}\right] \quad (1)$$

Furthermore, the value of Block error rate (BLER) is also calculated according to the Equation 2 for both downlink and uplink transmissions. This calculation depends of both the number of blocks in error, and the total number of blocks received with the mobile users within the geographic cell.

$$BLER = \frac{Number\ of\ Erroneous\ Blocks}{Total\ Number\ of\ Received\ Blocks} \quad (2)$$

Based on simulation with OPNET modeler simulator, in both the SNR and BLER measures, the macrocell has an insufficient range of transmission on its own, which could result to facing a decreasing quality of service on the network, as well as the shortage in coverage. Remembering the quality and coverage, as the two main objectives of network satisfaction, this shortage of the macro-based networks convinces the network designers to adjoin at least one more supportive network layer, even with smaller transmission power to the platform.

To overcome the shortage of the spectrum and coverage difficulties of the networks while considering interference problem, especially at the cell boundaries, supportive femto applications is implemented as the second layer/tier of the LTE network in addition to the fundamental macro layer and in cooperation with it, as is shown in Figure 9.

According to the above mentioned scenario, Figure 10 shows the internal network units, gateways and interfaces to link femtocell HeNodeBs to the Evolved Packet Core (EPC) unit. It could be observed that the additional gateways are added into the existing serving gateways to transport massages and demands from and to the femocell units to implement the required messaging for the explained interference coordination technique.

Figure 9. Two-tier deployment's units and backbones

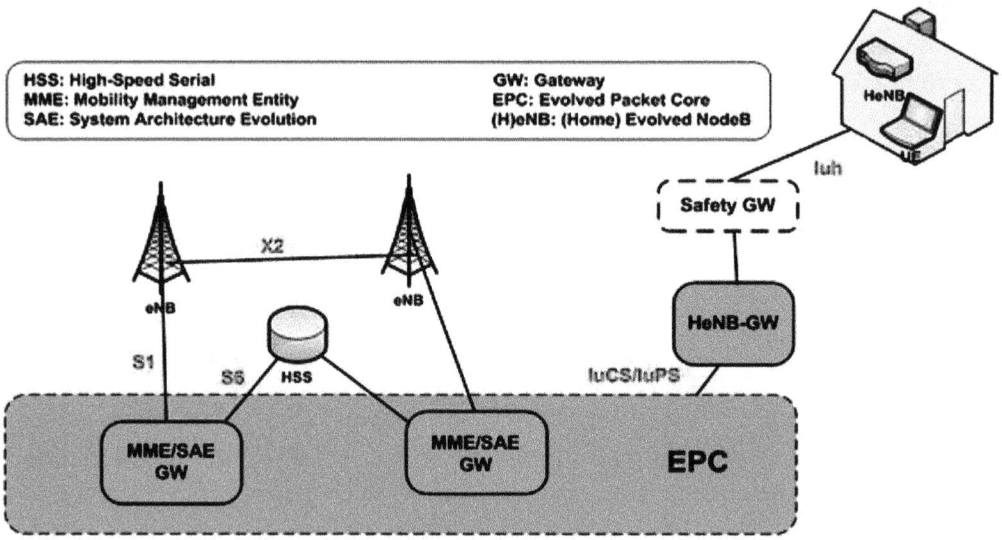

Figure 10. Internal interfaces in the two-tier deployment

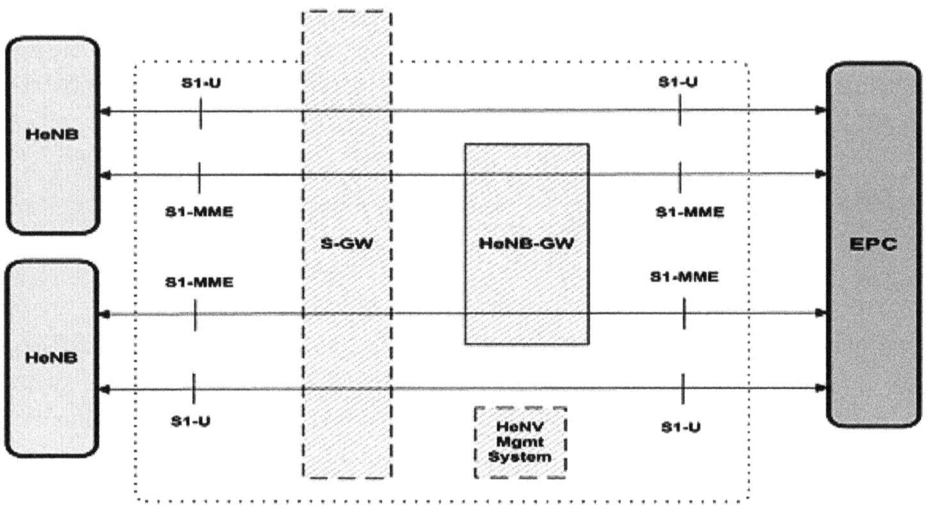

By applying the mentioned network management strategy, it could be totally observed that the signal coverage of the added numbers of femtocell routes on the macro-femto scenario improve the value of SNR by about 20%, which is proven to be due to the signal strength of the network, as well as the increased power to defeat the interference and background noises when the frequency channel is same. On the other hand, it could also be easily realized that the BLER value for the macro-femto scenario starts with a higher value due to the higher number of blocks in error at the beginning, but the UEs automatically go under the coverage of the femto base stations by approaching closer to the femto nodes, and therefore the overall trend of BLER is more gentle compared to the first scenario with macro-only policy. The value of the BLER is therefore improved by 12.5% comparing to the conventional architecture of macro-only systems. These improvements could confirm the positive effects of femto applications towards LTE-Advanced.

CONCLUSION

This chapter introduces the evolved features of LTE-Advanced systems, called self-organizing strategies, to deploy a ubiquitous network which supports the mobile network clients with both improved quality, as well as higher capacity, as a significant social step on mobile communications. The sociality aspect of this work is fulfilled through the user satisfaction objectives including the financial improvements. In the technical view, this work proposes using of femtocell deployment, as a low-power/short-range indoor application to produce heterogeneity of the new releases of the network. Considering mobility management, power consumption and specially interference problems of the advanced versions, alongside with the existing and added interfaces over the conventional LTE systems open new doors to take the utmost advantages of HetNet cooperation. As the results show, as a great instance in quality of service of the network, using of femtocell with a multi-layer structure could result 20% and 12.5% improvement in SNR and BLER respectively. This improvement could motivate the network researcher and mobile operators to keep stepping toward self-organization ideas within cooperative networks.

As the future work to this research, the authors of this chapter are currently focused on new technologies for 5G networks, i.e. SDN, NFV, IoT, M2M, Massive MIMO, D2D, Smart Networks, etc.

REFERENCES

Airvana Mobile Broadband. (2010). *Femtocell Technology.* Author.

Attar, A., Krishnamurthy, V., & Gharehshiran, O. N. (2011). Interference Management Using Cognitive Base-Stations for UMTS LTE. *IEEE Communications Magazine*, *49*(8), 152–159. doi:10.1109/MCOM.2011.5978429

Behjati, M., & Cosmas, J. (2012). Cognitive Spectrum Allocation and Interference Mitigation in 4G LTE-Advanced Networks. *5th Annual Research Student Conference (ResCon'12).* Brunel University.

Behjati, M., & Cosmas, J. (2012). *Cognitive-based Spectrum Flexibility Enhancements in LTE-Advanced Networks (DAMAVAND). Graduate School Poster Competition.* London, UK: Brunel University.

Behjati, M., & Cosmas, J. (2013). Multi-Layer Cell Deployment Strategy for Self-Organizing LTE-Advanced Networks. *IEEE International Wireless Communications and Mobile Computing Conference (IWCMC).* doi:10.1109/IWCMC.2013.6583663

Behjati, M., & Cosmas, J. (2013). Self-Organizing Network Interference Coordination for Future LTE-Advanced Networks. *IEEE International Symposium on Broadband Multimedia Systems and Broadcasting (BMSB)*. doi:10.1109/BMSB.2013.6621759

Behjati, M., Cosmas, J., & Nilavalan, R. (2013). Hierarchical Self-Organizing Strategy in LTE-Advanced Networks. *6th Annual Research Student Conference (ResCon'13)*. Brunel University.

Dahlman, E., Parkvall, S., & Sköld, J. (2013). *4G: LTE/LTE-Advanced for Mobile Broadband*. AP-Elsevier.

Góra, J., Pedersen, K. I., Szufarska, A., & Strzyz, S. (2010). Cell-Specific Uplink Power Control for Heterogeneous Networks in LTE. *IEEE 72th Vehicular Technology Conference Fall (VTC2010-Fall)*, 1-5.

Khandekar, A., Bhushan, N., Tingfang, J., & Vanghi, V. (2010). LTE-Advanced: Heterogeneous Networks. *European Wireless Conference (EW), IEEE*, 978-982. doi:10.1109/EW.2010.5483516

Mahmoud, Q. H. (2007). Front Matter. In *Cognitive Networks: Towards Self-Aware Networks*. Chichester, UK: John Wiley & Sons, Ltd. doi:10.1002/9780470515143.fmatter

Tabia, N., Gondran, A., Baala, O., & Caminada, A. (2011). Interference Model and Evolution in LTE Networks. *4th Joint IFIP Wireless and Mobile Networking Conference (WMNC)*, 26-28.

Wang, Z., Xiong, W., Dong, C., Wang, J., & Li, S. (2011). A Novel Downlink Power Control Scheme in LTE Heterogeneous Network. *IEEE International Conference on Computational Problem-Solving (ICCP)*, 241-245. doi:10.1109/ICCPS.2011.6089792

Yang, Y., Hu, H., Xu, J., & Mao, G. (2009). Relay Technologies for WiMAX and LTE-Advanced Mobile System. *IEEE Communications Magazine*, *47*(10), 100–105. doi:10.1109/MCOM.2009.5273815

KEY TERMS AND DEFINITIONS

Coordinated Multipoint (CoMP) Transmission: The internal cooperation between multiple base-stations to provide more consistent network coverage and support for mobile users(s).

Device to Device (D2D): Direct communication between two mobile users in cellular networks, without overpassing the base station or core network.

Dynamic Spectrum Allocation (DSA): A set of techniques to provide optimized spectrum allocation to the different network applications.

Internet of Things (IoT): The interworking of different objects or physical devices, embedded with network elements, in order to collect and exchange data.

Machine to Machine (M2M): Direct communication, either wired or wireless, between devices, using an appropriate communication channel.

Massive MIMO: A MIMO system with less number of terminals than the number of base station antennas within a network.

Multi-Input Multi-Output (MIMO): Use of multiple antennas at both the transmitter and receiver sides of the network to improve the network performance.

Network Functions Virtualization (NFV): A network architecture perception to virtualize network functions into building blocks with advanced capabilities.

Orthogonal Frequency Division Multiple Access (OFDMA): A Multi-user version of orthogonal frequency division multiplexing (OFDM) method, by assigning of different subcarriers to individual users.

Quality of Service (QoS): The overall performance of the network which is seen by the end users within the entire network.

Radio over Fibre (RoF): A network technology in which light is modulated by a radio signal and then transmitted over an optical fiber link for different applications.

Single Carrier Frequency Division Multiple Access (SC-FDMA): A FDMA scheme with single carrier to share the resources, which is mostly used in uplink communication.

Software-Defined Networking (SDN): An approach in network control and management via de-coupling the control plane from the data plane.

Compilation of References

Agarwal, R., & Motwani, M. (2009). Survey of clustering algorithms for MANET. *International Journal on Computer Science and Engineering*, *1*(2), 98–104.

Agila Bitsch Link, J. I., Schmitz, D., & Wehrle, K. (2011). GeoDTN: Geographic Routing in Disruption Tolerant Networks. *Proceedings of IEEE, Global Telecommunications Conference (GLOBECOM-2011)*, 1 - 5.

Agropedia. (2014, Feb 10). Retrieved Agropedia site from: http://agropedia.iitk.ac.in/

Ahmad, M., & Farooq, O. (2011). Chaos Based PN Sequence Generator for Cryptographic Applications. In *Proceedings of International Conference on Signal Processing and Communication Technologies*, (pp. 83-86). doi:10.1109/MSPCT.2011.6150443

Airvana Mobile Broadband. (2010). *Femtocell Technology.* Author.

Akbar, I., & Tranter, W. (2007). Dynamic spectrum allocation in cognitive radio using hidden Markov models: Poisson distributed case. *Proceedings of IEEE Southeast Conference*, 196–201. doi:10.1109/SECON.2007.342884

Akyildiz, I. F., Lo, B. F., & Balakrishnan, R. (2010, March). Cooperative Spectrum Sensing in Cognitive Radio Networks: A Survey. *Physical Communication*, *4*(1), 40–62. doi:10.1016/j.phycom.2010.12.003

Alibaygi, A., Karamidehkordi, M., & Karamidehkordi, E. (2011). Effectiveness of rural ICT centers: A perspective from west of Iran. *Procedia Computer Science, 3*, 1184–1188. Retrieved from http://www.sciencedirect.com/science/article/pii/S1877050910005661

Anastasios, M., Koutsouris, A., & Konstadinos, M. (2010). Information and Communication Technologies as Agricultural Extension Tools: A Survey among Farmers in West Macedonia, Greece. *Journal of Agricultural Education and Extension, 16*(3), 249–263. doi:10.1080/1389224X.2010.489767

Android Security Overview (ANDROID Security). (2014). Retrieved January 20, 2015, from http://source.android.com/devices/tech/security

Angell, I. O., & Smithson, S. (1991). *Information Systems Management: Opportunity and Risk.* London: Palgrave Macmillan. doi:10.1007/978-1-349-21555-3

Annonymous. (2014). *Data Tables: Planning Commission, Government of India.* Planning Commission of India. Retrieved from http://planningcommission.nic.in/data/datatable/

Arif, T. Y., & Sari, R. F. (2010). *Performance Comparison of Video Traffic Over WLAN IEEE 802.11e and IEEE 802.11n.* Paper presented at the UBICOMM 2010: The Fourth International Conference on Mobile Ubiquitous Computing, Systems, Services and Technologies, Florence, Italy. Retrieved from http://www.thinkmind.org/index.php?view=article&articleid=ubicomm_2010_13_30_10179

Armenta, A., Serrano, A., Cabrera, M., & Conte, R. (2012). The new digital divide: The confluence of broadband penetration, sustainable development, technology adoption and community participation. *Information Technology for Development*, *18*(4), 345–353. doi:10.1080/02681102.2011.625925

Arslan, H., & Yarkan, S. (2007). Binary time series approach to spectrum prediction for cognitive radio. *IEEE 66th Vehicular Technology Conference,* 1563 – 1567.

Attar, A., Krishnamurthy, V., & Gharehshiran, O. N. (2011). Interference Management Using Cognitive Base-Stations for UMTS LTE. *IEEE Communications Magazine*, *49*(8), 152–159. doi:10.1109/MCOM.2011.5978429

Bachrach, Y., Parnes, A., Procaccia, A. D., & Rosenschein, J. S. (2009). Gossip-based aggregation of trust in decentralized reputation systems. *Autonomous Agents and Multi-Agent Systems*, *19*(2), 153–172. doi:10.1007/s10458-008-9073-6

Baras, J. S., & Jiang, T. (2005, February). Managing trust in self-organized mobile ad hoc networks. In N. D. S. S. Workshop (Ed.), Extended abstract. Academic Press.

Bar, R. K., Mandal, J. K., & Singh, M. M. (2013). QoS of MANet Through Trust based AODV Routing Protocol by Exclusion of Black Hole Attack. *Procedia Technology*, *10*, 530–537. doi:10.1016/j.protcy.2013.12.392

Behjati, M., & Cosmas, J. (2012). Cognitive Spectrum Allocation and Interference Mitigation in 4G LTE-Advanced Networks. *5th Annual Research Student Conference (ResCon'12)*. Brunel University.

Behjati, M., & Cosmas, J. (2012). *Cognitive-based Spectrum Flexibility Enhancements in LTE-Advanced Networks (DAMAVAND). Graduate School Poster Competition*. London, UK: Brunel University.

Behjati, M., & Cosmas, J. (2013). Multi-Layer Cell Deployment Strategy for Self-Organizing LTE-Advanced Networks. *IEEE International Wireless Communications and Mobile Computing Conference (IWCMC)*. doi:10.1109/IWCMC.2013.6583663

Behjati, M., & Cosmas, J. (2013). Self-Organizing Network Interference Coordination for Future LTE-Advanced Networks. *IEEE International Symposium on Broadband Multimedia Systems and Broadcasting (BMSB)*. doi:10.1109/BMSB.2013.6621759

Behjati, M., Cosmas, J., & Nilavalan, R. (2013). Hierarchical Self-Organizing Strategy in LTE-Advanced Networks. *6th Annual Research Student Conference (ResCon'13)*. Brunel University.

Bernsen, J., & Manivannan, D. (2009). Unicast routing protocols for vehicular ad hoc networks: A critical comparison and classification. *Journal of Pervasive and Mobile Computing*, *5*(1), 1–18. doi:10.1016/j.pmcj.2008.09.001

Bernsen, J., & Manivannan, D. (2012). RIVER: A reliable inter-vehicular routing protocol for vehicular adhoc networks. *Journal of Computer Networks*, *56*(17), 3795–3807. doi:10.1016/j.comnet.2012.08.017

Bilal, S., Madani, S. A., & Khan, I. (2011). Enhanced junction selection mechanism for routing protocol in VANETs. *International Arab Journal of Information Technology*, *8*(4), 422–429.

Blake, S., Black, D., Carlson, M., Davies, E., Wang, Z., & Weiss, W. (1998). *An Architecture for Differentiated Services (RFC-2475)*. Academic Press.

Boba, M. S., Nor, S. M., & Nagar, S. A. (2014). A survey of unicast routing protocols based greedy Forwarding strategies for vehicular Ad–hoc networks in urban scenario. *Journal of Theoretical and Applied Information Technology*, *62*(1).

Boudhir, A. A., Mohamed, B., & Mohamed, B. A. (2010). New Technique of Wireless Sensor Networks Localization based on Energy Consumption. *International Journal of Computers and Applications*, *9*(11), 25–28. doi:10.5120/1436-1935

Bo, W., & Chengji, P. (2012). A New Optimal Design Method of Arbitrary Limited Period Spreading Sequences Based on Logistic Mapping. In *Proceedings of Fourth International Conference on Computational and Information Sciences (ICCIS)*, (pp 1269-1272).

Boyce, J. M., & Gaglianello, R. D. (1998). Packet Loss Effects on MPEG Video Sent Over the Public Internet. *Proceedings of the sixth ACM International Conference on Multimedia.*

Broderson, Wolisz, Cabric, Mishra, & Willkomm. (2004). *White paper: Corvus: A cognitive radio approach for usage of virtual unlicensed spectrum.* Academic Press.

Bulusu, N., Heidemann, J., & Estrin, D. (2000). GPS-less low-cost outdoor localization for very small devices. *IEEE Personal Communications*, *7*(5), 28–34. doi:10.1109/98.878533

Cao, Y., Sun, Z., Wang, N., & Cruickshank, H. (2013). Converge-and-diverge: A geographic routing for delay/disruption-tolerant networks using a delegation replication approach. *IEEE Transactions on Vehicular Technology*, *62*(5), 2339–2343. doi:10.1109/TVT.2013.2238958

Capkun, S., & Hubaux, J.-P. (2005). Secure positioning of wireless devices with application to sensor networks. *Proc. 24th Annual Joint Conference of the IEEE Computer and Communications Societies (INFOCOM'05).* doi:10.1109/INFCOM.2005.1498470

Chaithanya Manam, V. K., Mahendran, V., & Siva Ram Murthy, C. (2012). Message-driven based energy efficient routing in heterogeneous delay tolerant networks. *Proceeding of HP-MOSys'12, 1st ACM workshop on High performance mobile opportunistic systems*, 39 - 46. doi:10.1145/2386980.2386989

Chakrabarti, A. (2007). *Grid Computing Security.* Springer Berlin Heidelberg.

Chauhan, N. M. (2010). *Expectations of the Farmers from ICT in Agriculture.* Academic Press.

Cheng, H. T., Shan, H., & Zhuang, W. (2011). Infotainment and road safety service support in vehicular networking: From a communication perspective. *Journal of Mechanical Systems and Signal Processing*, *25*(6), 2020–2038. doi:10.1016/j.ymssp.2010.11.009

Chen, Z., & Qiu, R. C. (2010). Prediction of channel state for cognitive radio using higher-order hidden Markov model. *Proceedings of the IEEE Southeast Conference Concord*, 276 – 282. doi:10.1109/SECON.2010.5453870

Chikkerur, S., Sundaram, V., Reisslein, M., & Karam, L. J. (2011). Objective Video Quality Assessment Methods: A Classification, Review, and Performance Comparison. *IEEE Transactions on Broadcasting*, *57*(2), 165–182. doi:10.1109/TBC.2011.2104671

Cho, J. H., Swami, A., & Chen, R. (2011). A survey on trust management for mobile ad hoc networks. *IEEE Communications Surveys and Tutorials*, *13*(4), 562–583. doi:10.1109/SURV.2011.092110.00088

Cho, K.-H., & Ryu, M.-W. (2012). A Survey of Greedy Routing Protocols for Vehicular Ad Hoc Networks. *Smart Computing Review*, *2*(2), 125–137.

Chunli, D., Yuning, D., & Li, W. (2009). Autoregressive channel prediction model for cognitive radio. *5th International Conference on Wireless Communications, Networking and Mobile Computing*, 1 – 4.

Cisco Visual Networking Index. Forecast and Methodology, 2014-2019 White Paper. (2014). Retrieved from http://www.cisco.com/c/en/us/solutions/collateral/service-provider/ip-ngn-ip-next-generation-network/white_paper_c11-481360.html

Cisco visual networking index: Global mobile data traffic forecast update, 2015–2020. (2016). Retrieved from http://kom.aau.dk/group/16gr1022/Data_Stuff/Cisco_Mobile-growth.pdf

CNN Global News View (CNN). (2014). Retrieved January 20, 2015, from http://money.cnn.com/2014/02/28/technology/mobile/mobile-apps-internet/

Confalonieri, R., Foi, M., Casa, R., Aquaro, S., Tona, E., Peterle, M., & Acutis, M. et al. (2013). Development of an app for estimating leaf area index using a smartphone. Trueness and precision determination and comparison with other indirect methods. *Computers and Electronics in Agriculture*, *96*, 67–74. doi:10.1016/j.compag.2013.04.019

Cruselles, E., Soriano, M., & Melus, J. L. (1995). Uncorrelated PN sequences generator for spreading codes in CDMA systems. In *Proceedings of 6th IEEE International Symposium on Indoor and Mobile Radio Communications*, (vol. 3, pp.1335-1340). doi:10.1109/PIMRC.1995.477380

Cygwin: Get that Linux feeling - on Windows. (2010). Retrieved from https://www.cygwin.com/

D'souza, J., & Jose, J. (2010). Routing approaches in Delay Tolerant Networks: A Survey. Proceeding of International Journal of Computer Applications, 17, 9-15.

Da Cunha, F. D., Vilas, L. A., Viana, A. C., & Loureiro, A. A. F. (2014). *Data communication in VANETs: A Survey, Challenges and Applications*. Research Report.

Dahlman, E., Parkvall, S., & Sköld, J. (2013). *4G: LTE/LTE-Advanced for Mobile Broadband.* AP-Elsevier.

Damiani, E., De Capitani di Vimercati, S., Samarati, P., & Viviani, M. (2006). A WOWA-based aggregation technique on trust values connected to metadata. *Electronic Notes in Theoretical Computer Science*, *157*(3), 131–142. doi:10.1016/j.entcs.2005.09.036

Delgado, J. A., Kowalski, K., & Tebbe, C. (2013). The first Nitrogen Index app for mobile devices: Using portable technology for smart agricultural management. *Computers and Electronics in Agriculture*, *91*, 121–123. doi:10.1016/j.compag.2012.12.008

Demiryürek. (2010). Information systems and communication networks for agriculture and rural people. *Agricltural Economics – Czech, 56*(5), 209–214.

Detti, A., Bianchi, G., Pisa, C., Saverio, F., Loreti, P., Kellerer, W., . . . Widmer, J. (2009). *SVEF: An Open-Source Experimental Evaluation Framework for H.264 Scalable Video Streaming.* Paper presented at the IEEE Symposium on Computers and Communications, 2009 (ISCC 2009), Sousse.

Deutsch, M. (1962). Cooperation and trust: Some theoretical notes. *Nebraska Symposium on Motivation*. Nebraska University Press.

Dhubkarya & Sonam. (2009). High Quality Audio Coding At Low Bit Rate Using Wavelet And Wavelet Packet Transform. *Journal of Theoretical and Applied Information Technology, 6*(2), 194-200.

Dogaru, R. (2010). HCA101: A chaotic map based on cellular automata with binary synchronization properties. In *Proceedings of 8th International Conference on Communications,* (pp 41 – 44).

Elmirghani, J. M. H., & Cryan, R. A. (1995). Point-to-point and multi-user communication based on chaotic sequences. In *Proceedings of IEEE International Conference on Communications,* (pp 582 – 584).

Eschenauer, L., Gligor, V. D., & Baras, J. (2004, January). On trust establishment in mobile ad-hoc networks. In *Security Protocols* (pp. 47–66). Springer Berlin Heidelberg. doi:10.1007/978-3-540-39871-4_6

Fall, K. (2003). A Delay –Tolerant Network Architecture for challenged internets. *Proceeding of annual conference of the special Internet Group on Data Communication (ACM Siggomm'03)*, 27-34.

Faruki, P., Bharmal, A., Laxmi, V., Ganmoor, V., Gaur, M. S., Conti, M., & Rajarajan, M. (2015). Android Security: A Survey of Issues, Malware Penetration and Defenses. IEEE Communications Surveys and Tutorials.

Fedora 16 (32 bit). (2011). Retrieved from https://fedoraproject.org/

Feltekh, K., & Jemaa, Z. B. J. (2011). Chaos based spread spectrum watermarking algorithm. In *Proceedings of International Conference on Communications, Computing and Control Applications (CCCA),* (pp 1 – 5).

Ferreiro-Lage, J. A., Gestoso, C. P., Rubinos, O., & Agelet, F. A. (2009). Analysis of Unicast Routing Protocols for VANETs. *5th International Conference on Networking and Services (ICNS '09)*, 518-521. doi:10.1109/ICNS.2009.96

FFmpeg (GitHub). (2011). Retrieved from https://github.com/FFmpeg/FFmpeg

FFmpeg. (2011). Retrieved from https://www.ffmpeg.org/

Firdhous, M., Ghazali, O., & Hassan, S. (2011). *Trust and Trust Management in Cloud Computing–A Survey*. Internetworks Research Group, Universiti Utara Malaysia, Technical Report.

Frenkiel, R. H., Badrinath, B. R., Bores, J., & Yates, R. D. (2000). The infestations challenge: Balancing cost and ubiquity in delivering wireless data. *Proceeding of IEEE Personal Communications*, *7*(2), 66–71.

Function, L. (2009). *Lecture 5 Maximum Likelihood Method*. Academic Press.

Gandhi, R., Veeraraghavan, R., Toyama, K., & Ramprasad, V. (2007). Digital Green: Participatory video for agricultural extension. In *Information and Communication Technologies and Development, 2007. ICTD 2007. International Conference on* (pp. 1–10). IEEE. doi:10.1109/ICTD.2007.4937388

Ganesh, S., & Ranjani, S. N. (2013). Security Threats on Vehicular Ad Hoc Networks (VANET): A Review Paper. *International Journal of Electronics Communication and Computer Engineering, 4*(6).

Garbacz, C., & Jr, H. T. (2007). Demand for telecommunication services in developing countries. *Telecommunications Policy*, *31*(5), 276–289. doi:10.1016/j.telpol.2007.03.007

García, A., Erenas, M. M., Marinetto, E. D., Abad, C. A., de Orbe-Paya, I., Palma, A. J., & Capitán-Vallvey, L. F. (2011). Mobile phone platform as portable chemical analyzer. *Sensors and Actuators. B, Chemical*, *156*(1), 350–359. doi:10.1016/j.snb.2011.04.045

Gemayel, N., El Koslowski, S., & Jondral, F. K. (2009). *A low cost TDOA Localization System : Setup, Challenges and Results*. Academic Press.

Ghosh, C., Cordeiro, C., Agrawal, D. P., & Rao, M. B. (2009). Markov chain existence and hidden Markov models in spectrum sensing. *Proceedings of IEEE International Conference on Pervasive Computing and Communications*. doi:10.1109/PERCOM.2009.4912868

Gichamba, A., & Lukandu, I. (2012). A Model for designing M-Agriculture Applications for Dairy Farming. *The African Journal of Information Systems*, *4*(4). Retrieved from http://digitalcommons.kennesaw.edu/ajis/vol4/iss4/1

Giudici, F., & Pagani, E. (2005). Spatial and traffic-aware routing (STAR) for vehicular systems. Lecture Notes in Computer Science: Vol. 3726. High Performance Computing and Communications (pp. 77-86). Berlin: Springer. doi:10.1007/11557654_11

Góra, J., Pedersen, K. I., Szufarska, A., & Strzyz, S. (2010). Cell-Specific Uplink Power Control for Heterogeneous Networks in LTE. *IEEE 72th Vehicular Technology Conference Fall (VTC 2010-Fall),* 1-5.

Government of India. (2013). State of Indian Agriculture. Ministry of Indian Agriculture, Department of Agriculture & Cooperation, New Delhi.

Govindan, K., & Mohapatra, P. (2012). Trust computations and trust dynamics in mobile adhoc networks: A survey. *IEEE Communications Surveys and Tutorials*, *14*(2), 279–298. doi:10.1109/SURV.2011.042711.00083

Grandison, T., & Sloman, M. (2000). A survey of trust in internet applications. *IEEE Communications Surveys and Tutorials*, *3*(4), 2–16. doi:10.1109/COMST.2000.5340804

Grossglauser, M., & Tse, D. N. C. (2002). Mobility increases the capacity of adhoc wireless networks. *Proceeding of IEEE/ACM Transactions on Networking*, 10, 477-486. doi:10.1109/TNET.2002.801403

Gunjal Baisa, L., & Mali Suresh, N. (2011). Secured Color Image Watermarking Technique In DWT-DCT Domain. *International Journal of Computer Science, Engineering and Information Technology*, 36-44.

Gunjal, B. L., & Manthalkar, R. R. (2010). Discrete Wavelet Transform based Strongly Robust Watermarking Scheme for Information Hiding in Digital Images. *IEEE 3rd International Conference on Emerging Trends in Engineering and Technology*, 124 – 129.

Ham, F. M., Imana, E. Y., Ondi, A., Ford, R., Allen, W., & Reedy, M. (2009). Reputation Prediction in Mobile Ad Hoc Networks Using RBF Neural Networks. In Engineering Applications of Neural Networks (pp. 485-494). Springer Berlin Heidelberg. doi:10.1007/978-3-642-03969-0_46

Heidari-Bateni, G., & McGillem, C. D. (1994, March/April). A Chaotic Direct-Sequence Spread-Spectrum Communication System. *IEEE Transactions on Communications*, *42*(21314), 1524–1527. doi:10.1109/TCOMM.1994.582834

He, T., Huang, C., Blum, B. M., Stankovic, J. A., & Abdelzaher, T. (2003). Range free localization schemes for large scale sensor networks. In *Proc. 9th Annual International Conference on Mobile Computing and Networking, (MobiCom'03)*, (pp. 81–95). doi:10.1145/938985.938995

Htut, S. M., & Puthusserypady, S. (2003). A novel CDP DS/SS system with 2-dimensional Ikeda map chaotic sequence. In *Proceedings of 14th IEEE conference on Personal, Indoor and Mobile Radio Communications*, (vol. 3, pp 2734-2738). doi:10.1109/PIMRC.2003.1259238

Huang, J., & Nicol, D. (2009, April). A calculus of trust and its application to PKI and identity management. In *Proceedings of the 8th Symposium on Identity and Trust on the Internet* (pp. 23-37). ACM. doi:10.1145/1527017.1527021

Hwang, S. Y., Park, G. Y., Kim, D. H., & Jhang, K. S. (2010). Efficient Implementation of Pseudorandom Sequence Generator for High Speed Data Communications. *ETRI Journal*, *32*(2), 222–229. doi:10.4218/etrij.10.1409.0047

Hyvarinen, A. (1999). Fast and robust fixed-point algorithms for independent component analysis. *IEEE Transactions on Neural Networks*, *10*(3), 626–634. doi:10.1109/72.761722 PMID:18252563

IEEE. (1999). *IEEE Std 802.11 - Part 11, Wireless LAN Medium Access Control (MAC) and Physical Layer (PHY) Specifications*. IEEE.

IEEE. (2005). *IEEE Std 802.11e - Part 11 (2005): Wireless LAN Medium Access Control (MAC) and Physical Layer (PHY) specifications, Amendment 8: Medium Access Control (MAC) Quality of Service Enhancements*. IEEE.

IEEE. (2009). IEEE Std 802.11n - Part 11. Wireless LAN Medium Access Control (MAC) and Physical Layer (PHY) specifications, Amendment 5. *Enhancements for Higher Throughput*. IEEE.

IMAI. (n.d.) Retrived January 20, 2015, from http://www.artechhouse.com/uploads/public/documents/chapters/imai_520_CH03.pdf

Indra, V. A., & Murali, R. (2014). Routing Protocols for Vehicular Adhoc Networks (VANETs): A Review. *Journal of Emerging Trends in Computing and Information Sciences*, *5*(1).

Information and Communication Technology. (2011). In *Wikipedia*. Retrieved June 15, 2011, from http://en.wikipedia.org/wiki/Information_and_communication_technologies

Information System. (2011). In *Wikipedia*. Retrieved June 15, 2011, from http://en.wikipedia.org/wiki/Information_system

International Telecommunication Union (ITU). (2011a, Jan 6). *ICT services getting more affordable worldwide*. Retrieved from International Telecommunication Union website: http://www.itu.int/net/pressoffice/press_releases/2011/15.aspx#.Ulf2LdJkPE0

International Telecommunication Union (ITU). (2011b, Nov 12). *World Telecommunication /ICT Indicators Database*. Retrieved from International Telecommunication Union website: http://www.itu.int/ITUD/ict/statistics/#.Ulf2LdJkPE0

Internet Usage. (2014). Retrieved January 20, 2015, from http://dazeinfo.com/2014/01/23/smartphone-users-growth-mobile-internet-2014-2017/

Iqbal, Z., & Bjorklund, R. B. (2011). Colorimetric analysis of water and sand samples performed on a mobile phone. *Talanta*, *84*(4), 1118–1123. doi:10.1016/j.talanta.2011.03.016 PMID:21530787

Ivrigh, S. (2011). A novel framework for spectrum sensing in cognitive radio networks. *IEICE Transactions on Communications*, *E9-B*(9), 2600–2609.

Jain, L., Kumar, H., & Singla, R. K. (2011). A Review of Fuzzy Rule Promotion Techniques in Agriculture Information System. *International Journal of Computer Applications*, 55–60.

Jain, L., Kumar, H., & Singla, R. K. (2013). *Localization of Information Dissemination in Agriculture using Mobile Networks Changing Agricultural Scenario and Information Needs Use Multimedia in Agriculture Web and Mobile Applications Need for Mobile Phone based Communication in Agriculture Lack of*. Academic Press.

Jain, S., Fall, K., & Patra, R. (2004). Routing in a delay tolerant network. *Proceeding of ACM SIGCOMM*.

Jain, L., Kumar, H., & Singla, R. K. (2014). Assessing Mobile Technology Usage for Knowledge Dissemination among Farmers in Punjab. *Information Technology for Development*, 1–9. doi:10.1080/02681102.2013.874325

Jain, L., Kumar, H., Singla, R. K., Singh, P., & Singh, L. J. (2014). Error Reduction in Promoted Confidence Factor of a Rule Using Improved Fuzzy Rule Promotion Technique. *Cybernetics and Information Technologies*, *14*(1), 72–83. doi:10.2478/cait-2014-0006

Jawandhiya, P. M., Ghonge, M. M., Ali, M. S., & Deshpande, J. S. (2010). A survey of mobile ad hoc network attacks. *International Journal of Engineering Science and Technology*, *2*(9), 4063–4071.

Jerbi, M., Senouci, S.-M., Rasheed, T., & Ghamri-Doudane, Y. (2009). Towards efficient geographic routing in urban vehicular networks. *IEEE Transactions on Vehicular Technology*, *58*(9), 5048–5059. doi:10.1109/TVT.2009.2024341

Jian, S., Sangman, M., & Chung, L. (2008). Routing Protocols in Delay Tolerant Networks: A Comparative Survey. *Proceeding of 23rd International Technical Conference on Circuits /Systems, Computer and Communications (ITC-CSCC 2008)*, 1577-1580.

Jiang, T., & Qu, D. (2008). On minimum sensing error with spectrum sensing using counting rule in cognitive radio networks. *Proceedings on 4th Annual International Conference on Wireless Internet*, 1–9. doi:10.4108/ICST.WICON2008.4954

Jiang, X., & Zhou, Y. (2013). *Android Malware*. Springer.

Jiang, H., & Fu, C. (2008). A Chaos-Based High Quality PN Sequence Generator. In *Proceedings of International Conference on Intelligent Computation Technology and Automation (ICICTA),* (pp 60 – 64).

Jing, Y., Zhao, Z., Ahn, G. J., & Hu, H. (2014). Morpheus: Automatically Generating Heuristics to Detect Android Emulators. *Proceedings of the 30th Annual Computer Security Applications Conference(ACSAC).* doi:10.1145/2664243.2664250

Jones, E., & Paul, A. S. W. (2008). Routing Strategies for Delay – Tolerant Networks. *Journal of Computer Communication.*

Jonker, C. M., & Treur, J. (1999). Formal analysis of models for the dynamics of trust based on experiences. In *Multi-Agent System Engineering* (pp. 221–231). Springer Berlin Heidelberg. doi:10.1007/3-540-48437-X_18

Joseph Mitola, I. I. I. (n.d.). *An Integrated Agent Architecture for Software Defined Radio* (Dissertation). KTH/IT/AVH—00/01—SE.

JSVM Reference Software. (2013). Retrieved from http://www.hhi.fraunhofer.de/en/departments/video-coding-analytics/research-groups/image-video-coding/research-topics/svc-extension-of-h264avc/jsvm-reference-software.html

JSVM Software Manual. (2013). Retrieved from http://evalsvc.googlecode.com/files/SoftwareManual.doc

Karp, B., & Kung, H. T. (2000). GPSR: Greedy Perimeter Stateless Routing for Wireless Networks. *Proceedings of 6th annual International Conference on Mobile Computing and Networking*, 243-254. doi:10.1145/345910.345953

Ke, C.-H. (2006). *A prototype for H.264 evaluation framework using NS2.* Retrieved from http://140.116.72.80/~smallko/ns2/h264.htm

Ke, C.-H. (2012a). *How to do H.264 SVC transmission simulations?* Retrieved from http://csie.nqu.edu.tw/smallko/ns2/svc.htm

Ke, C.-H. (2013). *myEvalvid: An Evaluation Framework for MPEG video transmission in NS2 environment.* Retrieved from http://csie.nqu.edu.tw/smallko/ns2/myEvalvid.htm

Ke, C. H., Shieh, C. K., Hwang, W. S., & Ziviani, A. (2008). An Evaluation Framework for More Realistic Simulations of MPEG Video Transmission. *Journal of Information Science and Engineering*, *24*(2), 425–440.

Ke, C.-H. (2012b). myEvalSVC: An Integrated Simulation Framework for Evaluation of H.264/SVC Transmission. *Transactions on Internet and Information Systems (Seoul)*, *6*(1), 379–394. doi:10.3837/tiis.2012.01.021

Ke, C.-H., Shieh, C.-K., Hwang, W.-S., & Ziviani, A. (2008). An Evaluation Framework for More Realistic Simulations of MPEG Video Transmission. *Journal Of Information Science And Engineering*, *24*(2), 425–440.

Kempe, D., Dobra, A., & Gehrke, J. (2003, October). Gossip-based computation of aggregate information. In *Foundations of Computer Science, 2003. Proceedings. 44th Annual IEEE Symposium on* (pp. 482-491). IEEE. doi:10.1109/SFCS.2003.1238221

Khajavi, N. T., Sadeghi, S., & Sadough, S. M.-S. (2010). An improved blind spectrum sensing technique for cognitive radio systems. *5th International Symposium on Telecommunications*, 13-17.

Khan, A. R. (2012, January). Performance evaluation of DS-CDMA system using MATLAB. *International Journal of Advances in Engineering and Technology*, *2*(1), 269–28.

Khandekar, A., Bhushan, N., Tingfang, J., & Vanghi, V. (2010). LTE-Advanced: Heterogeneous Networks. *European Wireless Conference (EW), IEEE*, 978-982. doi:10.1109/EW.2010.5483516

Kirk, M., Steele, J., Delbe, C., Crow, L., Keeble, J., Fricke, C., ... Bulloch, G. (2011). *Connected Agriculture: The role of mobile in driving efficiency and sustainability in the food and agriculture value chain.* Retrieved from http://www.accenture.com/SiteCollectionDocuments/PDF/Accenture-Connected-Agriculture.pdf

Klaue, J., Rathke, B., & Wolisz, A. (2003). *EvalVid - A Framework for Video Transmission and Quality Evaluation.* Paper presented at the 13th International Conference, TOOLS 2003, Urbana, IL.

Klaue, J., Rathke, B., & Wolisz, A. (2003). EvalVid - A Framework for Video Transmission and Quality Evaluation. *Proc. of the 13th International Conference of modelingtechnques and tools for computer performenceevalution.*

Klaue, J., Rathke, B., & Wolisz, A. (2014). *EvalVid - A Video Quality Evaluation Tool-set.* Retrieved from http://www2.tkn.tu-berlin.de/research/evalvid/fw.html#bin

Kleinberg, J. (2000). Navigation in a small world. *Nature, 406*(6798), 845. doi:10.1038/35022643 PMID:10972276

Kolhe, S., Kamal, R., Saini, H. S., & Gupta, G. K. (2011). A web-based intelligent disease-diagnosis system using a new fuzzy-logic based approach for drawing the inferences in crops. *Computers and Electronics in Agriculture, 76*(1), 16–27. doi:10.1016/j.compag.2011.01.002

Kwon, Y. M., Mechitov, K., Sundresh, S., Kim, W., & Agha, G. (2005). Resilient localization for sensor networks in outdoor environments. In *Proc. 25th IEEE International Conference on Distributed Computing Systems (ICDCS'05),* (pp. 643–652). doi:10.1109/ICDCS.2005.68

Lalitha, N. V., Suresh, G., & Sailaja Dr, V. (2011). Improved Audio Watermarking Using DWT-SVD. *International Journal of Scientific & Engineering Research, 2-6,* 1–7.

Lantz, P. (2011). *Droidbox - android application sandbox.* Retrieved January 20, 2015, from http://code. google.com/p/droidbox/

Le, T. A., Nguyen, H., & Zhang, H. (2010). *EvalSVC - An evaluation platform for Scalable Video Coding transmission.* Paper presented at the IEEE 14th International Symposium on Consumer Electronics (ISCE 2010), Braunschweig. doi:10.1109/ISCE.2010.5523712

Lee, D., Chou, W. P., Yeh, S. H., Chen, P. J., & Chen, P. H. (2011). DNA detection using commercial mobile phones. *Biosensors & Bioelectronics, 26*(11), 4349–4354. doi:10.1016/j.bios.2011.04.036 PMID:21621401

Li, Shi, Chen, & Yan. (2011). Square Law Combining Double threshold Energy Detection in Nakagami Channel. *International Journal of Digital Content Technology and its Application, 5*(12).

Li, Y., Dong, Y. N., Zhang, H., Zhao, H. T., Shi, H. X., & Zhao, X. X. (2010). Spectrum usage prediction based on high-order markov model for cognitive radio networks. *IEEE 10th International Conference on Computer and Information Technology,* 2784 – 2788.

Liu, J., & Issarny, V. (2004). Enhanced reputation mechanism for mobile ad hoc networks. In *Trust management* (pp. 48–62). Springer Berlin Heidelberg. doi:10.1007/978-3-540-24747-0_5

Liu, L., & Sun, Q. (2006). A new SVD-DWT composite watermarking. *Proceeding 8th International Conference on Signal Processing,* 4.

Li, Y., & Wang, Z. (2011). Optimal relaying in heterogeneous delay tolerant networks. *IEEE International Conference on Communications ICC 2011,* 1 - 5. doi:10.1109/icc.2011.5963142

Lochert C., Mauve, M., Füßler, H., & Hartenstein, H. (2005). Geographic routing in city scenarios. *ACM SIGMOBILE Mobile Computing and Communications Review, 9*(1), 69-72.

Mahmoud, Q. H. (2007). Front Matter. In *Cognitive Networks: Towards Self-Aware Networks.* Chichester, UK: John Wiley & Sons, Ltd. doi:10.1002/9780470515143.fmatter

Ma, J., & Li, Y. (2007). Soft combination and detection for cooperative spectrum sensing in cognitive radio networks. *Proceeding of IEEE Global Telecommunication Conference*, 3139–3143. doi:10.1109/GLOCOM.2007.594

Malgenome. (n.d.). Retrieved 2014, from http://www.malgenomeproject.org/policy.html

Mandayam, N. (2002). Wireless Communication Technologies. In *Lecture Notes* (2nd ed.). Rutgers University. Retrieved from http//mathworld.wolfram.com/LogisticMap.html

Mangrulkar & Atique. (2012). Design and development of delay tolerant network: A case study. *Proceedings of IEEE International conference Nuicone 2012*, 1 - 5.

Mangrulkar, R. S., & Mohammad, A. (2010). Routing Protocol for Delay Tolerant Network: A Survey and Comparison. *Proceeding of IEEE International Conference on Communication, Computing and Control* during, 210-215.

Mao, G., Fidan, B., & Anderson, B. D. O. (2007). Wireless sensor network localization techniques. *Computer Networks*, *51*(10), 2529–2553. doi:10.1016/j.comnet.2006.11.018

Mapp, G., Shaikh, F., Aiash, M., Vanni, R. P., Augusto, M., & Moreira, E. (2009). Exploring efficient imperative hand over mechanisms for heterogeneous wireless networks. *Proceedings of the 12th International Conference on Network-Based Information Systems NBiS09*, 286 - 291.

Martoyo, I., Susanto, A., & Wijanto, E. (2010). Chaos codes vs Orthogonal codes for CDMA. In *Proceedings of IEEE 11th International Symposium on Spread Spectrum Techniques and Applications (ISITA)*, (pp 189 – 193).

Masud, M. Z., Sahib, S., Abdollah, M. F., Selamat, S. R., & Yusof, R. (2014). Android Malware Detection System Classification. Research Journal of Information Technology, 325-341.

Mathurapoj, A., & Pornavalai, C. (2009). Fuzz-Spray: Efficient Routing in Delay Tolerant ad-hoc network based on Fuzzy Mechanism. *Proceeding of FUZZ-IEEE 2009*, 104-109.

Mehta, P. (2003). Information Technology in Agriculture: Reaching the Unreached. *National Workshop on ICT for Agriculture and Rural Development*, Ahmedabad.

Meireles, R., Boban, M., Steenkiste, P., Tonguz, O. K., & Barros, J. A. (2010). Experimental study on the impact of vehicular obstructions in VANETs. *IEEE Vehicular Networking Conference (VNC)*, 338–345. doi:10.1109/VNC.2010.5698233

Mishra, S., Krishna, P. V., & Abraham, K. I. (2010). Adaptive Link State Routing and Intrusion Detection in Wireless Mesh Networks. Information Security, 4(4), 374-389.

Misra, Das, Khatua, & Obaidat. (2014). Qos-guaranteed bandwidth shifting and redistribution in mobile cloud environment. *IEEE Transaction on Cloud Computing, 11*(99), 1 - 10.

Mobile device protection and security threat measures. (2003). Retrieved February 21, 2017, from http://searchmobilecomputing.techtarget.com/guides/Mobile-device-protection-and-security-threat-measures

Mobile/Tablet Operating System Market Share (MOBILE Share). (2015). Retrieved January 20, 2015, from https://www.netmarketshare.com/operating-system-market-share.aspx?qprid=8&qpcustomd=1

Models, B. C. (2003). *ML Estimation, Binary Choice Models, (Ml).* Academic Press.

MP4Box. (2014). Retrieved from http://gpac.wp.mines-telecom.fr/mp4box/

MPEG-4 video trace file. (2010). Retrieved from http://www2.tkn.tu-berlin.de/research/trace/trace.html

MSU Video Quality Measurement tools. (2014). Retrieved from http://compression.ru/video/quality_measure/index_en.html

Mukherjee, A., Maiti, S., & Datta, A. (2014). Spectrum Sensing for Cognitive Radio Using Blind Source Separation and Hidden Markov Model. *IEEE Proceedings of Advanced Computing & Communication Technologies*, 409 – 414. doi:10.1109/ACCT.2014.63

Mukherjee, et al. (2015). Source Number detection and Realization of ROC for Spectrum sensing in Cognitive Radio. *Indian Journal of Science and Technology*, *27*.

Mukherjee, A. (2016). HML based Smart positioning of Fusion Center for Cooperative Communication in Cognitive Radio Networks. *IEEE Communications Letters*, *20*(11). doi:10.1109/LCOMM.2016.2602266

Mukherjee, A., & Datta, A. (2016). Vector Quantization based Power Allocation for Non-Ergodic Cognitive Radio Systems. *Journal of Engineering Science and Technology Review*, *9*(5), 85–87.

Mutagi, R. N. (April 1996). Pseudo-noise Sequences For Engineers. Electronics and Communication Engineering Journal, 79-87.

Nanda, S., & Arunachalam, S. (2010). Reaching the unreached (Vol. 7). Chennai: Jamsetji Tata National Virtual Academy (NVA), M S Swaminathan Research Foundation.

Network Simulator (NS2.29.3). (2011). Retrieved from http://sourceforge.net/projects/nsnam/files/allinone/ns-allinone-2.29/ns-allinone-2.29.3.tar.gz/download

Neuner, S., Veen, V. V. D., Lindorfer, M., Huber, M., Merzdovnik, G., Mulazzani, M., & Weippl, E. (2014). Enter Sandbox: Android Sandbox Comparison. *Proceedings of the 3rd IEEE Mobile Security Technologies Workshop (MoST).*

Nikolaidis, A., & Pitas, I. (2003). Asymptotically optimal detection for additive watermarking in the DCT and DWT domains. *IEEE Transactions on Image Processing*, *12*(5), 563–571. doi:10.1109/TIP.2003.810586 PMID:18237932

Nokia Life Tools. (n.d.). In *Wikipedia.* Retrieved February 10, 2014, from http://en.wikipedia.org/wiki/Nokia_Life_Tools

Noor, T. H., Sheng, Q. Z., Zeadally, S., & Yu, J. (2013). Trust management of services in cloud environments: Obstacles and solutions. *ACM Computing Surveys*, *46*(1), 12. doi:10.1145/2522968.2522980

Nsnam Web Pages. (n.d.). Retrieved February 27, 2017, from http://www.isi.edu/nsnam

Ou, C.-H. (2011). Wireless Sensor Networks A Networking Perspective. *IEEE Sensors Journal*, *11*(7), 1607–1616. doi:10.1109/JSEN.2010.2102748

Oureiro, A. N. A. F. L., Niversity, F. E. U., & Inas, O. F. M. (2007). *Localization systems for wireless sensor networks.* Academic Press.

Pal, A. (2010). Localization Algorithms in Wireless Sensor Networks: Current Approaches and Future Challenges. *Network Protocols and Algorithms*, *2*(1), 1607–1616. doi:10.5296/npa.v2i1.279

Pardo, P. (2009). Aggregation of trust for iterated belief revision in probabilistic logics. In *Scalable Uncertainty Management* (pp. 165–179). Springer Berlin Heidelberg. doi:10.1007/978-3-642-04388-8_14

Patch file for NAM. (n.d.). Retrieved from http://www.linuxquestions.org/questions/attachment.php?attachmentid=3326&d=1271149430

Patch file for Wireless Application. (2011). Retrieved from http://www.telematica.polito.it/fiore/ns2_wireless_update_patch.tgz

Patil, M. M., Shaha, U., Desai, U. R., & Merchant, S. N. (2011). Localization in wireless sensor networks using three masters. In *2005 IEEE International Conference on Personal Wireless Communications, 2005. ICPWC 2005.* (Vol. 11, pp. 384–388). IEEE. doi:10.1109/ICPWC.2005.1431371

Patwari, N., Hero, A. O., Perkins, M., Correal, N. S., & Dea, R. J. O. (2003). Article. *Wireless Sensor Networks*, *51*(8), 2137–2148.

Pawlikowski, K. (2003). *Do Not Trust All Simulation Studies of Telecommunication Networks.* Paper presented at the International Conference, ICOIN 2003, Cheju Island, Korea.

Peng, R., & Sichitiu, M. L. (n.d.). Robust, Probabilistic. *Constraint-Based Localization for Wireless Sensor Networks.*

Petsas, T., Voyatzis, G., Athanasopoulos, E., Polychronakis, M., & Ioannidis, S. (2014). Rage Against the Virtual Machine: Hindering Dynamic Analysis of Android Malware. *Proceedings of the Seventh European Workshop on System Security (EuroSec).* doi:10.1145/2592791.2592796

Pham, N. N., Youn, J., & Won, C. (2006). A Comparison of Wireless Sensor Network Routing Protocols on an Experimental Testbed. *IEEE International Conference on Sensor Networks, Ubiquitous, and Trustworthy Computing (SUTC'06)*, *2*, 276–281. doi:10.1109/SUTC.2006.1

Philip, H., Vennila, M. A., & Gomathy, M. (2003). Multimedia: an effective communication technology for extension | OpenAgri: An Open Access Agriculture Research Repository. In B. Jirli (Ed.), International Conference on Communication for Development in the Information Age: Extending the Benefits of Technology for All., Varanasi, India. Varanasi, India: Department of Extension Education, Institute of Agricultural Sciences, Banaras Hindu University. Retrieved from http://agropedia.iitk.ac.in/openaccess/?q=content/multimedia-effective-communication-technology-extension

Pirzada, A. A., & McDonald, C. (2004, January). Establishing trust in pure ad-hoc networks. In *Proceedings of the 27th Australasian conference on Computer science-Volume 26* (pp. 47-54). Australian Computer Society, Inc.

Qiu, R. C., & Chen, Z. (2010). Prediction of channel state for cognitive radio using higher-order hidden Markov model. *Proceedings of the IEEE Southeast Conference*, 276 – 282.

Quan, Z., Cui, S., Poor, H., & Sayed, A. (2008). Collaborative wideband sensing for cognitive radios. *IEEE Signal Processing Magazine*, *25*(6), 63–70. doi:10.1109/MSP.2008.929296

Radenkovic & Grundy. (2012). Efficient and adaptive congestion control for heterogeneous delay-tolerant networks. *Journal of Adhoc Networks, 10*(7), 1322 - 1345.

Raj, S. (2013). e-Agriculture Prototype for Knowledge Facilitation among Tribal Farmers of North-East India: Innovations, Impact and Lessons. *Journal of Agricultural Education and Extension*, *19*(2), 113–131. doi:10.1080/1389224X.2012.718247

Ramakrishnan, K., Floyd, S., & Black, D. (2001). *The Addition of Explicit Congestion Notification (ECN) to IP (RFC-3168).* Retrieved from.

Ramamritham, K. (2004). aAqua : A Database-backended Multilingual. *Multimedia Community Forum*, 1–4.

Ramamritham, K., Bahuman, A., Duttagupta, S., Bahuman, C., & Balasundaram, S. (2005). *Innovative ICT Tools for Information Provision in Agricultural Extension (December 2005).* Retrieved from http://www.cse.iitb.ac.in/~krithi/papers/ ICTD2006.pdf

Ranjan, P., & Ahirwar, K. K. (2011). Comparative Study of VANET and MANET Routing Protocols. *Proceedings of International Conference on Advanced Computing and Communication Technologies (ACCT 2011).*

Rappaport, T. S. (2002). *Wireless Communications* (2nd ed.). Principles and Practice, Prentice- Hall India.

Recommendation ITU-T J.144 (03/2004): Objective perceptual video quality measurement techniques for digital cable television in the presence of a full reference. (2004). (pp. 1-156).

Recommendation ITU-T P.910 (04/2008): Subjective video quality assessment methods for multimedia applications. (2008). (pp. 1-42).

Reddy, P. K., & Ankaiah, R. (n.d.). A framework of information technology-based agriculture information dissemination system to improve crop productivity. *Current Science, 88*(12), 1905–1913. Retrieved from http://tejas.serc.iisc.ernet.in/~currsci/ jun252005/1905.pdf

Rehman, Khan, Zia, & Zheng. (2013). Vehicular Ad-Hoc Networks (VANETs) - An Overview and Challenges. *Journal of Wireless Networking and Communications, 3*(3), 29-38. DOI: 10.5923/j.jwnc.20130303.02

Rice, J. A. (1995). Mathematical Statistics and Data Analysis (2nd ed.). Duxbury Press.

Ries, M., Nemethova, O., & Rupp, M. (2007). *Motion Based Reference-Free Quality Estimation for H.264/AVC Video Streaming* Paper presented at the 2nd International Symposium on Wireless Pervasive Computing, 2007. ISWPC '07, San Juan.

Rossholm, A., & Lovstroem, B. (2008). *A new low complex reference free video quality predictor.* Paper presented at the IEEE 10th Workshop on Multimedia Signal Processing, 2008, Cairns, Qld.

Ruohomaa, S., Kutvonen, L., & Koutrouli, E. (2007, April). Reputation management survey. In *Availability, Reliability and Security, 2007. ARES 2007. The Second International Conference on* (pp. 103-111). IEEE. doi:10.1109/ARES.2007.123

Ruohomaa, S., & Kutvonen, L. (2005). Trust management survey. In *Trust Management* (pp. 77–92). Springer Berlin Heidelberg. doi:10.1007/11429760_6

Sahoo, S. (2015). Advanced Energy Sensing Techniques Implemented Through Source Number Detection for Spectrum Sensing in Cognitive Radio. In *Computational Vision and Robotics* (pp. 179–187). Springer. doi:10.1007/978-81-322-2196-8_21

Sandulescu & Nadjm-Tehrani. (2008). Opportunistic DTN Routing with Window-aware Adaptive Replication. *Proceeding of AINTEC'* 08'.

Santos, F. (2008). Localization in Wireless Sensor Networks. *ACM Journal,* 1–19.

Saravanan, R. (2008). *Agricultural extension: worldwide innovations*. New Delhi: New India Publishing Agency.

Saravanan, R. (2010). *Icts for@ Agricultural Extension: Global Experiments, Innovations and Expereience*. New Delhi: New India Publishing Agency.

Seet, B. C., Liu, G., Lee, B. S., Foh, C. H., Wong, K. J., & Lee, K. K. (2004). A-star: A mobile ad hoc routing strategy for metropolis vehicular communications. *Proceeding NETWORKING, 3042*, 989-999. doi:10.1007/978-3-540-24693-0_81

SE-Linux. Validating Security-Enhanced Linux. (2013). Retrieved January 20, 2015, from. http://source.android.com/devices/tech/security/se-linux.html

Sen, S. A., Clark, J., & Tapiador, J. (2010). Security threats in mobile Ad hoc networks. Security of Self-Organizing Networks: MANET, WSN, WMN, VANET. Taylor & Francis. doi:10.1201/EBK1439819197-9

Shah, R. C., Roy, S., & Brunette, W. (2003). Data Mules: Modeling three tier architecture for sparse sensor networks. Proceeding of Sensor Network Protocols and Applications, 30-41.

Shahid, M., Rossholm, A., Lövström, B., & Zepernick, H.-J. (2014). No-reference image and video quality assessment: A classification and review of recent approaches. *EURASIP Journal on Image and Video Processing*, *1*(40), 1–32. doi:10.1186/1687-5281-2014-40

Sherchan, W., Nepal, S., & Paris, C. (2013). A survey of trust in social networks. *ACM Computing Surveys*, *45*(4), 47. doi:10.1145/2501654.2501661

Shih, D. H., Lin, B., Chiang, H. S., & Shih, M. H. (2008). Security aspects of mobile phone virus: A critical survey. *Industrial Management & Data Systems*, *108*(4), 478–494. doi:10.1108/02635570810868344

Silver, M. S., Markus, M. L., & Beath, C. M. (1995). The Information Technology Interaction Model: A Foundation for the MBA Core Course. *Management Information Systems Quarterly*, *19*(3), 361–390. doi:10.2307/249600

Singh, D., Naveen, C., & Madhuri, A. (2012). Scrambled Hybrid Wavelet-Singular Domain Digital Image Watermarking. *CSI 6th International Conference on Software Engineering (CONSEG)*, 219-224.

Singh, Naveen, & Madhuri. (2012). Spatial and Frequency Domain for Grey level Digital Images. *IJCA*, 16-20.

Singh, P. (2004). Indian Agricultural Development in Changing Scenario Past, Present and Future. *Journal of the Indian Society of Agricultural Statistics*, *58*, 37–49. Retrieved from http://isas.org.in/jsp/volume/vol58/panjabsingh.pdf

Singh, S. K. (2008). The diffusion of mobile phones in India. *Telecommunications Policy*, *32*(9-10), 642–651. doi:10.1016/j.telpol.2008.07.005

Skupin, R., Hellge, C., & Schierl, T., & Wiegand, T. (2011). Packet level video quality evaluation of extensive H.264/AVC and SVC transmission simulation. *Journal of Internet Services and Applications*, *2*(2), 129–138. doi:10.1007/s13174-011-0025-z

Song, S., Hwang, K., Zhou, R., & Kwok, Y. K. (2005). Trusted P2P transactions with fuzzy reputation aggregation. *IEEE Internet Computing*, *9*(6), 24–34. doi:10.1109/MIC.2005.136

SourceForge. NS2 downloader stats. (2014). Retrieved from http://sourceforge.net/projects/nsnam/files/stats/timeline

Spreitzenbarth, M., Schreck, T., Echtler, F., Arp, D., & Hoffmann, J. (2014). Mobile-Sandbox: combining static and dynamic analysis with machine-learning techniques. International Journal of Information Security.

Spyropoulos, T., Psounis, K., & Raghvendra, C. S. (2003). Spray and Wait Efficient routing in intermittently connected Networks. Proceeding of Mobile Computer and Communication Review, 7(3).

Spyropoulos, T., Psounis, K., & Raghvendra, C. S. (2007) Spray and Focus Efficient Mobility Assisted Tolerant Networks. *Proceeding of workshop on PerCom apos*, 79-85.

Su, Zeng, & Xiao. (2014). Key Technologies for SON in Next Generation Radio Access Networks. *Journal of Communication and Computer, 11*, 448-459. doi: 00510.17265/1548-7709/2014.05

Subramanian, S., & Ramachandran, B. (2012). *Trust Based Scheme for QoS Assurance in Mobile Ad-Hoc Networks.* arXiv preprint arXiv:1202.1664

Sumriddetchkajorn, S., Chaitavon, K., & Intaravanne, Y. (2013). Mobile device-based self-referencing colorimeter for monitoring chlorine concentration in water. *Sensors and Actuators. B, Chemical*, *182*, 592–597. doi:10.1016/j.snb.2013.03.080

Suryanarayana, G., & Taylor, R. N. (2004). *A survey of trust management and resource discovery technologies in peer-to-peer applications.* Academic Press.

SVC. Original YUV sequences in 4CIF and CIF resolution. (2014). Retrieved from ftp://ftp.tnt.uni-hannover.de/pub/svc/testsequences/

SVEF. Scalable Video-streaming Evaluation Framework. (2012). Retrieved from http://svef.netgroup.uniroma2.it/

Sverdlov, A., Dexter, S., & Eskicioglu, A. M. (2005). Robust DCT-SVD domain image watermarking for copyright protection: embedding data in all frequencies. *Multimedia Computing and Networking Conference*, 16-20.

Tabia, N., Gondran, A., Baala, O., & Caminada, A. (2011). Interference Model and Evolution in LTE Networks. *4th Joint IFIP Wireless and Mobile Networking Conference (WMNC)*, 26-28.

Taneja, S., & Kush, A. (2010). A Survey of routing protocols in mobile ad hoc networks. International Journal of Innovation. *Management and Technology*, *1*(3), 2010–0248.

Tangil, G. S., Tapiador, J. E., Lopez, P., & Blasco, J. (2014). Dendroid: A Text Mining Approach to Analyzing and Classifying Code Structures in Android Malware Families. *Expert Systems with Applications*, *41*(4), 1104–1117. doi:10.1016/j.eswa.2013.07.106

Tang, K. W., & Tang, W. K. S. (2005). A Chaos-based Secure Voice Communication System. In *Proceedings of IEEE International Conference on Industrial Technology*, (pp. 571-576).

Teguig, S., & Le Nir, V. (2012). Data fusion schemes for cooperative spectrum sensing in cognitive radio networks. *Communications and Information Systems Conference*, 1–7.

The Economist. (2011). Beyond the PC. *The Economist*. Retrieved November 12, 2012, from http://www.economist.com/node/21531109

The Network Simulator - ns-2. (2009). Retrieved from http://www.isi.edu/nsnam/ns/

Tian, J., Han, L., & Rothermel, K. (2003). Spatially aware packet routing for mobile ad hoc inter-vehicle radio networks. *Proceedings IEEE Intelligent Transportation Systems*, *2*, 1546–1551.

Ting & Ling. (1999). Hidden digital watermarks in Image. *IEEE Transactions on Image Processing, 8*(1), 58-68.

TRAI. (2014). *Telecom Subscription Data as on 31st January, 2014*. Retrieved from http://trai.gov.in/Content/PressDetails/2135_0.aspx

Tranter, W. H., Shanmugan, K. S., Rappaport, T. S., & Kosbar, K. L. (2003). *Principles of Communication Systems Simulation with Wireless Applications* (1st ed.). Prentice- Hall.

Types of Wireless Communication Technologies (WIRELESS Types). (n.d.). Retrieved January 20, 2015, from http://www.ehow.com/info_12157177_types-wireless-communication-technologies.html

UNESCAP. (2011). *Statistical Yearbook for Asia and the Pacific 2011*. Retrieved from http://www.unescap.org/stat/data/syb2011/

Unnikrishnan & Veeravalli. (2007). *Cooperative Spectrum Sensing and Detection for Cognitive Radio*. IEEE GLOBCOM.

Vahdat, A., & Becker, D. (2000). Epidemic Routing for partially connected adhoc networks. Tech Report CS- 200006, Duke University, April 2000.

van Schyndel, R. G., Tirkel, A. Z., & Osborne, C. F. (1994). A Digital Watermark. *Proceedings of the IEEE International Conference on Image Processing*, 2, 86-90. doi:10.1109/ICIP.1994.413536

Vander Stoep, J. (2009). Design and Implementation of Reliable Localization Algorithms using Received Signal Strength. *IEEE Sensors Journal*, 1–47.

VCEG. J. V. T. J. o. I. I. M. I.-T. (2007). JSVM 9.8 Software. Geneva, Switzerland.

Velloso, P. B., Laufer, R. P., de O Cunha, D., Duarte, O. C. M. B., & Pujolle, G. (2010). Trust management in mobile ad hoc networks using a scalable maturity-based model. *Network and Service Management, IEEE Transactions on, 7*(3), 172-185.

Venu Madhav, & Sarma. (2012). Energy efficient cluster routing protocol for heterogeneous wireless sensor networks. *LNICST*, (108), 452 - 455.

Vidas, T., & Christin, N. (2014). Evading Android Runtime Analysis via Sandbox Detection. *Proceedings of the 9th ACM Symposium on Information, Computer and Communications Security (ASIACCS) Virustotal.* Retrieved January 20, 2015, from www.virustotal.com

Video Library and Tools - NSL. (Raw Sequences from the H.264/SVC Group). (2015). Retrieved from https://cs-nsl-wiki.cs.surrey.sfu.ca/wiki/Video_Library_and_Tools

Video Sequences, Y. U. V. (2014). Retrieved from http://trace.eas.asu.edu/yuv/

Visual Networking Index (VNI). (n.d.). Retrieved January 20, 2015, from http://www.cisco.com/c/en/us/solutions/service-provider/visual-networking-index-vni/index.html#~mobilevni

VMware Workstation. (2011). Retrieved from http://www.vmware.com/in/products/workstation

Vu, K. M. (2011). ICT as a Source of Economic Growth in the Information Age: Empirical Evidence from the 19962005 Period. *Telecommunications Policy, 35*(4), 357–372. doi:10.1016/j.telpol.2011.02.008

Wang, C.-Y., & Wei, H.-Y. (2013). *WMNLAB IEEE NS-2 802.11n Module.* Retrieved from http://wmnlab.ee.ntu.edu.tw/a/ns-2.29-11n.tar.gz

Wang, Y., & Singh, M. P. (2006, July). Trust representation and aggregation in a distributed agent system. In AAAI (Vol. 6, pp. 1425-1430). AAAI.

Wang, Y., Tao, Y., Yu, P., Xu, F., & Lü, J. (2007). A trust evolution model for P2P networks. In Autonomic and Trusted Computing (pp. 216-225). Springer Berlin Heidelberg.

Wang, J., Ghosh, R. K., & Das, S. K. (2010). A survey on sensor localization. *Journal of Control Theory and Applications, 8*(1), 2–11. doi:10.1007/s11768-010-9187-7

Wang, Q., Dai, H.-N., & Zhao, Q. (2013). Eavesdropping Security in Wireless Ad Hoc Networks with Directional Antennas. *Proceedings of IEEE WOCC.* doi:10.1109/WOCC.2013.6676462

Wang, Z., Xiong, W., Dong, C., Wang, J., & Li, S. (2011). A Novel Downlink Power Control Scheme in LTE Heterogeneous Network. *IEEE International Conference on Computational Problem-Solving (ICCP)*, 241-245. doi:10.1109/ICCPS.2011.6089792

World Bank. (2011). *ICT in Agriculture: Connecting Smallholders to Knowledge, Networks, and Institutions.* World Bank.

Wu, B., Chen, J., Wu, J., & Cardei, M. (2007). A survey of attacks and countermeasures in mobile ad hoc networks. In Wireless Network Security (pp. 103-135). Springer US. doi:10.1007/978-0-387-33112-6_5

Xiao, F. (2000). *DCT-based Video Quality Evaluation.* Retrieved from http://compression.ru/video/quality_measure/vqm.pdf

Xiao, Lu, & Zhang. (2006). Performance Enhancement of DS/SS System Using Hybrid Chaotic Spreading Sequence. In *Proceedings of International Conference on Communication Technology*, (pp 1 - 5, 27-30). doi:10.1109/ICCT.2006.341981

Xu, G., Lu, Y., He, J., & Hu, N. (2008). Primary users detect for multiple antenna cognitive radio based on blind source separation. *8th International Symposium on Intelligent Information Technology workshops*, 777 – 780. doi:10.1109/IITA.Workshops.2008.112

Yadav, K. (2011, April 4). *Impact Assessment of ICT – enabled Knowledge Sharing Agri-portals in Uttarakhand.* Govind Ballabh Pant University of Agriculture & Technology, Pant Nagar, Uttrakhand. Retrieved from http://oar.icrisat.org/id/eprint/5845

Yang, Y., Hu, H., Xu, J., & Mao, G. (2009). Relay Technologies for WiMAX and LTE-Advanced Mobile System. *IEEE Communications Magazine*, *47*(10), 100–105. doi:10.1109/MCOM.2009.5273815

Yarkan, S., & Arslan, H. (2007). Binary time series approach to spectrum prediction for cognitive radio. *IEEE 66th Vehicular Technology Conference*, 1563 - 1567. doi:10.1109/VETECF.2007.332

Yu, B., Xu, C.-Z., & Guo, M. (2012). Adaptive Forwarding Delay Control for VANET Data Aggregation. *IEEE Transactions on Parallel and Distributed Systems*, *23*(1).

Yucek, T., & Arslan, H. (2009). A Survey of Spectrum Sensing Algorithms for Cognitive Radio Applications. *IEEE Communications Surveys and Tutorials*, *11*(1), 116–130. doi:10.1109/SURV.2009.090109

YUV test sequences. (2014). Retrieved from http://videocoders.com/yuv.html

Zhang, Zhao Li, Shi, Zhao, & Dong. (2010). Spectrum usage prediction based on high-order markov model for cognitive radio networks. *IEEE 10th International Conference on Computer and Information Technology (CIT)*, 2784 – 2788.

Zhang, Q., Foh, C. H., Seet, B.-C., & Fong, A. C. M. (2010). Location estimation in wireless sensor networks using spring-relaxation technique. *Sensors (Basel, Switzerland)*, *10*(5), 5171–5192. doi:10.3390/s100505171 PMID:22363204

Zhang, Q., Yu, T., & Irwin, K. (2004, November). A Classification Scheme for Trust Functions in Reputation-Based Trust Management. In *ISWC Workshop on Trust, Security, and Reputation on the Semantic Web.*

Zhauniarovich, Y. (2014). *Android Security (and Not).* Retrieved January 20, 2015, from Iwww.zhauniarovich.com/files/asani/asani.pdf

About the Contributors

Dharm Singh is a fellow member of Institution of Engineers (India), 25 years professional experience and currently teaching and guiding B.E., ME and PhD students in the Department of Computer Science & Engineering, College of Technology and Engineering, Maharana Pratap University of Agriculture and Technology (MPUAT), Udaipur.. Served as Incharge Head, Department of Computer Science and Engineering, MPUAT, Udaipur. Serving as Incharge University Intranet since last nine years. Honorary working as Convenor and Chairman Special Interest Group Wireless Networks (SIG-WNs), Computer Society of India, Board member of Computer Engineering Division Board, The Institution of Engineers (India), Secretory IEI ULC and Chairman CSI Udaipur Chapter. Received more than 13 Awards and Honours including "Eminent Engineering Personality Award" in recognition of eminence and contribution to the profession of Computer Engineering from The Institution of Engineers (India), Significant Contribution Award –National level from Computer Society of India, the largest association of IT professionals in India, Best Technical Staff Award and Appreciation Award by Maharana Pratap University of Agriculture & Technology, Udaipur. Written 16 books (3 texts books as sole author and edited 13 books including conference proceedings in the fields of Computer Science and Engineering and published more than 100 papers in referred Journals and Conferences. More than Twenty (20) Workshops, National workshops, National Seminars and conferences, In-country workshop and IEEE International Conference were organized as Organizing Chair and Secretary. Conference attended in Hong Kong and Taiwan and academic Visit at UK and Dubai in 2011. An upright affiliation of various engineering societies like: Fellow of The Institution of Engineers (India), Senior Member IEEE, Life Senior Member, International Association of Computer Science and Information Technology (IACSIT), Senior Member, Computer Society of India (CSI),) and Life member, The Indian Science Congress Association (Engineering Science).

* * *

Ahmed Abo-Elhassab is a Researcher at Electronics and Communication Eng. Department, Ain-Shams University, Cairo, Egypt, and Network Operation Center Engineer at TEDATA Company, Cairo, Egypt.

Madhuri Agarwal did her master in Computer Science and Engineering form College of Technology and Engineering, Udaipur. She published more than five papers and presented papers in various conference including IEEE conference. Presently she is working in Cognizant as Software Engineer.

Mohammad Atique is presently working as Professor in Post Graduate Department of Computer Science & Engineering Sant Gadge Baba Amravati University, Amravati. He has more than 25 years of Teaching and administrative experience. He is Life Member of ISTE, New Delhi, Fellow IE Kolkata, Fellow IETE New Delhi and Senior Life Member of CSI, Mumbai. His has more than 40 publications in National and International Journals to his credit. He is also reviewer of reputed National and International Journals. He is recognized PhD supervisor for RTM Nagpur and SGBAU Amravati University. Currently 5 students have been awarded Phd under him and 3 students are perusing PhD under him. His area of interest include Soft Computing, Real-time System, Ad hoc Network, Delay tolerant network, data mining and Machine Intelligent etc. He has chaired many sessions and also delivered keynote speeches in National and International conference, STTPs. He has received research grant from various funding agencies like UGC,AICTE and DST in India.

Mohammadreza Behjati received the BSc (Hons) degree in Electrical and Electronic Engineering in 2009 and the MSc degree in Wireless Communication Systems in 2011, as well as the PhD in Electrical Engineering and Electronics, Wireless Communications, from Brunel University London in 2015. He is also the winner of the Walduck Prize for Research Impact in year 2015. Dr. Behjati has been working as a Postdoctoral Researcher at 5G Research Centre (5GRC), Loughborough University, UK. His current research activities and interests include the novel techniques and solutions for the next generations of heterogeneous cellular networks, Massive MIMO systems, 5G networks and the related areas.

Lal Chand Bishnoi, BE(CSE), ME(CSE), Ph.D(CSE), MIEEE, MACM, MCSI, MIEI, is an Assistant Director (T&P Head), Govt. Polytechnic College, Bikaner, India. He did his PhD from Uttrakhand Technical University, Dehradun. He published more than 10 referred papers in Journals and conferences. He is having more than 22 years academic experience.

John Cosmas is currently a Professor of multimedia systems with the School of Engineering and Design, Brunel University London, UK. He co-leads the Wireless Networks and Communications Research Centre, and is the Course Director of MSc Advanced Multimedia Design and 3D Technologies. He has participated in 11 EU-IST and two EPSRC funded research projects since 1986 and has led three of these (CISMUNDUS, PLUTO, 3-D MURALE). His current research interests include the development of multimedia systems applied to future of broadcasting, the future of Internet and 3-D multimedia video/graphics design, and the synergies between these technologies. Prof. Cosmas is an Associate editor of the IEEE Transactions on Broadcasting.

Sherine M. Abd El-Kader is a Professor, Computers & Systems Dept., at the Electronics Research Institute (ERI). She has published more than 30 papers, 5 book chapters. Her topics of interest are: Wi-MAX, Wi-Fi, IP Mobility, QoS, Wireless sensors Networks, Ad-Hoc Networking, realtime traffics, Bluetooth, and IPv6. Also she is a technical reviewer for many international Journals. She is also a technical member at both the ERI projects committee and at the telecommunication networks committee, Egyptian Organization for Standardization & Quality since February 2007 till 2011, She is currently the head of Technology Innovation Support Center (TISC) at ERI.

Salwa Elramly, IEEE Senior Member. She obtained the BSc. Degree 1967, MSc. Degree 1972 from Faculty of Engineering, Ain Shams University, Egypt & PhD degree 1976 from Nancy University, France. She is now Professor Emeritus with the Electronics and Communications Engineering Department, Faculty of Engineering, Ain Shams University; where she was the Head of the Department (2004-2006). Her research field of interest is Wireless Communication Systems and Signal Processing, Language Engineering, Coding, Encryption, and Radars. She is the IEEE Signal Processing Chapter chair in Egypt. She was awarded Ain Shams Award of Appreciation in Engineering Sciences (2010), Award of Excellence from the Society of Communications Engineers (2009) & Award of Excellence from the Egyptian Society of Language Engineering. She shared in the establishment of the Egyptian Society of Language Engineering in 1996, and is now its President. She published more than hundred scientific papers and supervised more than 100 MSc and PhD thesis.

Lokesh Jain is Assistant Professor in School of Electrical Engineering and Information Technology, College of Agricultural Engineering & Technology, Punjab Agricultural University, Ludhiana, India. His field of interest includes information and communication systems, mobile and web based application development. He has overall experience of 12 years in teaching and research. He has joined as Scientist in Indian Council of Agricultural research and has been posted at Central Institute for research on Cotton Technology, Mumbai in year 2003. In 2005, he joined Punjab Agricultural University, Ludhiana and is serving there till now. He is actively involved in the development and implementation of various software applications for the university.

Harish Kumar is working as Associate Professor in Computer Science & Engineering at University Institute of Engineering & Technology, Panjab University, Chandigarh, India. He has completed his Ph.D. from Panjab University, M.Tech. from Punjab Agriculture University. He also got in-service Executive training from Indian School of Business (ISB), Hyderabad as well as Indian Institute of Management (IIM), Lucknow. He had worked in the IT industry prior to adopting academia as a profession. He was consultant to develop an online education portal currently handling 2+ million users and connecting them with 55000+ teachers across the world. He is Academic Coordinator for MHRD and World Bank funded TEQIP-II project. He has also executed sabbatical project at Infosys Technologies as a member of their development team in PED division. His areas of interest include Mobile Ad hoc networks, Traffic Profiling, Next Generation Networks, Software Test Estimation, Open source software, and e-learning. He has more than 55 publications to his credit. He has also delivered several lectures on various topics during workshops/ conferences/UGC refresher courses at national as well as international forums. He is life member of Computer Society of India, Indian Society for Technical Education and International Association for Engineers.

R. S. Mangrulkar has done his PhD in Computer Science and Engineering from SGBAU Amravati in 2016. He has done his B.E (Comp sci. & Engg) from B.N.C.O.E Pusad in 2001 and M.Tech (CSE) from National Institute of Technology, Rourkela, Orissa in 2008. Currently he is working as Associate Professor and Head, department of Comp. Engineering, Bapurao Deshmukh College of Engineering Sevagram. He has total 15 years of teaching experience to UG and PG level. He has total 14 research publication in international journal and 28 research paper in National and International Conferences. He is reviver of IJCA, Elsevier and AIRCSE journal. He evaluated 9 dissertation report of M.E / M.Tech candidate in RTMNU Nagpur and SRTMNU Nanded. He received gold medal on securing Merit position in SGB Amravati University. He also delivered 28 guest lecture / Expert lectures in Research oriented programs at SGGS Nanded, PIET Nagpur, RCERT Chandrapur and SKNCOE Pune, PVPIT Pune, COE Ambojogai, MGM Nanded, Anjuman COE Nagpur and BDCE Sevagram on "Latex and Its Importance in Manuscript Preparation" and "Network Simulator Programming" and trained almost 700 students of PG and Faculty Members. He also chaired 25 sessions in various national level conference and technical symposium. He also received certification of appreciation from DIG Special Crime Branch Pune and Supretendant of Police and broadcasting media gives wide publicity for the project work guided by him on the topic "Face Recognition System". He received certification of achievement from IIT Kanpur for ACM Programming Context. He is life member of ISTE,CSI,IACSIT and IAENG. He is highly motivated towards research and always promotes use of recent technology for presenting and promoting open source research. His area of interest is Adhoc Network specially design of Routing Protocols. He also received 3.5 lakhs grant under Research Promotion Scheme of AICTE for the project "Secured Energy Efficient Routing Protocol for Delay Tolerant Hybrid Network". Under his leadership, Department of Computer Engineering also received grant of Rs. 1.29 lacs from UGC, AICTE and CSIR for conduction of STTP/Workshop. He is active member of BOS, RTMNU and handling responsibility of promoting new syllabus to B.E. Computer Engineering.

Amrit Mukherjee, currently a Research Scholar in KIIT University, working on sensing algorithms in cognitive radio and its corresponding simulations. He has worked in C-DAC as International Faculty for ICT in Armenia and Ghana respectively. He has completed his M.Tech from Dr.Babasaheb Ambedkar Technological University, Raigad and B.E from Amravati University Maharashtra on 2011 and 2009.He has submitted hi Ph.D thesis in 2016, and the research work is based on Spectrum sensing for cooperative communication in Cognitive Radio.

Monika Rani has completed her BTech and MTech from Punjab Technical University. Currently she is pursuing PhD from Panjab University, Chandigarh. She is also working as System Manager at Panjab University, Chandigarh. Earlier, she worked as Programmer in PGIMER and EDP Supervisor in EPFO, GoI. She has research interests in Computer Networks, Information Retrieval etc.

Arif Sari, born October 21, 1986 in Nicosia (Northern Cyprus), is a full-time instructor of the department of Management Information Systems at the European University of Lefke, North Cyprus. He received his BS degree - in Computer Information Systems and MBA degree - (2008 and 2010) at European University of Lefke, and Ph.D. degree (2013) in Management Information Systems at The American University. He is an IEEE, ACM, and IEICE Member since Sept. 2012 and has published various papers, book chapters, participated in variety of conferences in the fields of Network Security, Network Simulation, Computer Applications, Mobile Networks, Information Communication Technologies, Mobile Network Security and Mobile Security Systems. His research interests include Network Security, Mobile and Wireless Networks, Network Simulations, Security and Cryptography and Radio Frequency Identification Systems. He is reviewer of many international journals and refereed conferences.

Kandarpa Kumar Sarma (S'04-M'09, SM '12) is presently with the department of Electronics and Communication Technology, Gauhati University, India. He completed his MTech degree in Digital Signal Processing from IIT Guwahati, India, in 2005 and PhD degree from the same institution. His areas of interest include applications of soft-computational tools, mobile communication, pattern recognition and language technology. He is a senior member of IEEE (USA), Fellow IETE (India) and has authored several books and research papers.

Charu Singh doing his Bachelor Degree in Electronics and Communication Engineering, Vellore. She was also associated solar car taxi developed by Namibia University of Science and Technology and presented her presentation in the President house, Republic of Namibia. She also published a paper in ACM conference proceedings.

Sarbjeet Singh received his B.Tech degree in Computer Science and Engineering from Punjab Technical University, Jallandhar, Punjab, India in 2001 and M.E. degree is Software Engineering from Thapar University, Patiala, India in 2003. He also received Ph.D degree in Computer Science and Engineering from Thapar University, Patiala, India in 2009, working on grid security systems architecture. Currently he is Associate Professor in Computer Science and Engineering at UIET, Panjab University, Chandigarh, India. He has more than 30 research publications in international journals and conferences to his credit. His research interests include parallel and distributed systems, distributed security architectures, distributed services like grid, cloud and web services, privacy and trust related issues in distributed environments. Dr. Singh is a life member of Computer Society of India and Indian Society for Technical Education.

Deb Sunder Swami is presently with Department of Electronics and Communication Engineering, Royal Group Of Institutions, Guwahati, India. He completed his Bachelor in Engineering from Visvesvaraya Technological University, Karnataka, India and Masters in Engineering from Gauhati University, India in 2014.His areas of interest include Wireless Communication, Embedded systems and Computer Architecture. He has 3 research papers to his credit so far.

Meenakshi Tripathi is presently Assistant Professor, Department of Computer Engineering, MNIT, Jaipur (India). She completed her B.E. (Computer Engineering) from IET, Alwar (India) in 2003 and M.Tech. (Computer Science Engineering) from Banasthali Vidyapith (India) in 2005. She has completed her Ph.D. from MNIT, Jaipur (India). Her research interests include Wireless Sensor Network Security and Wireless Network Routing Protocols.

Index

www.ingramcontent.com/pod-product-compliance
Ingram Content Group UK Ltd.
Pitfield, Milton Keynes, MK11 3LW, UK
UKHW050951290926
15470UKWH00005B/1753

9 781522 52342